CLASSIC CARS AND

AUTOMOBILE ENGINEERING

Volume 5

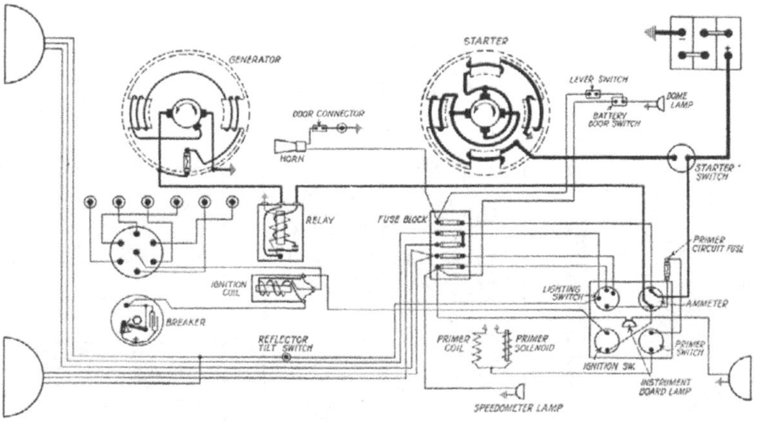

WIRING DIAGRAMS, DATA SHEETS, QUESTIONS AND ANSWERS

Classic Cars and Automobile Engineering:
Volume 5

Wiring Diagrams, Data Sheets, Questions and Answers

Restored by Mark Bussler

Automobile Engineering

A General Reference Work

FOR REPAIR MEN, CHAUFFEURS, AND OWNERS; COVERING THE CONSTRUCTION, CARE, AND REPAIR OF PLEASURE CARS, COMMERCIAL CARS, AND MOTORCYCLES, WITH SPECIAL ATTENTION TO IGNITION, STARTING, AND LIGHTING SYSTEMS, GARAGE EQUIPMENT, WELDING, FORD CONSTRUCTION AND REPAIR, AND OTHER REPAIR METHODS

Prepared by a Staff of

AUTOMOBILE EXPERTS, CONSULTING ENGINEERS, AND DESIGNERS OF THE HIGHEST PROFESSIONAL STANDING

Illustrated with over Fifteen Hundred Engravings

FIVE VOLUMES

STUDEBAKER BIG SIX, FIVE-PASSENGER SPORT PHAETON

Table of Contents

VOLUME V

Electrical Equipment

By Charles B. Hayward   *Revised by Tom C. Plumridge*

BALLOON TIRE—FRONT WHEEL

READING WIRING DIAGRAMS AND AUXILIARY ELECTRICAL EQUIPMENT

EXPLANATION OF WIRING DIAGRAMS

The first requisite in the work of electrical trouble shooting and repair work is the ability to trace a circuit on a wiring diagram. The wiring diagram is very useful both for rewiring work and for checking up correct connections in trouble shooting.

In the making of a wiring diagram, it would be a difficult task to write in the names of the different parts of the several pieces of electrical equipment. The manufacturers have compiled a number of symbols which are used to show these different parts. In Figs. 1 to 12 inclusive, are shown the symbols that are used in the Delco installation. While these symbols are standard for Delco only, they may be considered standard for all since they bear a close resemblance. These symbols should be carefully studied so that the different parts of the circuits can be easily recognized on sight. Switches, contact points, resistances, etc., are points at which trouble occur. If the repair man can find the part on the wiring diagram and trace the circuit on the diagram to the part, he can easily find the circuit on the car.

Current Direction. The plus and minus, or positive and negative signs, + positive, — negative, scarcely call for any extended explanation. They indicate the direction in which the current flows. It is of the utmost importance, where the manufacturers' directions are to connect certain apparatus with a given wire to the plus, or positive, side, and another wire to the negative, that these instructions be followed explicity. Otherwise, the apparatus either will refuse to work or it may be damaged, as in the case of a storage battery on which the connections have been reversed. Wherever it is necessary that the current flow through a piece of apparatus in a certain direction, the manufacturer stamps plus and minus signs at the terminals.

WELDING MANIFOLD TUBES IN AN AUTOMOBILE PLANT
Courtesy of Oxweld Acetylene Company, Chicago, Illinois

Battery; Generator. A battery, regardless of its type, is always shown by alternate heavy and light lines, as indicated in Fig. 1, each pair of lines representing a cell, so that the number of cells in the battery may be told at a glance. Other sources of current, such as generators, are indicated by a conventional sign consisting of a circle with two short heavy lines tangent to its circumference at opposite points and usually at an angle to the horizontal, as shown in Fig. 2. The origin of this sign will be apparent in its resemblance to the end view of a commutator with a pair of brushes bearing on it. This sign is also used to indicate a motor, in which case the letter **M** is inserted in the circle.

Coils. Coils which are wound on an iron core are generally indicated by a conventional sign consisting of a few loops of wire, as in Fig. 3, but this is only the case where such a coil occurs at a place in the circuit where there might be a chance of confusion in identifying it. Where there is no possibility of confusion—as in the case of the windings of a generator or motor, ignition coils, and the like—the sign shown in Fig. 4 is often used. Where the lines are heavy, a coarse wire, such as is employed for series windings of generators or motors, or the primary winding of an ignition coil, is intended.

Resistance. Resistance in a circuit is usually shown by an arbitrary sign, Fig. 5, similar in outline to a piece of the cast-iron grid frequently used in charging resistances, though sometimes shown as a coil and marked "resistance".

Grounds. The sign of a ground connection is the inverted pyramid of short lines, Fig. 6, and indicates that the circuit is grounded. This may be either by a wire directly connected at some point with the frame, as in the case of the storage battery, or it may be through an internal ground connection in the apparatus itself, as in the lamps and sometimes the generator or motor, the connection being made simply by fastening them in place. In any case, the sign indicates that the circuit is completed through a ground.

Contacts. There are a number of signs employed to indicate contact points, switches, and the like, and, where they are not of an arbitrary character, such as Fig. 7, which shows contact such as used in switches, distributors, etc., and Fig. 8, which indicates a lighting switch (Delco diagrams), they usually will be found to bear sufficient resemblance to the apparatus itself to make their identification easy.

Induction Coil. Fine lines indicate a generator shunt winding, the secondary of an ignition coil, or the coil of a relay or cut-out. The primary and secondary windings of an induction coil as used for ignition are indicated by a fine and a coarse coil sign, as in Fig. 9.

Condenser. A condenser with its overlapping plates is shown in Fig. 10.

Crossed Wires. To show wires that cross one another without making connection, a straight-line cross is made at that point to show that the wires do not touch, as in Fig. 11, while wires that are connected are shown by a black dot at the junction.

General and Special Usage. While these signs are not universally used in exactly the form shown here, their employment is very general and in the majority of cases, such as the positive and negative, battery, ground, generator, induction-coil windings, and coil signs, they are never changed. In some instances special signs are employed, such as that shown in Fig. 12, which indicates the motor commutator of the Delco single-unit machine or *dynamotor*, and shows the special brush lifting switch. Incandescent lamps are almost always indicated by small circles, though the lamp itself is sometimes drawn in. As a matter of fact, very little system is followed by different makers in making these wiring diagrams. In an effort to simplify its reading to the uninitiated, a diagram will sometimes picture most of the apparatus in such form that it will be recognized from its resemblance to the original, including the battery, generator, lamps, and the like, using only signs for showing coils and ground connections; others go to the opposite extreme and show nothing but signs.

Tracing the Circuit. The correct point at which the tracing of the circuit should begin is at the source of supply, which in the automobile is the battery. Irregardless of whether it is the positive or the negative terminal that is connected to the frame of the car, the current always flows from the battery through the positive and returns through the negative. Examples of both of these connections are shown in the diagrams. A line, which represents a wire in the system, is followed to the apparatus to which the connection is to be made. If, in the tracing of the circuit, a resistance unit is found on the diagram, there must be a resistance in the circuit on the car. By tracing the circuit, the repair man finds the position of the apparatus in each of the circuits.

In tracing the headlight circuit on the wiring diagram of the Dodge car, Diagram F, the positive terminal of the battery is attached to the frame of the car. The current flows from the positive terminal of the battery through the frame to the ground connection of the lamp and out by way of the wire to the switch terminal *H*. At this point the switch is closed and the current cannot complete the circuit. Open the switch and the current will flow out at the point *B* to the ammeter, across the ammeter to the battery side of the starting switch and then to the negative terminal of the battery, completing the circuit. It will be found that both headlights are attached to the terminal *H*. The battery side of the switch for the headlights is point *B*. If, in tracing the circuit on a car, it is found that the headlights are attached to the point marked *IG*, the repair man would know that the connection was not correct and he would know exactly how to make the proper connection.

In tracing the ignition primary circuit on the same diagram, start from the positive terminal of the battery. The current flows to the grounded side of the contact points in the breaker housing and, when the points come together, passes through the wire and the primary winding of the coil, out through the resistance and to the ignition and lighting switch marked *IG*. If this switch is open, the current returns to the negative side of the battery by way of the point *B* on the ignition and lighting switch and through the wire to the ammeter. It finally passes across the ammeter to the battery side of the starting switch and to the battery. In tracing this primary circuit, the proper switch connections have been found and the location of the resistance unit. If there was trouble in the primary current to the extent that there was no current found at the contact points, the repair man would know where to look for the resistance unit to see whether it was broken or burnt out. He would also know where to look for other connections to see if everything was correct in the primary circuit or system. It will be seen that the current in this system always returns to the battery by way of the ammeter and starting switch and the supply is by way of the frame connection.

The Delco system, as installed on the Roamer car, Diagram C, shows that the battery negative terminal is connected to the frame. This is a true ground-return system and the current supply is furnished by way of a wire and returns by way of the frame. In tracing the ignition primary circuit, the current flows to the battery side of the starting switch from the positive side of the battery to the ammeter and across to the main point marked *6*. When the switch is open, the current flows out of the switch, at *5*, to the resistance unit, through the primary winding of the coil, across the timing contact points to the grounded point and back to the battery by way of the ground connection which, in this case, is the metal part of the car. In tracing the circuit to the headlights for the dim position, the flow of the current is from the positive side of the battery to the starter switch, through the ammeter to the main switch at point *6*, through the switch connections and out at point *3*, through the dimmer resistance back to point *2* of the switch. It then goes to the headlight bulbs and out from the lamps to the battery by way of the ground connection. All the current except the starter current passes through the ammeter.

TYPICAL ELECTRICAL SYSTEMS

The six sample diagrams here given are typical of the wiring systems used on the present-day car and embrace the Remy, Delco, Atwater-Kent and Westinghouse. The single and double units are also shown in these diagrams. If the operation of one system is understood, the other systems of the same manufacture and the tests for troubles will be understood. The explanation on the operation of the systems mentioned should be sufficient to give the repair man a good working knowledge and enable him to make satisfactory repairs in any other system. The operation and control of all parts of the electrical equipment as installed on the automobile have been explained and discussed in previous sections. The reader should be able to follow the points mentioned in connection with the sample diagrams.

Diagram A. Diagram A shows the wiring for the Remy system as used on the Studebaker Models *EG* and *EH*. The ignition and lighting system in this installation is Remy throughout. A Remy lighting system can usually be recognized by the thermostatic control

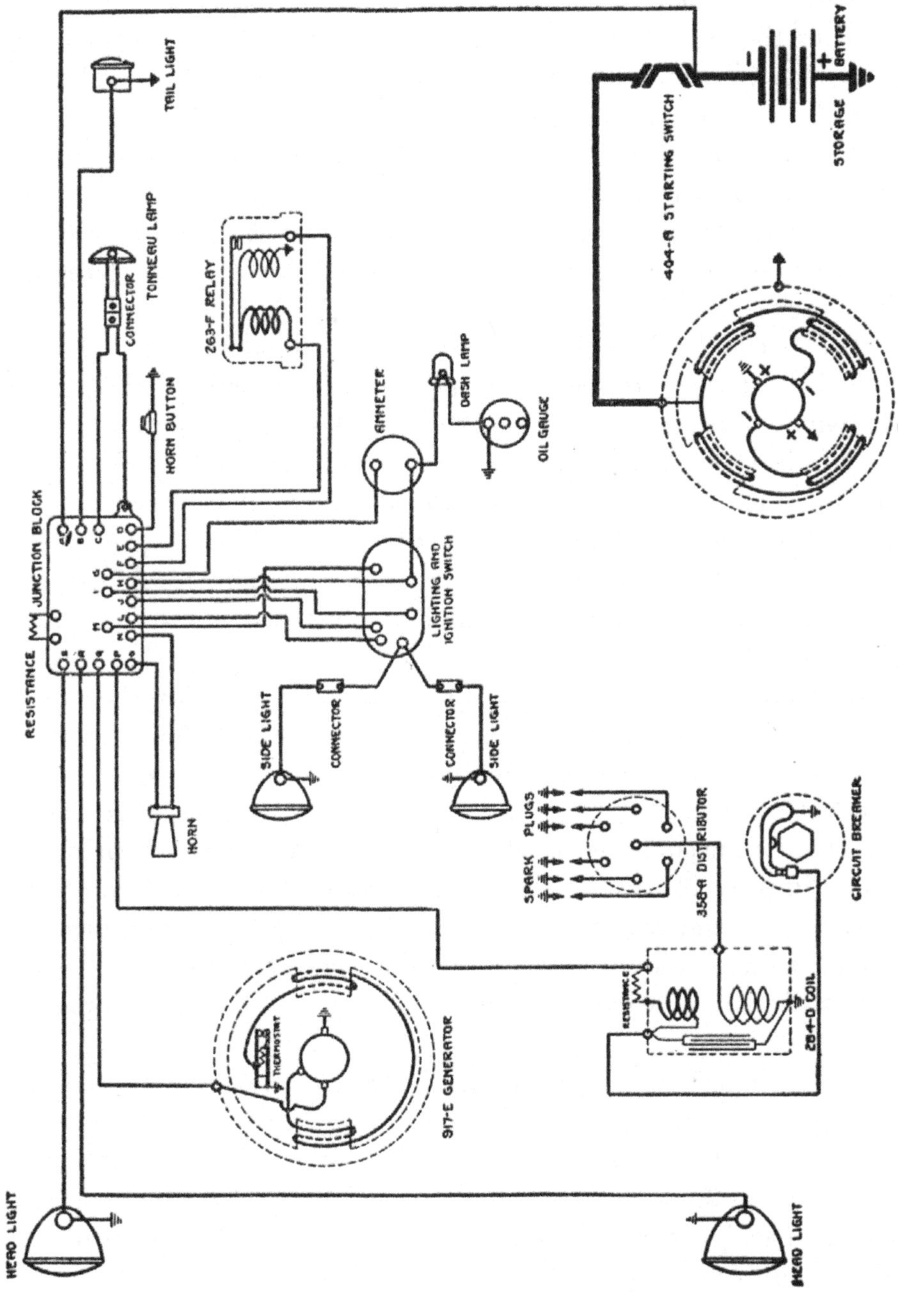

Diagram A. Studebaker. Remy Lighting and Ignition

installed in the generator. The operation of this control is fully discussed in the lesson on "Generators." This type of control is typical of the Remy and is a patented feature. The system is a two-unit installation with the thermostatic control used in conjunction with a third brush. The relay is separate from the generator and is usually fastened to the outside of the generator rather than built into the generator itself.

In the ignition system it will be noted that the resistance unit as well as the condenser is built into the coil or, at least, located in connection with it.

The battery, which is fastened to the frame by the battery terminal, is of the three-cell type indicating a six-volt system, and the ground connections show it to be a single-wire system.

It is interesting to note that the current supply for all units of this installation is taken through a junction block instead of directly through the switch as in other systems. When trouble is experienced in this system, the fuses in the junction block should be the first point of examination and a blown fuse is the indicator for the circuit in which there is trouble.

Diagram B. Diagram B is the wiring diagram for the two-unit Delco installation as used on the Oldsmobile. The generator control is by the third-brush method. There is no relay in this generator circuit and the battery and generator connection is made when the switch is turned to the "on" position. A circuit breaker is found in place of the relay and is located at the switch to protect the battery should any excess current start to flow through any of the lighting circuits. In addition to the regular side and head lamps, additional low candlepower lamps for city driving and parking are used.

In the ignition, a very marked change from the preceding Diagram A is found. The condenser and resistance units have been placed in the breaker housing instead of at the coil. A double set of breaker contacts are used and are arranged for synchronous timing. The battery to frame connection is negative instead of positive and a six-volt single-wire system is again used.

In tracing the circuit in this diagram beginning at the battery terminal, opposite to the one connected to the frame, the circuit is regular as far as the starting switch. A lead goes to the ammeter

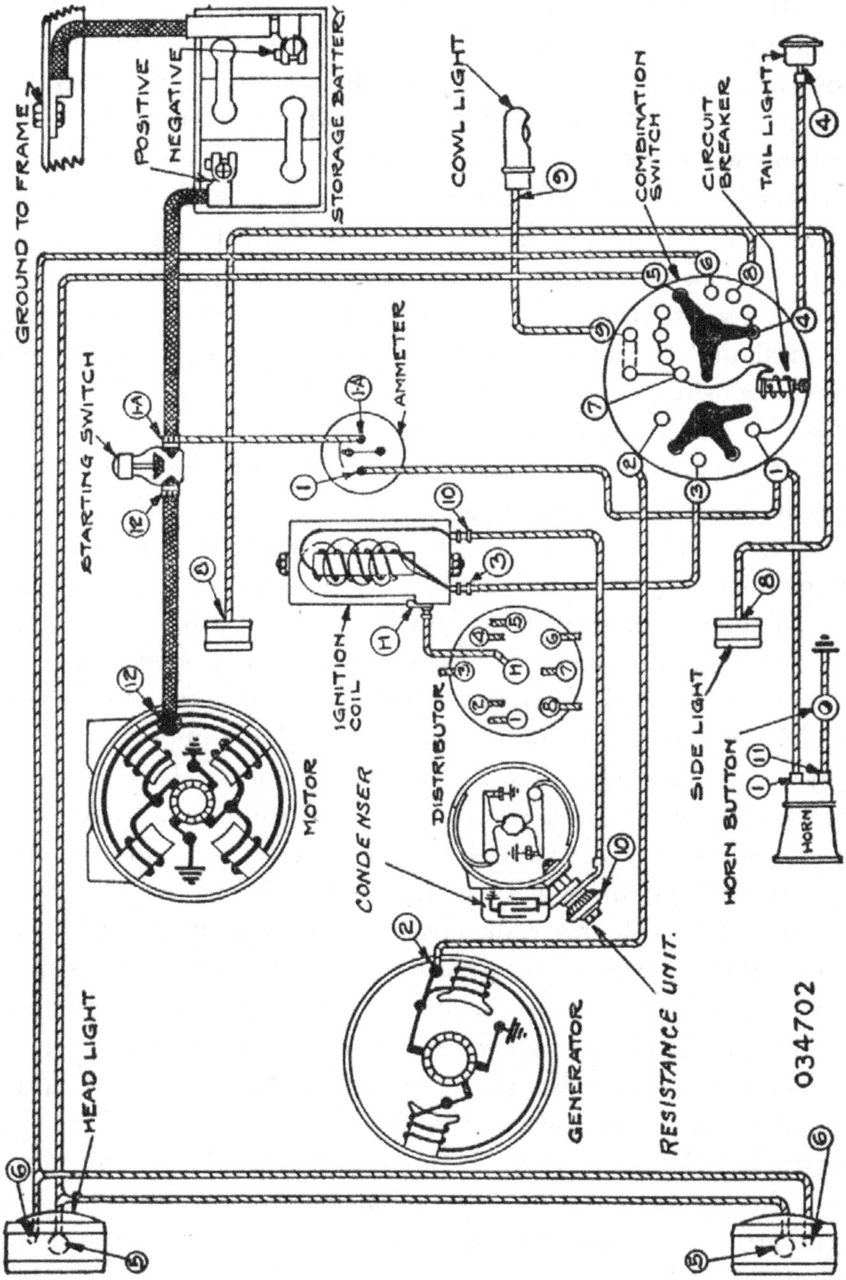

Diagram B. Oldsmobile. Delco Two-Unit

for ignition, lighting, and signal purposes. When the starting switch is closed, the battery circuit is completed through the starter fields and armature to the ground. With such short and direct connections and with the low resistance of the starting motor it is very evident that there can be little excess battery current for lights and ignition when the starting-motor circuit is complete especially if the battery is low. Low battery power accounts for the frequent lack of brilliancy from the lights and sometimes for the failure of the ignition when the starting motor is in use on any automobile.

A lead goes to the ignition and lighting switch from the ammeter. It should be noted that the horn circuit is so arranged that the current consumed by it is recorded by the ammeter. It cannot be broken by either the ignition or lighting switches. We do, however, find that the circuit breaker is usually so arranged that the horn circuit is completed through it.

From the circuit breaker, the current can be traced to the various contact points on the lighting switch (marked combination switch). The cowl or dash light can be operated from the contact points at will. The current can go no further without moving the switch which can be done by rotating it in a clockwise direction. The idle finger will come into contact with the supply taps. The finger resting on contact *5* will rest on *6* completing the current for the low candlepower lights, commonly called dimmers. The finger resting on contact *4* will soon be connected to the same contact and the tail light circuit will be completed. By turning the switch in a counter clockwise direction past the "off" position, the finger resting on contact *5* will be brought into connection with the supply taps. The finger resting on *4* will rest on contact *8* and the idle finger will connect the tail light circuit. At this point the side lights will be connected and by rotating it still farther the head lamps will be connected instead of the side lamps.

On the ignition side, there is but one throw, that being in the clockwise direction, one finger coming in contact with the top by which the primary ignition circuit is completed through the primary winding of the coil to the breaker contacts and ground. The other two fingers complete the circuit between the generator and battery. It will be noticed that the ignition and lighting circuits are so placed as to "float" on the generator to the battery line. What actually happens is that the ignition and lighting circuits are first supplied and the surplus current output of the generator is driven through the ammeter to the battery, giving a lower charging reading when lights are on than when they are off. The circuit breaker is necessary should short cuts occur anywhere past this point since all the generator output, as well as the battery output, would flow directly through the short, being limited only by the capacity of the wires which would soon fuse and cause trouble. When the engine is not running and the ignition switch is turned on, there is nothing to prevent the battery from discharging through the generator, which would consume about 5 amperes. It is therefore very important that the ignition switch be closed when the engine is not running. This is usually taken care of by the relay but as none is used in this installation the warning given is very important.

Diagram C. Diagram C is the diagram for the Delco system as installed on the Roamer. This is also a two-unit system with the third brush regulation. The relay is in the generator and battery circuit. The operation of the relay has been explained in the section on generator control.

A cut-out relay is used to protect the battery and close the generator when a sufficient charging rate has been reached. The horn circuit is so arranged that the current consumed by it does not pass through the ammeter.

A circuit breaker is located on the lighting and ignition switch and, in addition, there is a dimmer resistance in conjunction with it, which is brought into the head lamps circuit when the switch is rotated to its proper position. The dash and tail lights are so arranged as to burn at full capacity irregardless of whether the lighting switch is in the dim or bright position. No side lights are used. The ignition wiring shows the resistance located at the ignition coil and the condenser located with the breaker contacts. A six-volt battery is used and the one negative terminal is connected to the frame.

Diagram D. Diagram D is the wiring diagram for that used on the Pan-American car. The starting and lighting system in this car is the Westinghouse, while the ignition is Atwater-Kent. The installation is of the two-unit type and the generator control is by the third-brush method. The relay is built into the generator and

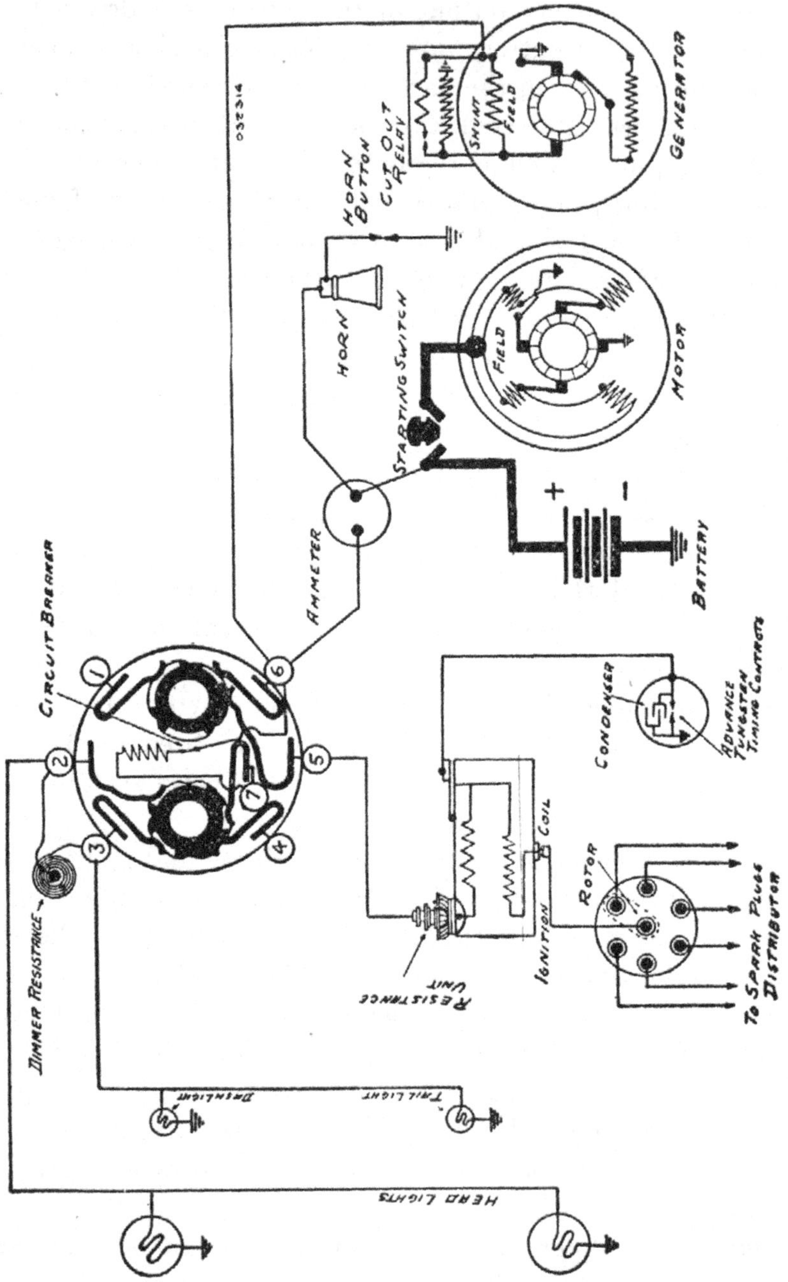

Diagram C. Roamer. Delco Two-Unit

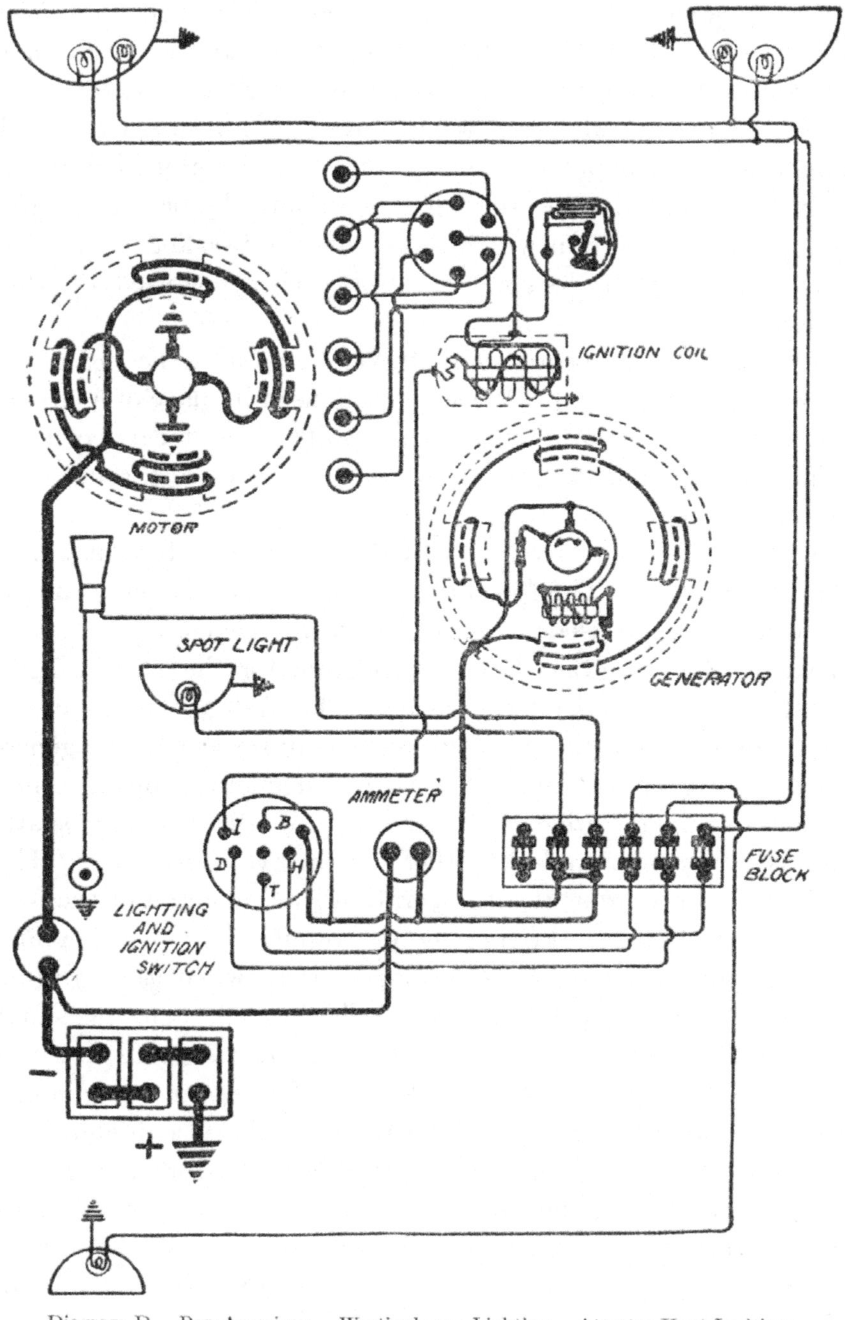

Diagram D. Pan-American. Westinghouse Lighting. Atwater-Kent Ignition

the latter is protected against damage should the relay become inactive or by some connection becoming loose or broken by a fuse being inserted in the field winding. Should any trouble occur that would cause a heavy load to be thrown on the generator, this fuse will melt or blow out allowing no current to flow through the field winding which would prevent generation. The ignition, being the Atwater-Kent, is of the open-circuit type. Note the peculiar shape of the part in the diagram which represents the breaker unit. The Atwater-Kent ignition system can always be recognized by this formation.

The condenser is located with the contacts and the resistance in connection with the coil. A fuse block is used in place of a circuit breaker and the spot lamp and horn circuits are not connected to the ignition and lighting switch but are connected through the ammeter.

The battery is a six-volt type with the positive terminal connected to the frame. The four-brush series-wound starter seems to be the common type used in most electrical systems.

Diagram E. Diagram E is the diagram for the Delco single-unit system as installed on the Essex. The system on this car is entirely different from the previous systems in respect to the generator and starting-motor units. The Delco single unit is one in which two commutators are used on the same end of the armature shaft: one for the generator and the other for the starting motor. The ground brushes are brought into operation by the action of the foot pedal thus closing the starting motor circuit. The ignition and starting motor circuits are so arranged that when the ignition switch is closed the battery current will flow through the shunt winding of the generator, causing the armature to rotate slowly—the driving connection to the engine being through an overrunning clutch. As the starting pedal is operated, it slides a double idler gear into mesh with a gear on this rotating armature and finally into the gear cut on the face of the flywheel. The purpose of rotating the armature is to make the engaging of these gears positive and easy. As soon as the gears have been fully engaged the foot pedal brings the brushes into contact with the commutator and the starting action takes place. The rotating armature causes a clicking noise at the generator clutch which can be heard when the

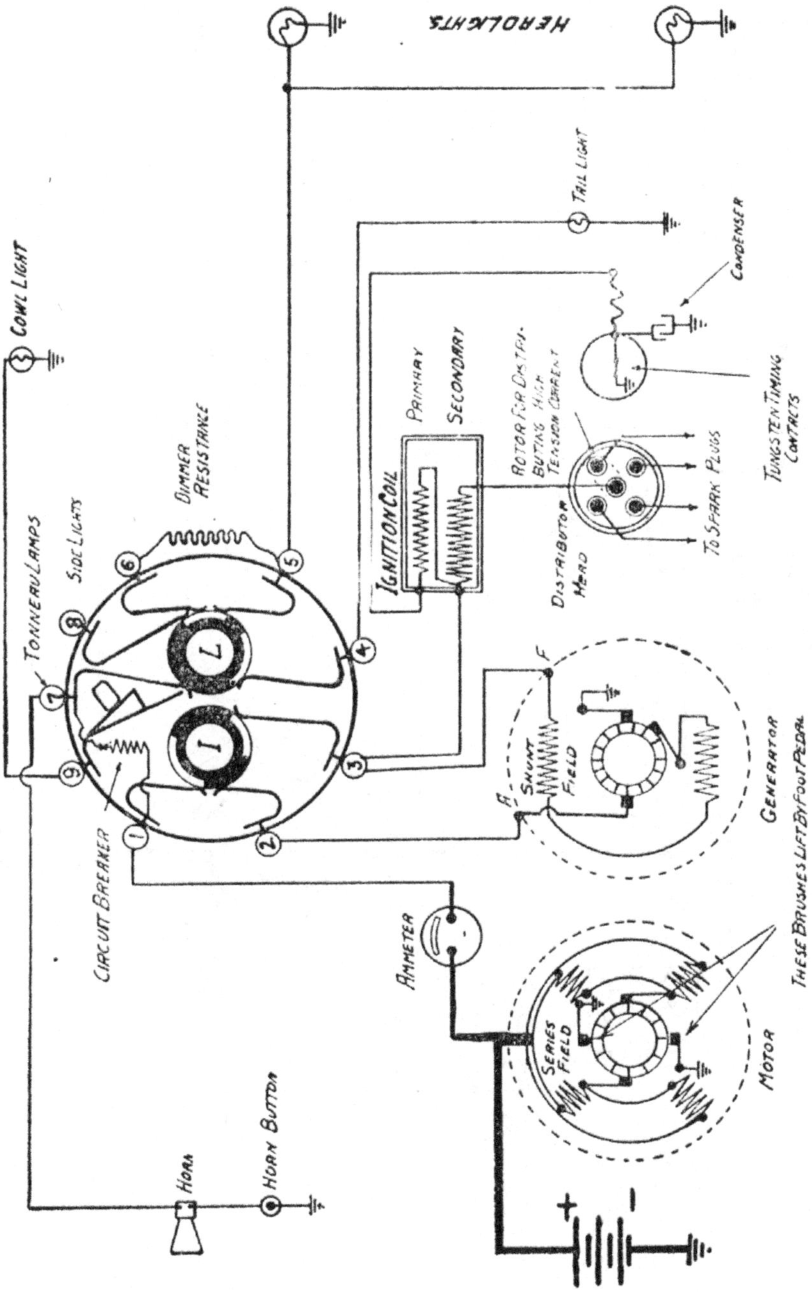

Diagram E. Delco Single-Unit

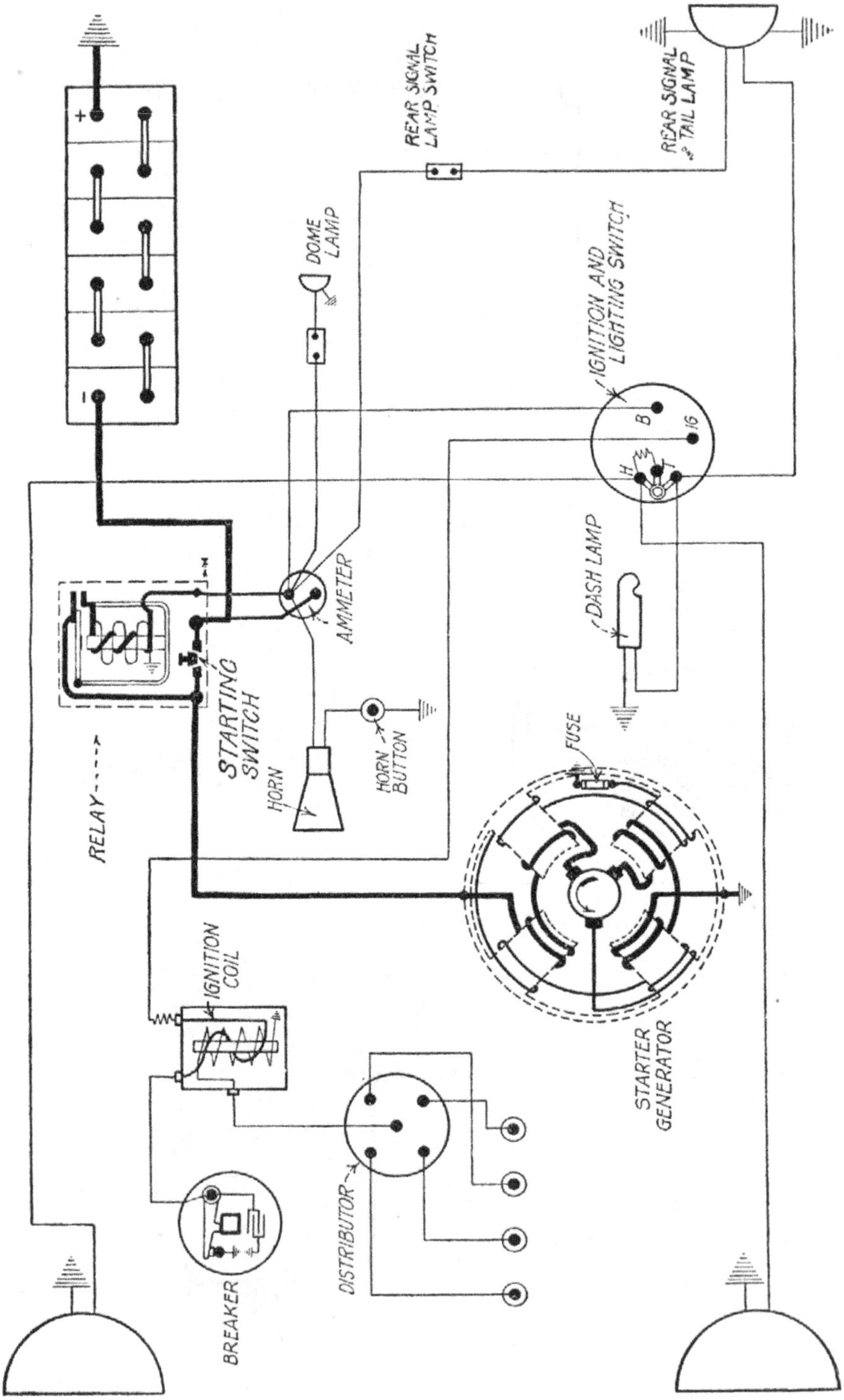

Diagram F. Dodge (1923-24). North-East System

ignition switch is turned on and often when the engine is permitted to idle below the rotating speed of the armature.

In the ignition wiring, the resistance and the condenser are located with the breaker contacts. A circuit breaker and dimmer resistance are located on the switch and all circuits, except that of the starting motor, are connected through the ammeter.

Diagram F. Diagram F is the wiring diagram for the North-East system as installed on the Dodge car. It illustrates another single unit in which the armature has only one commutator on the armature shaft and which is used for both starting motor and generator. The field winding is of the differential-wound type. It consists of a series winding, which is the starter motor winding, and the shunt winding, which is the generator winding. When the unit is operating as a starter motor, the two windings operate together, that is, the series winding is helped by the shunt winding. When the unit is operating, the shunt winding acts as a bucking coil for regulating purposes. A fuse is installed in the field winding which burns out if the charging rate rises above 10 amperes. A relay, which is combined with a starting switch is placed in the generator and battery circuit. The current taken for ignition, lights, and horn is shown on the ammeter. It will be noticed that the system is of the 12-volt type, while in the previous systems it was a 6-volt type. The condenser is located with the points and the resistance at the coil in the ignition system. The dimmer resistance, for use in connection with the head lights, is located at the switch. A 12-volt battery is used and the positive terminal is connected to the frame.

While in these systems there is a little different arrangement of the several units, the principle of operation and the tests for troubles are practically the same. If the previous systems have been studied carefully, all systems will be understood. For instance, if the operation of one ignition coil or the operation of one relay or the test for one type of field winding is understood in one, they will be understood in all.

Two typical magneto ignition circuits are shown in Fig. 13, which is the Bosch, and in Fig. 14, which is the Dixie.

Size of Conductors. The influence of the factor of resistance makes plain the reason for using wires of different sizes for the various circuits of the ignition starting and lighting systems of the automobile. If an ample flow of compressed air is desired for power purposes, a liberal outlet must be provided, while if only a small spray is required, as for cleaning purposes, a small-bore tube will suffice. If we try to employ the small-tube line for power purposes, we shall not gain the desired result because its resistance is so great that it will not permit a sufficient flow of air. For the

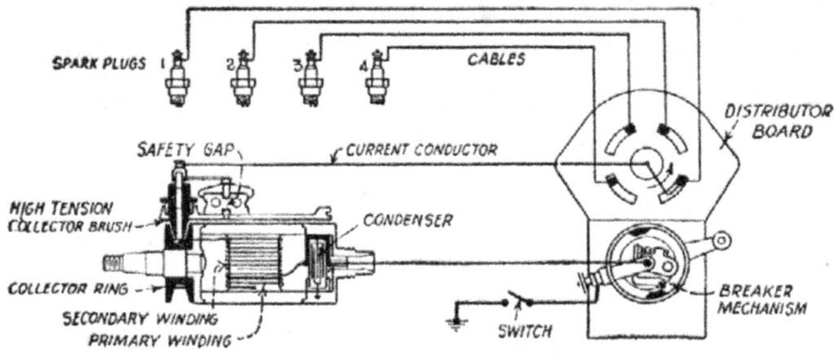

Fig. 13. Bosch Magneto Ignition Circuit

same reason a conductor of much larger diameter and, therefore, of correspondingly low resistance must be employed to handle the

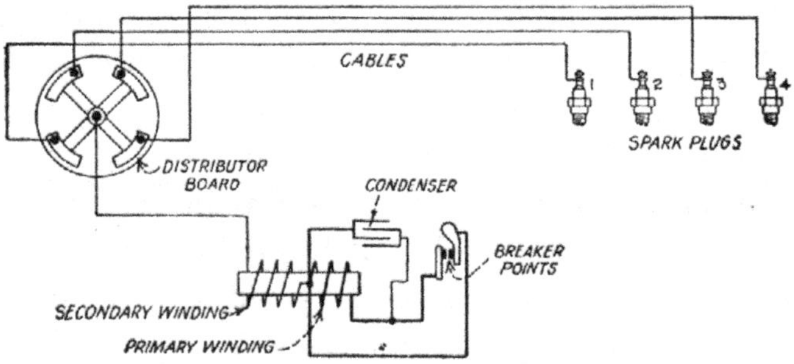

Fig. 14. Dixie Magneto Ignition Circuit

heavy current necessary to operate the electric starting motor, than is needed for the comparatively small current which is demanded by the ignition system.

Whether it is mechanical or electrical in its nature, the power necessary to overcome resistance is liberated in the form of heat. Mechanical resistance is friction and its presence between moving bodies always generates heat. Electrical resistance may, for the purpose of illustration, be termed internal or molecular friction, and it also results in heat. The extent of the rise in temperature of a conductor or wire, depends entirely upon the proportion that its size and, consequently, its current-carrying ability bear to the amount of current that is sent through it. Roughly speaking, if a wire is three-fourths the size it should be to carry the starting current, it will become uncomfortably warm to the hand after the motor has been operated several times in succession. If it is only one-half the size it should be, continuous operation of the starting motor for a few minutes will doubtless burn off most of the insulation. Further reducing its size would cause the wire to become so hot as to set fire to the insulation the moment the current was turned on, and any great decrease in diameter would result in the immediate fusing of the wire itself. The wire would literally "burn up" and in a flash.

It would not be practical to attempt to conduct live steam at high pressure through a cardboard tube. Nor is it any more so to attempt to send a heavy current through "any old piece of wire." Electric lighting and starting systems as they exist on cars today are of all degrees of merit. The cars themselves have reached a stage of reliability where their useful life is now on the average from five to ten years or more. Consequently, there are a great many cars in service equipped with electric systems that were brought out several years ago. These are the cars on which the repair man will get a great deal of his early experience, and he need not take it for granted that just because the electric systems have worked for a certain length of time they were properly designed at the outset. Overheated conductors not only indicate excessive resistance caused by small wires or poor joints, but they also indicate a waste of power that is being drawn from the battery and dissipated in the air. The utilization of this energy or rather the prevention of its transformation into heat would mean all the difference between poor and good operation between an efficient and a wasteful system.

Since the amount of current received by the electrical unit is in direct proportion to the size of the wire or conductor, it is necessary to see that the correct size wire is installed when and where replacements are made. Suppose that a new wire must be installed on the starting motor. If the wire is too small, the machine will not get enough current to make it operate correctly. The following is a rule by which the correct size of the wire may be found. Test the starting motor with a high-reading ammeter (scale should read to at least 300 amperes) after having made certain by hydrometer and voltage tests that the storage battery is fully charged. Carefully note ammeter reading exactly at instant of closing switch, to determine maximum current flow. Measure the length of cable between the battery and the starting motor, i.e., both sides of starting switch. Then maximum starting current times 10.7 times number of feet of cable used, divided by .25 will give the cross-section of the wire in circular mils. For example, assume that the starting motor required a maximum of 300 amperes momentarily to break away the engine, and five feet of cable are employed for the connections. Then

$$\frac{300\times10.7\times5}{25}=64{,}200 \text{ circular mils}$$

By referring to Table I, which gives the various size wires in circular mils and their equivalent in gage sizes, it will be noted that the closest approach to this is No. 2 cable, which is 66,373 circular mils, so that the largest size cable would have to be used. If the starting cable used on an old system which does not show particularly good efficiency is much smaller than this, it would probably be an advantage to replace it with larger cable, assuming, of course, that every other part of the system is in good condition and working properly. Table II shows the carrying capacity of various sized wires.

AUXILIARY ELECTRICAL EQUIPMENT

Electrical signaling devices are becoming more popular with regard to the noise signal and the stoppping and turning light which is placed at the rear of the car.

Electric Horns. The use of a storage battery which is of sufficient capacity for starting purposes and which is kept constantly charged by the lighting generator has made it possible to employ numerous auxiliary electrical devices. The electrical horn is the chief of these, and it has to a very large extent displaced warning devices of every other class. Two different types of electric horns are used, in both of which the sound is produced by the vibrations of a sheet-metal diaphragm several inches in diameter. The only

TABLE I

American Wire Gage (B. & S.)

No.	Diameter in Mils	Diameter in Mm.	Circular Mils	Ohms per 1000 Ft.	No.	Diameter in Mils	Diameter in Mm.	Circular Mils	Ohms per 1000 Ft.
0000	460.00	11.684	211600.0	.051	19	35.89	.912	1288.0	8.617
000	409.64	10.405	167805.0	.064	20	31.96	.812	1021.5	10.566
00	364.80	9.266	133079.4	.081	21	28.46	.723	810.1	13.323
0	324.95	8.254	105592.5	.102	22	25.35	.644	642.7	16.799
1	289.30	7.348	83694.2	.129	23	22.57	.573	509.5	21.185
2	257.63	6.544	66373.0	.163	24	20.10	.511	404.0	26.713
3	229.42	5.827	52634.0	.205	25	17.90	.455	320.4	33.684
4	204.31	5.189	41742.0	.259	26	15.94	.405	254.0	42.477
5	181.94	4.621	33102.0	.326	27	14.19	.361	201.5	53.563
6	162.02	4.115	26250.5	.411	28	12.64	.321	159.8	67.542
7	144.28	3.665	20816.0	.519	29	11.26	.286	126.7	85.170
8	128.49	3.264	16509.0	.654	30	10.03	.255	100.5	107.391
9	114.43	2.907	13094.0	.824	31	8.93	.277	79.7	135.402
10	101.89	2.588	10381.0	1.040	32	7.95	.202	63.2	170.765
11	90.74	2.305	8234.0	1.311	33	7.08	.108	50.1	215.312
12	80.81	2.053	6529.9	1.653	34	6.30	.160	39.7	271.583
13	71.96	1.828	5178.4	2.084	35	5.61	.143	31.5	342.433
14	64.08	1.628	4106.8	2.628	36	5.00	.127	25.0	431.712
15	57.07	1.450	3256.7	3.314	37	4.45	.113	19.8	544.287
16	50.82	1.291	2582.9	4.179	38	3.96	.101	15.7	686.511
17	45.26	1.150	2048.2	5.269	39	3.53	.090	12.5	865.046
18	40.30	1.024	1624.1	6.645	40	3.14	.080	9.9	1091.865

TABLE II

Carrying Capacity of Wires

B. & S. Gage	Circular Mils	Rubber Insulation Amperes	Other Insulation Amperes
18	1,624	3	5
16	2,583	6	8
14	4,107	12	16
12	6,530	17	23
10	10,380	24	32
8	16,510	33	46
6	26,250	46	65
5	33,100	54	77
4	41,740	65	92
3	52,630	76	110
2	66,370	90	131
1	83,690	107	156
0	105,500	127	185
00	133,100	150	220
000	167,800	177	262
0000	211,600	210	312

difference between the two forms lies in the method of causing this diaphragm to vibrate, one employing a small electric motor and the other a simple electric magnet. Fig. 15, which is a phantom view of the operating mechanism of a Klaxon horn, shows the first type. On the upper end of the armature shaft of the electric motor is fastened a toothed wheel which strikes the button in the center of the diaphragm and sets it vibrating at the rate of several thousand times per minute, giving rise to the raucous squawk which has come to be identified with automobile warning signals. A variation of the Klaxon horn is one in which the armature is in a horizontal position with the notched ratchet which rubs against a diaphragm as in the Klaxon vertical type. As shown in Fig. 16, which illustrates a section of the Apollo horn, this type is nothing more nor less than an ordinary buzzer on an enlarged scale. The armature of the electromagnet vibrates at high speed producing a sound by taps on the rod attached to the diaphragm.

Fig. 15. Phantom View Klaxon Horn

Care of the Electric Horn. As the operation of the electric horn is based upon exactly the same principles as the essentials of the

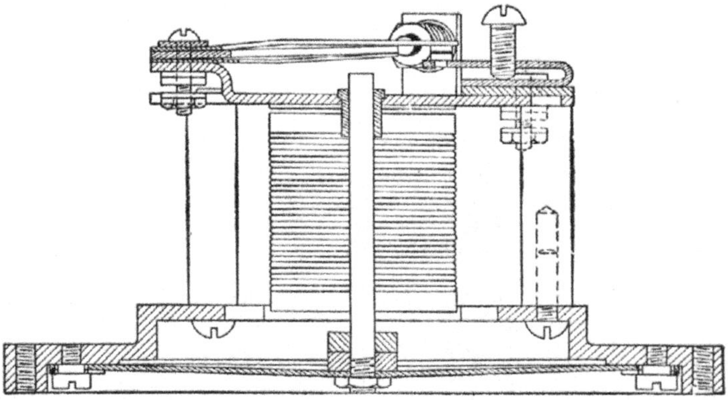

Fig. 16. Mechanism of Apollo Electric Horn (*Horseless Age*)

starting and lighting systems, the instructions given for the care and adjustment of the latter will apply to it as well. In the case of the motor-driven type of horn, the commutator and brushes of the motor will require attention from time to time. Failure to operate may be due to a broken connection at the horn or at the battery; ground in the circuit between it and the battery; brushes not bearing properly on the commutator; or an excess of oil and dirt on the latter. If the motor runs properly, but the horn produces either no sound or a very weak sound, the trouble will be due to the poor contact of the toothed wheel with the button on the diaphragm. This button is made glass hard to obviate wear at that point, but, in time, replacement of either the button or the toothed wheel or of both may be necessary.

While the brushes and commutator should be kept free of oil, the oiling of the small bearings of the horn should not be neglected. The best oil for this purpose is a thin grade of sewing machine oil. A few drops occasionally will be found sufficient to keep the armature turning freely in its bearings.

The attention required by the vibrating type of horn, of which there are many thousands in use, is very similar to that described for the battery cut-out and the voltage regulator. The contact points will require cleaning, truing up, and adjustment at intervals, and the spring may also need occasional attention. Failure to operate may be caused by a loose connection or break in the circuit, or by a lack of adjustment which causes the contacts to be held apart so that no current can pass through the winding of the electromagnet. A weak sound from this horn will result either from insufficient current or from lack of adjustment.

LIGHTING

For automobile headlights, side lamps, tail lamps, and general illumination, electric lighting has superseded all other systems. In the best electric-lighting systems the current is supplied by a dynamo driven constantly by the engine, with a storage battery auxiliary.

Incandescent Lamps. *Tungsten and Other Filaments.* Incandescent lamps are usually provided with tungsten filaments. These filaments are much shorter and much stronger than in standard lamps, a condition that is further contributed to by the necessities of low voltage and high amperage, which require short and thick rather than long and thin filaments. A good tungsten lamp will afford 1 candle power of illumination for each 1.2 watts of current.

Mazda Type. Fig. 17 shows the standard types of lamps generally used. These are Westinghouse Mazda lamps for 6 volts, those at the left being 15 c.p. headlight lamps; the next two, 6 c.p.

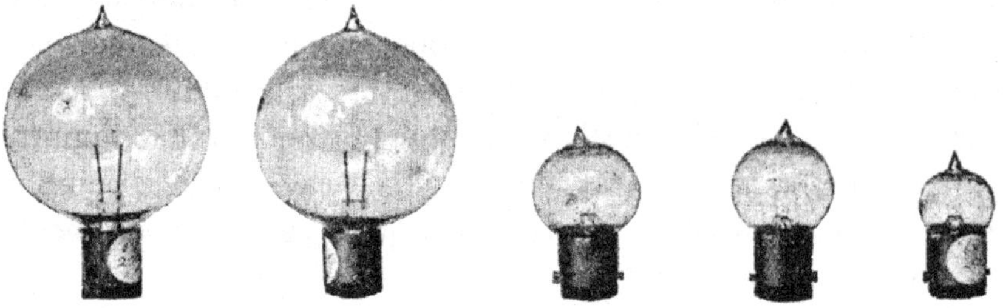

Fig. 17. Westinghouse Lamps—Head, Side, and Tail

side-light lamps; and the smallest one is a 2 c.p. size designed for the tail light, meter light, and for interior lighting of closed cars. At 6 volts, the 15 c.p. lamps require 2.5 amperes, the 6 c.p. side lights 1.25 amperes, or where 4 c.p. lamps are employed—a better size for the purpose—.85 ampere; the 2 c.p. lamps take .42 ampere. The larger lamps have the filament in the form of a spiral coil occupying the

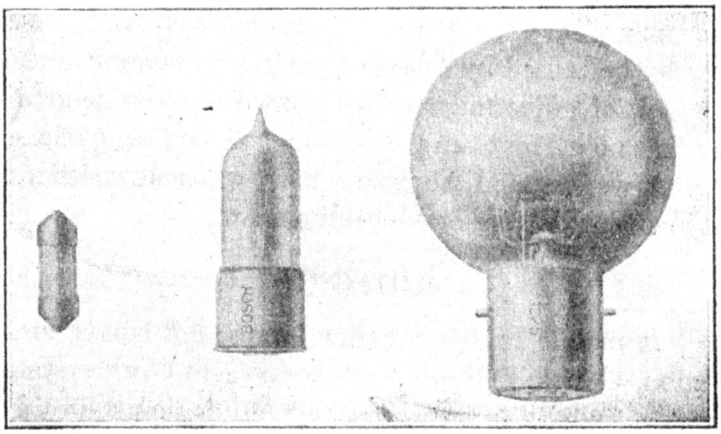

Fig. 18. Bosch Type Automobile Incandescent Bulbs

minimum space so that the whole source of light can be placed at the exact focus point of the paraboloidal reflector.

When lamp bulbs are bought for replacement, care should be taken to see that they are the correct type of bulb in regard to the voltage, amperage, and base contact. The lamps can be purchased with the single-contact and double-contact base but the two types are not interchangeable. In other words, the single-contact base bulbs will not burn in the double contact socket or vice versa.

Bosch Type. Fig. 18 shows the Bosch lamps, which are of special form. The headlight lamp at the right is of 25 c.p. and has the filament stretched horizontally across wire supports, while the side lamps of 8 c.p. have a loop of corrugated wire, and the tail lamp, of tubular form, a single filament running straight across it. Tail lamps are usually in series with the instrument lamp so that failure of the latter to light also indicates a failure of the tail lamp.

Lamp Voltages. When Edison was asked how he came to hit upon 110 volts as the standard for incandescent lighting, he said he "just guessed it." Evidently the 6-volt standard came about in pretty much the same way. It is not practicable to operate small lamps at a high voltage, as the lamp of that type requires a long slender filament. Many manufacturers of starting apparatus have deemed it necessary to employ a higher voltage, but the lamps are usually run at 6 volts, so that the batteries employed are accordingly some multiple of 3, as 6, 9, or 12 cells, giving 6, 12, 18, or 24 volts. Where more than three cells are used, this necessitates operating the lamps from a part of the battery, which is not advantageous, as it involves discharging the battery unevenly. As a battery capable of delivering current at 12 volts weighs and costs about 35 per cent more than one giving current at 6 volts and the attention required is greater, the lower voltage is generally favored.

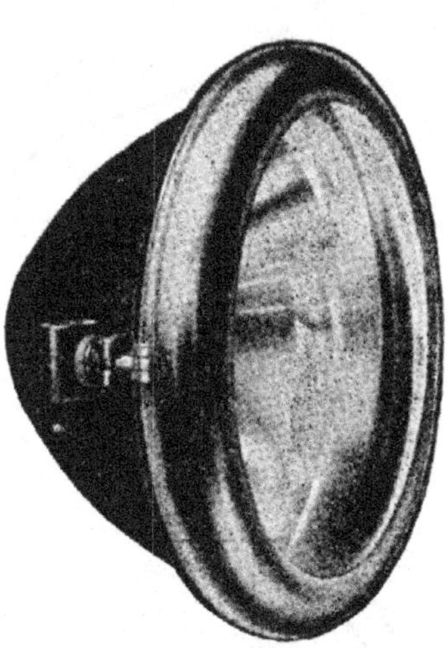

Fig. 19. Typical Electric Automobile Headlight

A recent addition to the lighting equipment on the automobile is the signal light placed at the rear of the car. Some of these lights not only have the stop signal but right-hand and left-hand turn direction signals as well. They are operated by a pull switch which is attached to the foot or service-brake pedal. Where the service and emergency brake operate together, care should be taken in installing the lights or when the emergency brake is set it will pull the switch and the signal light will burn continuously and discharge

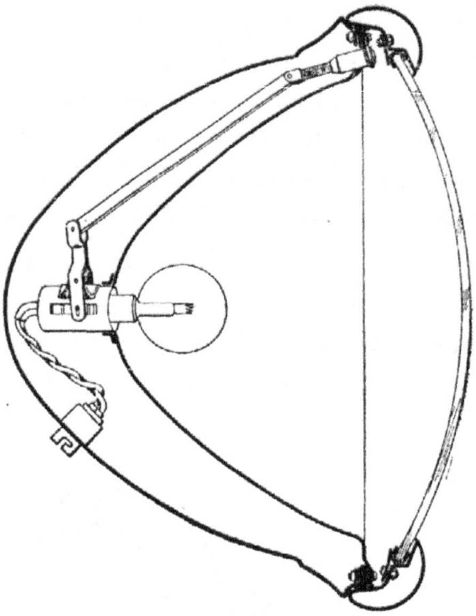

Fig. 20. Section of Fig. 19

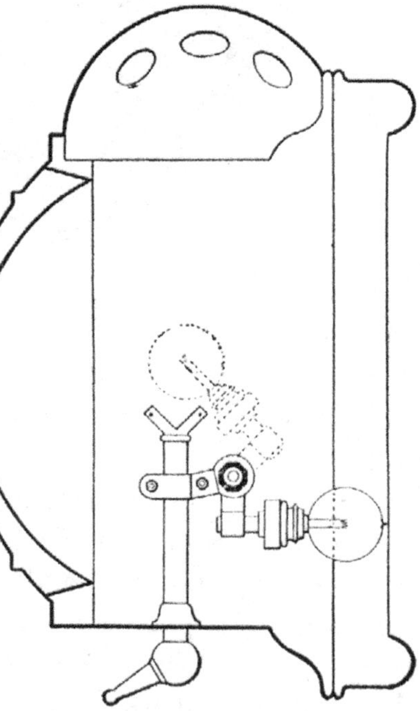

Fig. 21. Early Type of Reflector

the battery. If there is trouble in the stop-light system, it can usually be traced to the switch mechanism. The pull wire may have been broken and the lamp will not light. The pull-off spring may become weak, preventing the operation of the switch and cause the lamp to burn continuously. When the lamp is being installed, it is best to make connection to some current supply point outside of the battery. If attached to the battery, the wires will soon corrode and break off at the battery terminals.

Reflectors. Much attention has been directed to the problem of defining the best type of reflectors for automobile headlights, and the conditions of lighting by acetylene gas have been determined to be very different from those involved when electric lighting is used.

Parabolic Type. A typical electric headlight for automobile use is that illustrated in Fig. 19. The plain form affords a minimum tendency to catch dirt and mud and greatly simplifies cleaning. The position of the lamp is adjusted to give correct focus, as this is essential to give a properly projected beam of light ahead on the road.

Comparison of Parabolic with Lens Type. The reflector in the foregoing lamp is of the deeply parabolic metal type, illustrated in Fig. 20. The advantage of this type of reflector is that it intercepts a much larger proportion of the light rays from the lamp than the lens-mirror type of reflector, Fig. 21.

Types for Various Locations. Fig. 22 -*a*, -*b*, -*c*, -*d*, and -*i* show the usual types of lamps employed. These are, in the order given, an outside side lamp, flush-type side lamp, two types of electric tail lamps, and a cowl or dash lamp for illuminating the instruments, such as the ammeter, oil telltale, and the like. Fig. 22-*f* shows a magnetic trouble-hunting lamp, the base of which attaches itself to any metal part of the chassis.

Headlight Glare. The increased efficiency of electric headlights has brought with it, in far more aggravated degree, the blinding

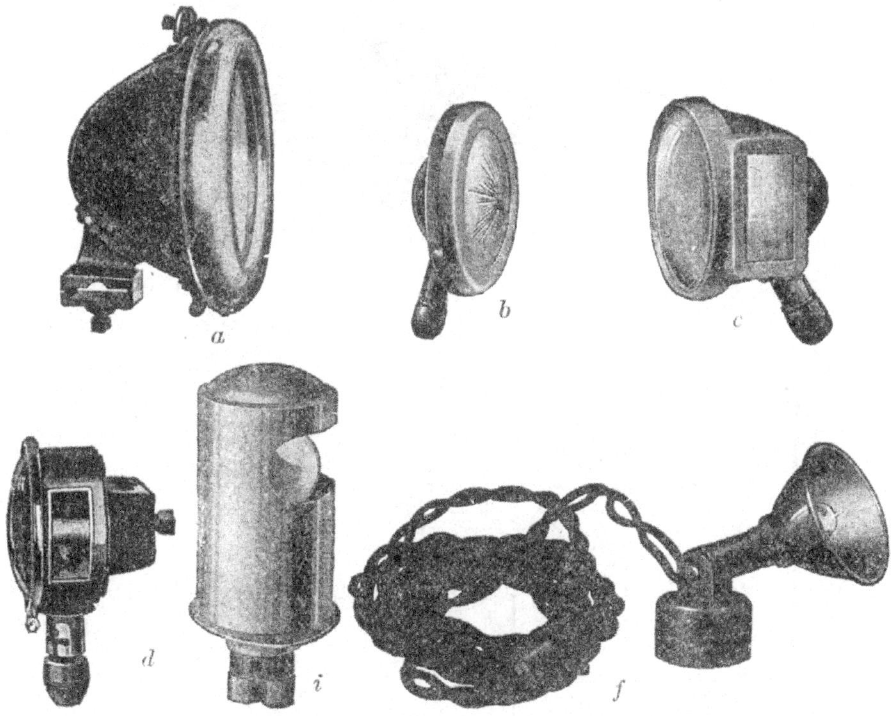

Fig. 22. Types of Side, Dash, Tail, and Trouble-Hunting Lamps

glare first experienced with the acetylene lamps. Originally, strong headlights bothered pedestrians; but since the introduction of electric lighting, they have been objected to most strongly by automobilists themselves, because to the driver of an automobile, the blinding glare from the headlights of an approaching car means not only annoyance but danger. Acuteness of vision is wholly destroyed for a period of thirty seconds or more during which only a slow-down to a walking pace will insure absolute safety to the automobilist, as a pedestrian or the usual black and lampless buggy are practically invisible.

Dimming Devices. Owing to the fact that glare and illumination are so closely related and that there is no objection to glare on

Fig. 23. Type of Headlight Dimming Switch

deserted country roads where the necessity for road illumination is greatest, permanently dimmed lights are naturally not practicable.

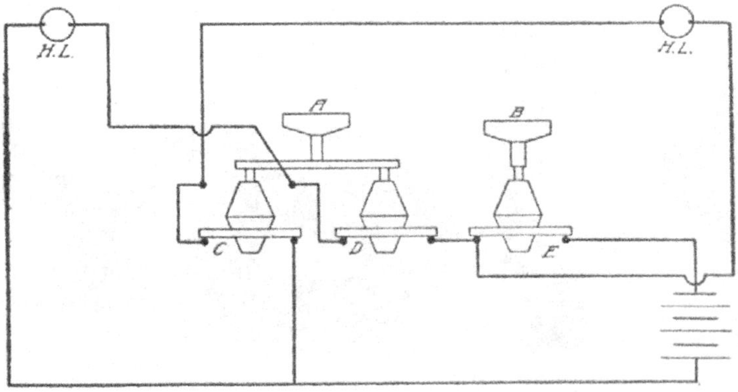

Fig. 24. Wiring Diagram of Parallel Control for Headlights
Courtesy of Horseless Age, New York City

What is required is a device under the control of the driver, so that either the full illuminating power of the head lamps or a subdued or dispersed light, free from glare, may be had as required.

A great many fundamentally different devices have been offered as a solution of the problem. While differing radically, practically

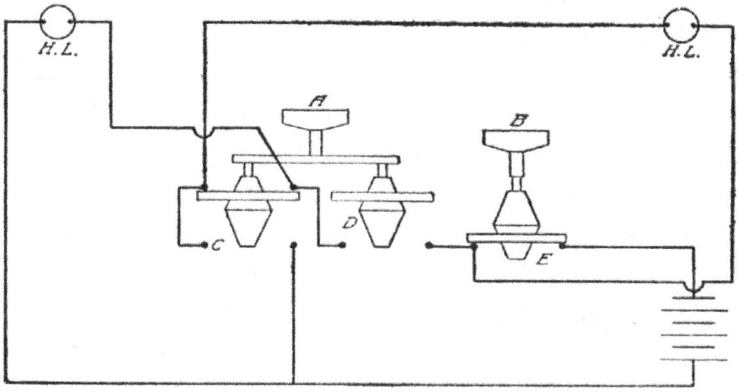

Fig. 25. Wiring Diagram of Series Control for Dimming Headlights
Courtesy of Horseless Age, New York City

all of them may be classed under two heads, i.e., electrical and mechanical.

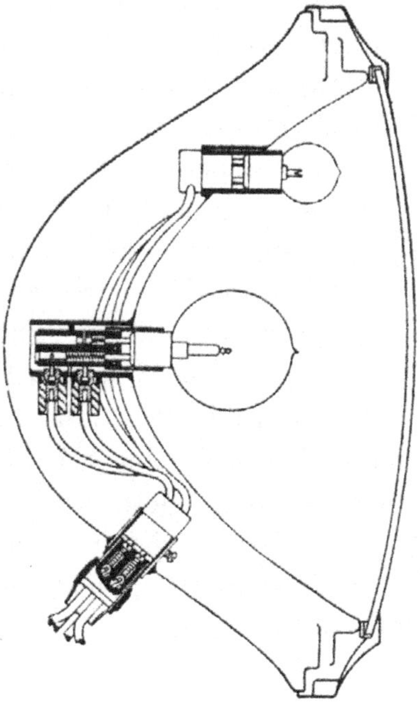

Fig. 26. Section of Hall Double Headlight

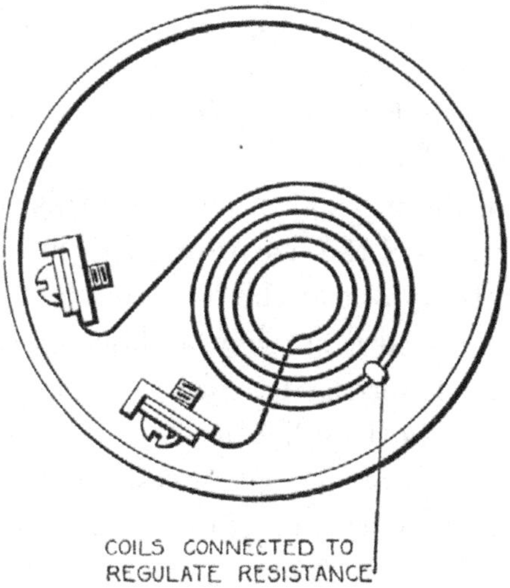

Fig. 27. Variable Dimming Resistance

Electrical Devices. One of the simplest of this class that has met with considerable favor is nothing more nor less than a *resistance* that may be inserted in the circuit of the headlights by turning a small switch, mounted either on the steering wheel or in some other easily accessible location. This cuts the voltage down and causes the lamps to burn a dull red, instead of the filaments being the dazzling white reached at full incandescence. A dimmer of this type is shown in Fig. 23. An equally simple and practical device is a switch to throw the headlights into series for a dim light and back into parallel again when full illumination is desired. With the series connection, the current must pass through both lamps successively and each bulb thus receives but half the voltage and, as even a comparatively slight drop in voltage causes the efficiency of an incandescent lamp to fall off very markedly, the same result is attained. It is equivalent to burning a 6-volt lamp on a 3-volt current. With the normal, or parallel, connection, the current flows through each lamp separately, and both receive the full voltage of the battery so that they burn at full brilliance. A switch of this kind is marketed by the Cutler-Hammer Company. Fig. 24 illustrates the connection for parallel arrangement, or full illumination, switches *D* and *B* being closed and the button *A* pulled out to make *C* contact with its lower set of connections. Fig. 25 shows the connections for series burning, effected by pulling out button *B* and pushing in *A*, this closing switch *E*, opening *D*, and contacting *C* with the upper connections.

The use of *two bulbs in each headlight* is also commonly resorted to, the method of effecting this being shown in Fig. 26. The second bulb is of the size ordinarily employed for side lights and is, moreover, entirely out of the focus of the reflector, so that the diminished light produced is entirely without glare and is mostly diverted downward.

Inserting a resistance in the headlight circuit is also done to dim the headlights. Fig. 27 shows a method of adjusting the intensity of the dimmers. The coils are connected together by a piece of wire as shown. This cuts out some of the resistance and causes the lights to burn brighter. Any desired intensity may be secured by experimenting with the position of this jumper.

NEW CADILLAC TWO-PASSENGER STANDARD COUPE

WIRING DIAGRAMS AND DATA SHEETS

INTRODUCTION

Since a great number of abbreviations are used on each of the data pages, a short explanation regarding the abbreviations will be in order and make perfectly clear the material which it is intended these data pages should convey. The first part dealing with the name of the car, model and year, or years, in which it was produced, is entirely clear, as is also the material with reference to the make of starting, generating and ignition systems.

Under the heading of "Generators," regulation may be of several kinds. The majority of cars, however, use the standard *third brush type* of regulation. Several makes of cars use methods of regulation other than the third brush type, such as the *relay regulation* type and that known as the *"inherent"* or *"bucking coil"* type.

The maximum charging rate is that with a normal setting of the regulator and taken when the generator is cold. The maximum armature speed at which the maximum charging rate is given is also shown on the data sheet.

A reference to the speed at which the relay points close is given. This speed is armature speed and where this cannot be given, the miles per hour of the car is given at the point of relay point closing. The air gap is always measured with the contact points closed.

On a large number of cars a *relay* is not used, the circuit being closed by the *ignition switch* when the switch is turned to the "on" position.

The *direction of rotation* of the starter is always determined from the *commutator end* and is thus recorded in all cases on the wiring diagram.

The standard method of indicating *firing orders* is used, and where given in connection with the V-type engine, the letters "R" and "L" refer to the *right* and *left* sides of the engine as determined from the *driver's position.* This corresponds in all cases to the material furnished by manufacturers. Some abbreviations are used with reference to *ignition timing.* The letters "TDC" refer to *top dead center* whenever used and when such a phrase as "TDC at retard" is used, it means that the *ignition advance* and *retard control* is set in *retard* position and the *contacts* are just starting to open when the *piston* of the firing cylinder is at *top dead center.* Methods of securing this position are carefully outlined. In some cases, the *number of degrees* or *inches* before *top dead center* are given, and this position can be determined from the marks on the *flywheel.* Wherever letters such as "IN-OP-6" appear, it means that these letters are stamped on the *rim* of the *flywheel* and that they must be brought in line with an *indicator* when the contacts start to open, the spark control being set at retard. In some cases, the letters "UD" are stamped on the rim of the flywheel. Also, "45" and "RET."

The bore and stroke of the engine are always given in inches and the taxable horsepower is also given on the data sheet. This material is useful when making out applications for State taxes and license plates. The *valve tappet clearance* is given in *thousandths* of an inch and should be *measured* only when the engine is *hot.*

In most cases, the make of the carburetor is given. In one case, a Zenith-Duplex carburetor is used and is indicated by "Zenith-Dup."

The width and diameter of the piston ring is given and the gap distance is also shown. This material is very useful when ordering oversized rings and also judging the amount of wear which has taken place on the piston rings where renewal of the rings is considered.

The *capacity* of the *oiling system* is given in *quarts* and the *type* indicated is based upon the standard nomenclature of *lubricating systems.* "Press" indicates a system where the *oil* is lifted from the *oil sump* by the *pump* and *forced up* to the *main bearings,* the overflow from these bearings running into oil troughs, which supply the *connecting rods* and other parts of the engine. The *pump* may be of either the *gear, vane* or *plunger* type. In some cases, the pressure is even carried to the *camshaft* and *piston pin bearings.* A recent modern tendency is to extend this pressure to the *hollow rocker arm* bearings on the overhead valve motors. Wherever the

WIRING DIAGRAMS AND DATA SHEETS

CHART OF ABBREVIATIONS

GENERAL

Amps.—Amperes
V.—Volts
C. P.—Candle Power
Bat. to Frame Con.—Battery to Frame Connection
U. S. L.—United States Lighting
R. & L. of Firing Order—Right and left determined from **driver's seat**
Max. Chg. Rate—Maximum Charging Rate
R. P. M.—Revolutions per Minute
M. P. H.—Miles per Hour
Thermo—Thermosyphon Cooling
Zenith-Dup.—Zenith-Duplex

IGNITION

T. D. C. at Retard—Top Dead Center; Spark Control Retarded
½ Advance at T. D. C.—Spark Control ½ Advanced, Piston at T. D. C.
12° past T. D. C.—Piston 12° past T. D. C. Spark Control Retarded
2½″ before T. D. C.—Piston 2½″, on Flywheel, before T. D. C.
"Ret" at Retard—Marks "Ret" on Flywheel
1″ B. H. C. on F. W.—"T. D. C." on Flywheel 1″ before T. D. C.
"Spark" at Rt. Sup.—"Spark" on Flywheel at Right Engine Support
"IN-OP-6" at Retard } Marks on Flywheel at Indicator, Spark Control Re-
"UD" at Retard } tarded
"45" at T. D. C.—Adv.—Marks on Flywheel at Indicator, Spark Control Advanced

LUBRICATION SYSTEM

Cap.—Capacity
Qt.—Quart
Press.—Pressure to Main Bearings
Splash.—Pressure Circulating Splash
Splash—Connecting Rod Dips

CLUTCH

Disc—Multiple disc, either wet or dry
Cone—Cone
R-Cone—Reverse Cone
Plate—Single or Double Plate Type

TRANSMISSION

D. G. & M.—Detroit Gear and Machine Company

REAR AXLE

Semi—Semi-Floating
¾ Flt.—Three-quarters Floating
Full Flt.—Full Floating

BRAKES

Front—Front Wheels
Rear—Rear Wheels
Hand—Hand Brake
Trans.—Transmission Brake

abbreviation "Splash. Press." appears, it indicates that the "Circulating System" is used, in which case the oil is merely circulated to the oil troughs, either by means of a *pump* or by the *flywheel*, no effort being made to maintain a pressure at any of the bearings.

The *size* and *type* of *clutch* is indicated, the size of the disc being shown in *inches* for the outside diameter and thickness. The word "disc" refers to the *multiple disc type* and may be either *dry* or *running* in oil. The word "plate" refers to the *single plate* type or *double plate* type of clutch. The word "cone" refers to the *cone* clutch and when preceded by the letter "R" means that it is of the *reversed cone* type. The *gear ratio* has to do with the *final drive* and, in all cases, the number given indicates the number of revolutions that the motor must make in order to turn the rear wheels one complete revolution when in high gear. A great many manufacturers give a choice between two gear ratios, the *lower gear ratio* generally being used with the *closed* and *heavier* models.

Wherever four-wheel brakes are used, the thickness, length, and width, of the brake lining will be given for each of the brakes. Under the heading of "Front" will be the dimensions of the brake lining given for front-wheel brakes; under the heading of "Rear" will be the dimensions for the brake linings given for brakes on the rear wheel and the dimensions for the "hand brakes" is given under that title. Where transmission brakes are used, the word "Transmission" will be written in either under hand brake or rear brake. Where there are no dimensions shown, the information is not available or the type of brake above the vacant space is not fitted on the car in question.

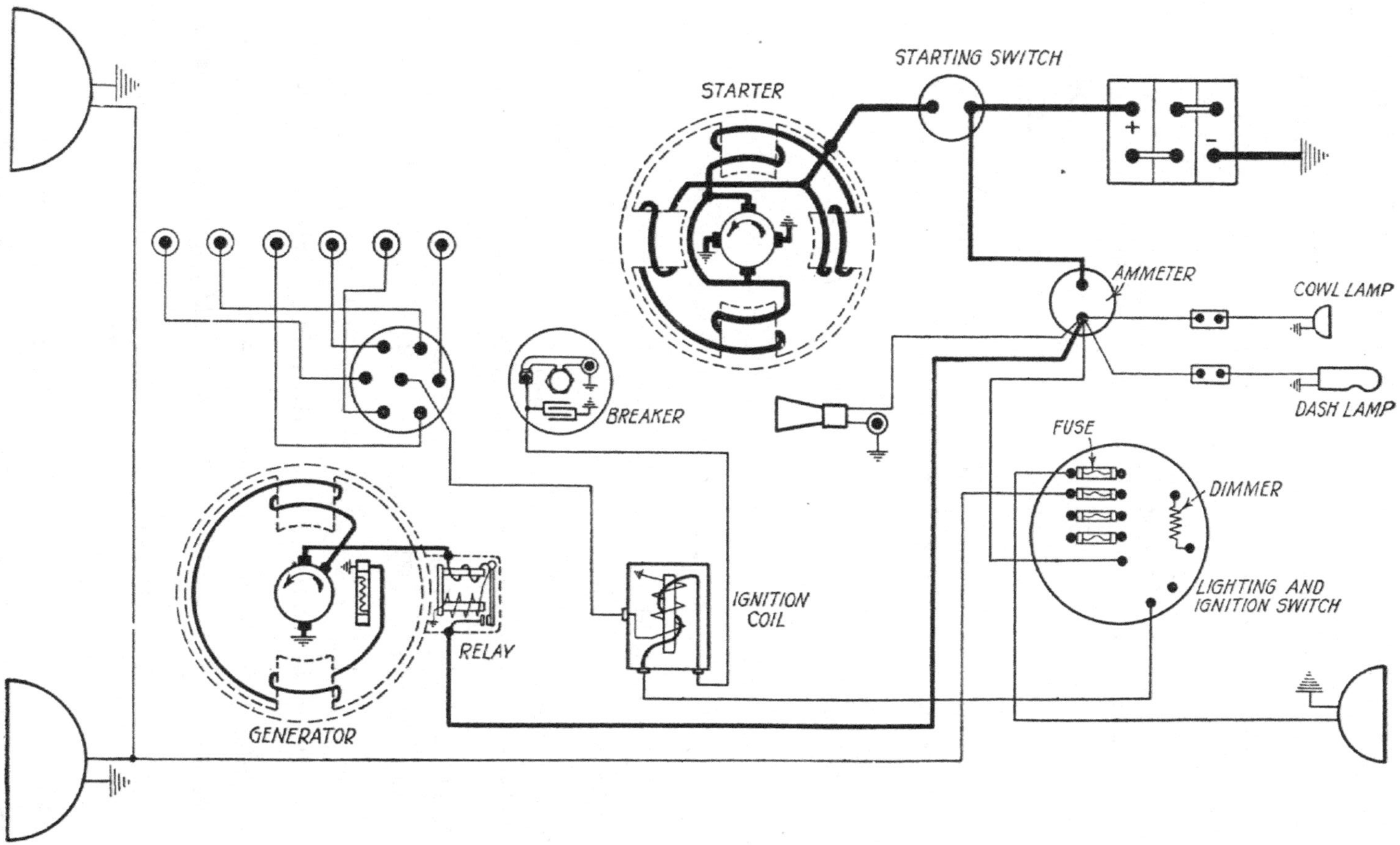

APPERSON WIRING DIAGRAM, 1923-1924, MODEL 6-23, REMY EQUIPMENT

APPERSON Model 6 Year 1923-24

Remy Starter & Generator Remy Ignition

GENERATOR Model	Regulation	Max. Chg. rate
No. 917 - L	Third Brush	20 Amps.

Amps. Cut-in	Volts Warm Test	R.P.M.	Amps. Cut-in	Volts Cold Test	R.P.M.
0	7.2	675	7	7.0-7.3	850
18-20	8.4	1900	9-12	7.7	1800

RELAY Model	Closes	Opens	Contact Gap	Air Gap
No. 265 - A	10 M P H	8 M P H	.010"-.016"	.015"

STARTING Motor Model	Drive	Rotation
No. 720 - L	Bendix	Anti-Clockwise

BATTERY Prest-O-L Type	Capacity	Bat. to Frame Con.
613 S. H. C.	6 Volts 100 Amps.	Negative

IGNITION Coil Model	DISTRIBUTOR Model	Breaker Contact Gap
No. 284 - L	No. 626 - H	.020" - .025"

FIRING ORDER	Ignition Timing	SPARK PLUG
1-4-2-6-3-5	T.D.C. at Retard	Size 7/8" Gap .025

LIGHTING	Headlamps	Dash & Tail	Side Lamps
Single Contact	21 C.P.	2 C.P.	2 C.P

ENGINE — Cylinder — Valve tappet clearance

Bore $3\frac{1}{8}$ Stroke $4\frac{1}{4}$ No. 6 1 Head Inlet .004 Exhaust .005

CARBURETOR	COOLING	PISTON RING	OILING SYSTEM
Stromberg - 1"	Thermo	3 1/8 X 3/16	Cap. 8qt. Type Force

CLUTCH	REAR AXLE	Gear Ratio	TRANSMISSION
Plate	3/4 Flt.	5.10	Own

BRAKES — Position — Brake lining width & length

Rear wheel Trans. or wheel W. Service 2 X 44 Hand $1\frac{3}{4}$ X 39

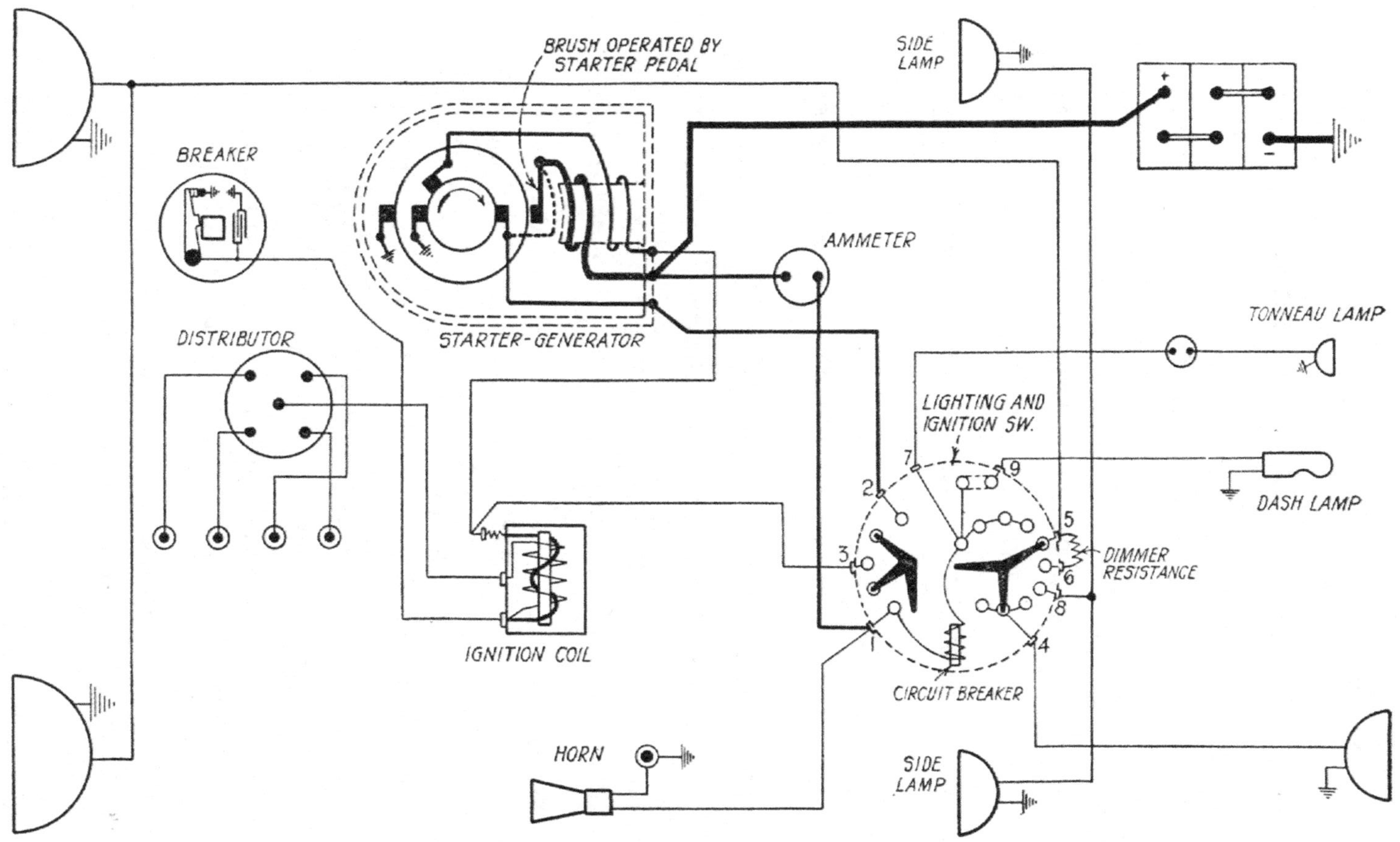

BUICK WIRING DIAGRAM, 1923-1924, MODEL 4, DELCO EQUIPMENT

B U I C K Model Four Year 1923-24

Delco Starter & Generator Delco Ignition

GENERATOR Model			Regulation		Max. Chg. rate
No. 251			Third Brush		18 Amps.
Amps. Cut-in	Volts Warm Test	R.P.M.	Amps. Cut-in	Volts Cold Test	R.P.M.
0		600	8-9		825
14-16		1500	12-15		1800

RELAY NONE USED. CONTACT MADE BY SWITCH.

STARTING Motor Model	Drive	Rotation
No. 251	Gear	Anti-Clockwise

BATTERY Exide Type	Capacity	Bat. to Frame Con.
3 - X C - 13 - 1	6 Volts 90 Amps.	Negative

IGNITION Coil Model	DISTRIBUTOR Model	Breaker Contact Gap
No. 2176	No. 251	.022" - .027"

FIRING ORDER	Ignition Timing	SPARK PLUG
1 - 3 - 4 - 2	"7°" at Retard	Size 7/8 Gap .030"

LIGHTING	Headlamps	Dash & Tail	Side Lamps
Single Contact	21 C.P.	2 C.P.	5 C.P.

ENGINE — Cylinder — Valve tappet clearance

Bore $3\frac{3}{8}$ Stroke $4\frac{3}{4}$ No. 4 I Head Inlet .009 Exhaust .009

CARBURETOR	COOLING	PISTON RING	OILING SYSTEM
Marvel 1 1/4"	Pump	3 3/8 X 3/16	Cap 4qt. Type Force

CLUTCH	REAR AXLE	Gear Ratio	TRANSMISSION
Disc 7 5/8	Semi	4.70	Own

BRAKES	Position	Brake lining width & length
Four wheel	Trans. or wheel W	Service $1\frac{3}{4}$ X 34 Hand $1\frac{3}{8}$ X $35\frac{1}{2}$

B U I C K Model Mast. 6 Year 1925

Delco Starter & Generator Delco Ignition

Regulation — Max. Chg. rate and speed

3rd Brush — 14-16 Amps. 20 M.P.H.

RELAY Air Gap — Contact Gap — Cut-in R.P.M.

None

BATTERY Exide Type 3 X 3 - 15-1 Volts 6 Amps 100

Bat. to Frame Con. Neg. CONTACT BREAKER Gap .020

Firing Order 1-4-2-6-3-5 Ignition Timing 7° Mark FL. Ret.

SPARK PLUG — ENGINE — Taxable Hp.

Size 7/8" Gap .020" Bore 3 3/8" Stroke 4 3/4" — 27

INTAKE VALVE TIMING — EXHAUST VALVE TIMING

Open 50 B.T.C. Close 68.2 A.T.C. Open 76 B.B.C. Close 32.2 A.T.C.

VALVE CLEARANCE Hot Intake .006" Exhaust .006"

CARBURETOR — COOLING SYSTEM — OILING SYSTEM

Marvel Type P Cap 4½ gal. Type Pres. Cap 5½ qts.

PISTON RING: Width 3/16" Diam. 3 3/8" Gap .004"-.012"

CLUTCH Disc 7 5"/8 X 5"/32 GEAR RATIO 4.7 AXLE F.F.

BRAKES

Front: 38 5"/8 X 2" X 3"/16 — Rear: 38 2"/8 X 2" X 3"/16 — Hand: 40 7"/16 X 1 5"/8 X 5"/32

Lighting — Headlights — Dash & Tail — Side Lamps

Single Contact 21 C.P. 3 C.P. 3 C.P.

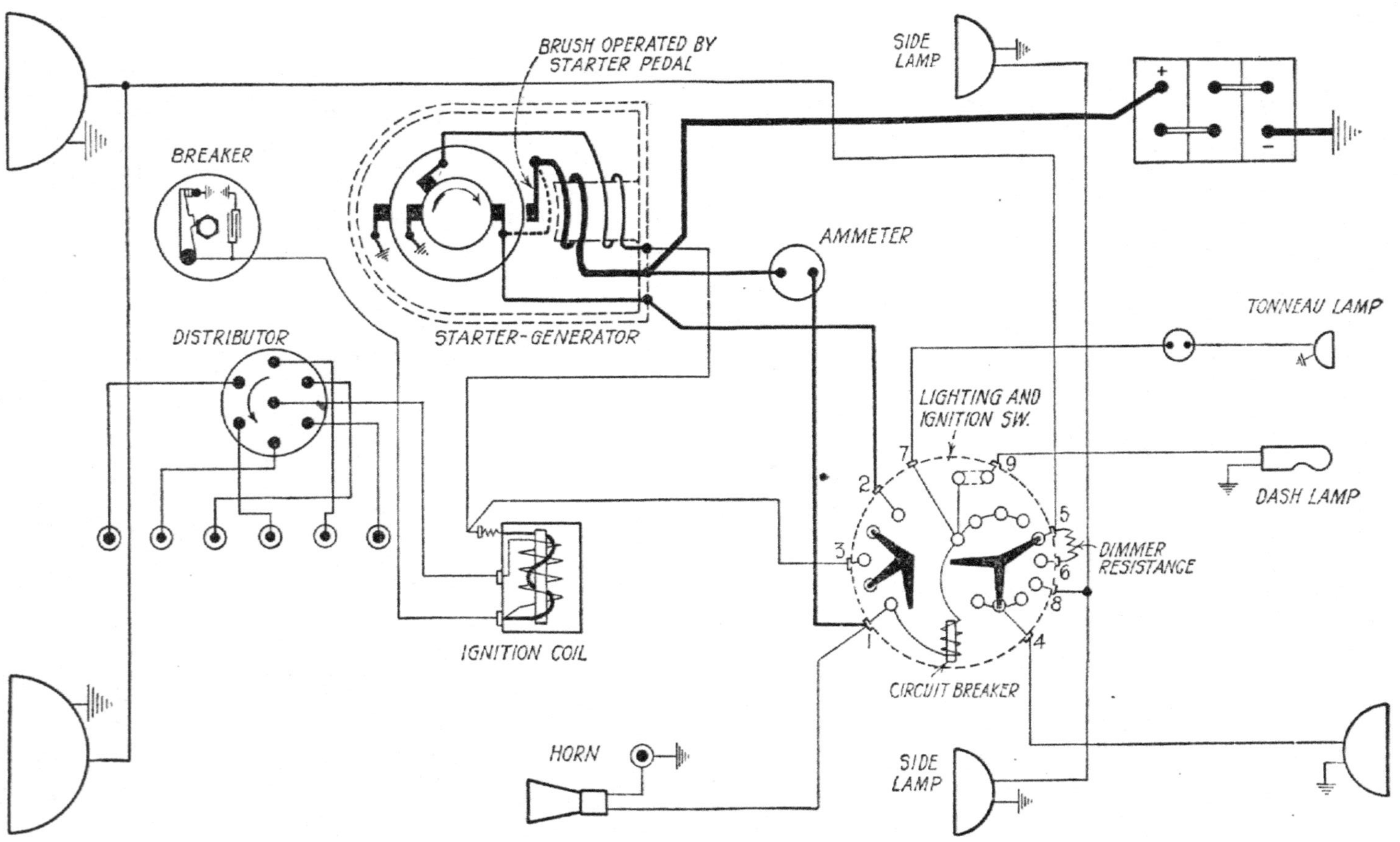

BUICK WIRING DIAGRAM, 1925, STANDARD SIX AND MASTER SIX, MODEL 1924-6-41 TO 1925-57

B U I C K Model Stan. 6 Year 1925

Delco Starter & Generator Delco Ignition

Regulation	Max. Chg. rate and speed
3rd Brush	20 Amps. 1500

RELAY Air Gap	Contact Gap	Cut-in R.P.M.
None		

BATTERY Exide Type 3-X3-13-1 Volts 6 Amps 90

Bat. to Frame Con. Neg. CONTACT BREAKER Gap .020"

Firing Order 1-4-2-6-3-5 Ignition Timing 7° Mark F L.Ret.

SPARK PLUG	ENGINE	Taxable Hp.
Size 7/8" Gap .020"	Bore 3" Stroke 4 1/4"	22

INTAKE VALVE TIMING	EXHAUST VALVE TIMING
Open 50 B.T.C. Close 68.2 A.B.C.	Open 76.8 B.B.C. Close 32.2 A.T.C.

VALVE CLEARANCE Hot Intake .008" Exhaust .008"

CARBURETOR	COOLING SYSTEM	OILING SYSTEM
Marvel	Type P Cap 3 gal.	Type Pres. Cap 4½ qts.

PISTON RING: Width 1/8" Diam. 3" Gap .004"-.012"

CLUTCH Disc 7 5"/8 X 5"/32 GEAR RATIO 4.9 AXLE 3/4 FL

BRAKES

Front	Rear	Hand
33 7"/8 X 1 3"/4 X 3"/16	33 7"/8 X 1 3"/4 X 3"/16	35 9"/16 X 1 3"/8 X 3"/16

Lighting	Headlights	Dash & Tail	Side Lamps
Single Contact	21 C.P.	2 C.P.	2 C.P.

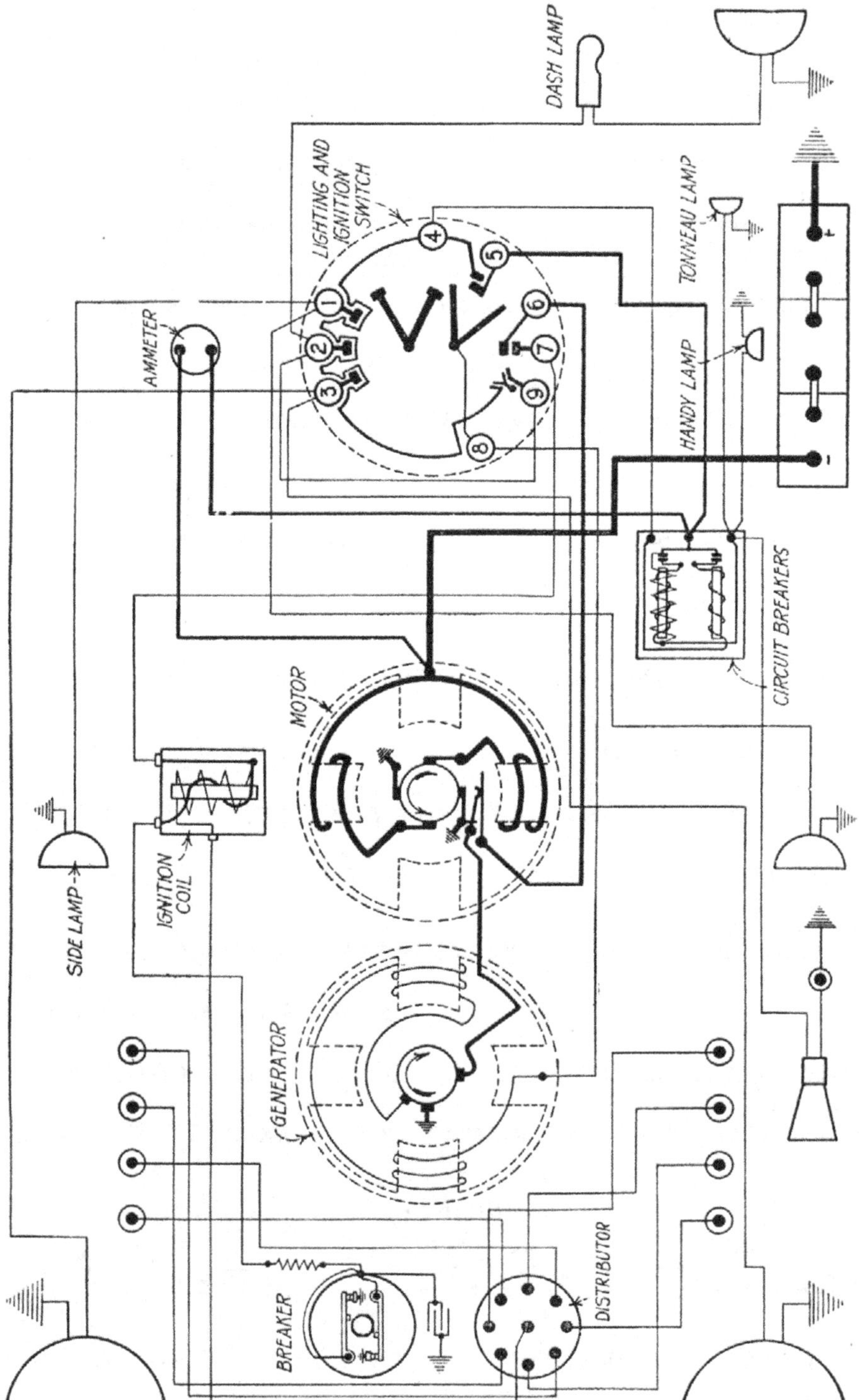

CADILLAC WIRING DIAGRAM, 1923, MODEL V-61, DELCO EQUIPMENT

C.A.D.I.L.L.A.C. Model V - 61 Year 1923

Delco Starter & Generator Delco Ignition

GENERATOR Model	Regulation	Max. Chg. rate
No. M. G. 98	Third Brush	17-20 Amps.

Amps. Cut-in	Volts Warm Test	R.P.M.	Amps. Cut-in	Volts Cold Test	R.P.M.
0	- - -	420	2	- - -	500
17-20	- - -	1500	15	- - -	1000

RELAY NONE USED. CONTACT MADE BY SWITCH.

STARTING Motor Model	Drive	Rotation
No. M. G. 98	Gears	Anti-Clockwise

BATTERY Exide Type Capacity Bat. to Frame Con.

3-LXRV-15-2 6 Volts 130 Amps. Positive

IGNITION Coil Model	DISTRIBUTOR Model	Breaker Contact Gap
No. 2115	No. 5239	.020"

FIRING ORDER	Ignition Timing	SPARK PLUG
1L-2R-3L-1R-4L-3R-2L-4R	1 3/4" before TDC	Size 7/8 Gap .023

LIGHTING	Headlamps	Dash & Tail	Side Lamps
Single Contact	6-8V-21 C.P.	3-4 V - 2 C.P.	6-8V-4 C.P.

ENGINE Cylinder Valve tappet clearance

Bore $3\frac{1}{8}$ Stroke $5\frac{1}{8}$ No. 8 L Head Inlet .004 Exhaust .006

CARBURETOR	COOLING	PISTON RING	OILING SYSTEM
Own 2"	Pump	3 1/8 X 3/16 Cap 7	Type F&S

CLUTCH	REAR AXLE	Gear Ratio	TRANSMISSION
Disc 7 3/4	Full	4.50 5.00	Own

BRAKES Position Brake lining width & length

Rear wheel Trans. or wheel W Service $2\frac{1}{2}$ X 54 Hand $2\frac{1}{2}$ X 51

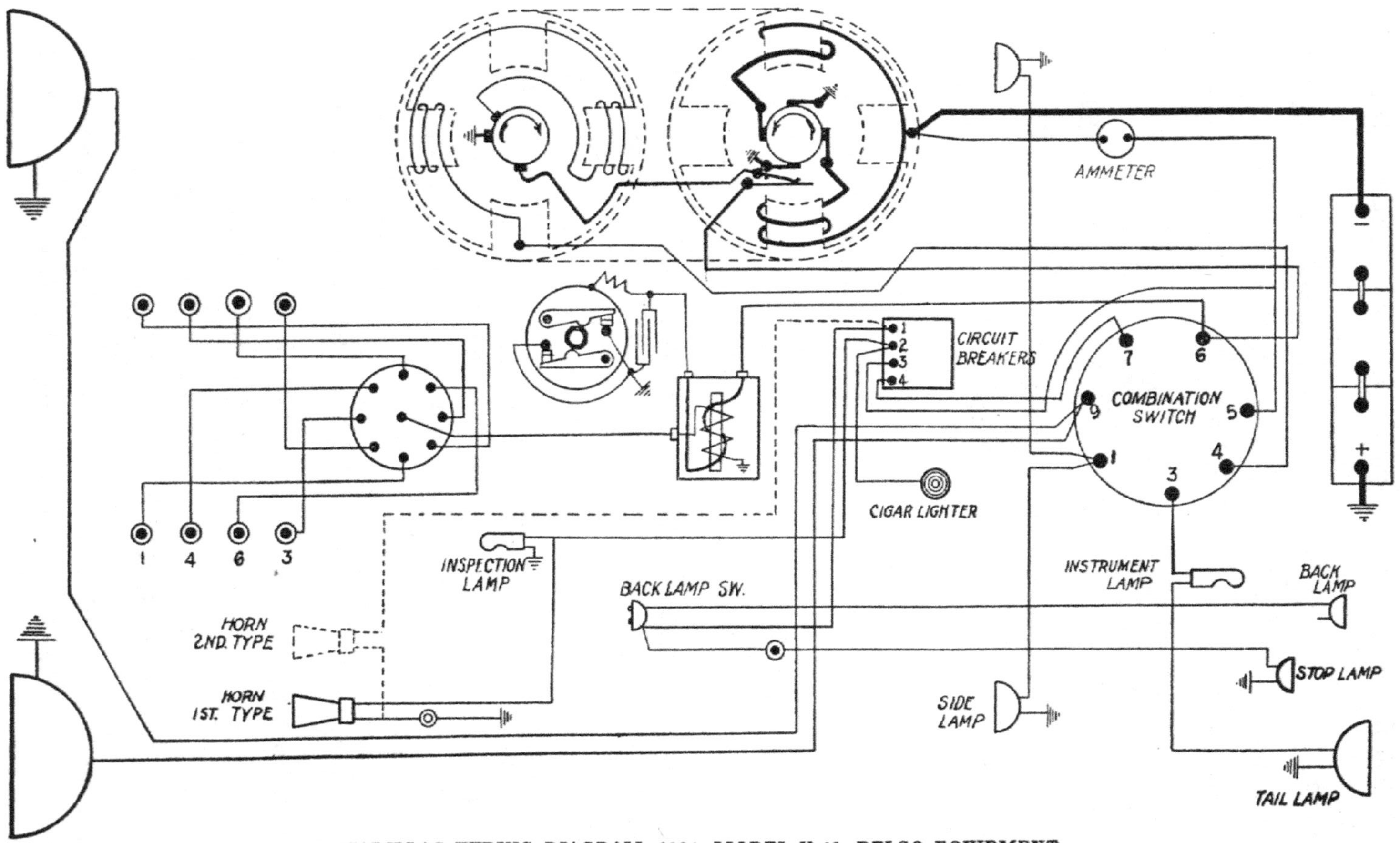

CADILLAC WIRING DIAGRAM, 1924, MODEL V-63, DELCO EQUIPMENT

C A D I L L A C Model V - 63 Year 1924

Delco Starter & Generator Delco Ignition

GENERATOR Model	Regulation	Max. Chg. rate
No. 98	Third Brush	19 Amps.

Amps. Cut-in	Volts Warm Test	R.P.M.	Amps. Cut-in	Volts Cold Test	R.P.M.
0	- - -	420	3	- - -	450
17-19	- - -	1500	17-19	- - -	1300

RELAY NONE USED. CONTACT MADE BY SWITCH.

STARTING Motor Model	Drive	Rotation
No. 98	Gear	Clockwise

BATTERY Exide Type Capacity Bat. to Frame Con.

3-LXRV-15-2 6 Volts 12 Amps. Positive

IGNITION Coil Model	DISTRIBUTOR Model	Breaker Contact Gap
No. 2179	No. 5239	.015" - .020"

FIRING ORDER	Ignition Timing	SPARK PLUG
1L-4R-4L-2L-3R-3L-2R-1R	T.D.C. at Retard	Size 7/8" Gap .023

LIGHTING	Headlamps	Dash & Tail	Side Lamps
Single Contact	6 V 21 C.P.	3 V - 2 C.P.	6 V - 2 C.P.

ENGINE Cylinder Valve tappet clearance

Bore $3\frac{1}{8}$ Stroke $5\frac{1}{8}$ No. 8 L Head Inlet .004 Exhaust .006

CARBURETOR	COOLING	PISTON RING	OILING SYSTEM
Own 2"	Pump	3 1/8 X 3/16	Cap. 7 qt. Type F&S

CLUTCH	REAR AXLE	Gear Ratio	TRANSMISSION
Disc 7 3/4	Full	4.50 5.00	Own

BRAKES Position Brake lining width & length

Four wheel Trans. or wheel W. Service $2\frac{1}{2}$ X 54 Hand $2\frac{1}{2}$ X 51

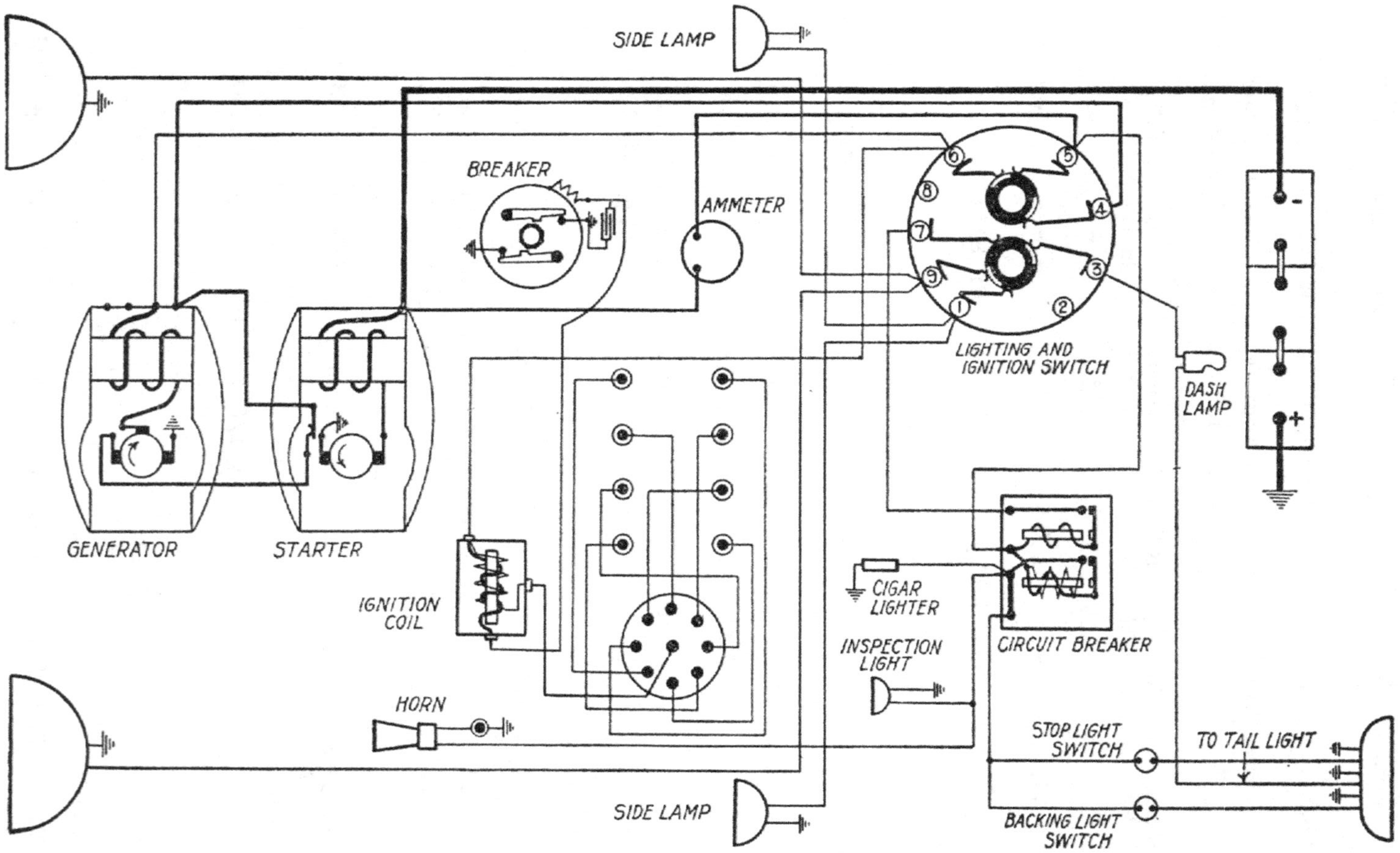

CADILLAC WIRING DIAGRAM, 1925, V-63, DELCO SYSTEM

C A D I L L A C Model V - 63 Year 1925

Delco Starter & Generator Delco Ignition

Regulation	Max. Chg. rate and speed
3rd Brush	17-19 Amps. 1500

RELAY Air Gap	Contact Gap	Cut-in R.P.M.
None		

BATTERY Exide Type 3-LXRV-15-2G Volts 6 Amps 130

Bat. to Frame Con. Pos. CONTACT BREAKER Gap .018"

Firing Order 1L-4R-4L-2L-3R-3L-2R-1R Ignition Timing 1 21/32" P.B.T.C. adv.

SPARK PLUG	ENGINE		Taxable Hp.
Size ___ Gap .023	Bore 3 1/8"	Stroke 5 1/8"	31

INTAKE VALVE TIMING		EXHAUST VALVE TIMING	
Open T.D.C.	Close 46°.7' A.B.C.	Open 46°.7' B.B.C.	Close T.D.C.

VALVE CLEARANCE Cold Intake .004" Exhaust .004"

CARBURETOR	COOLING SYSTEM		OILING SYSTEM	
Own	Type P	Cap 6 gal.	Type Pres.	Cap 7 qts.

PISTON RING: Width 3/16" Diam. 3 1/8" Gap

CLUTCH Disc GEAR RATIO 4.5 AXLE F F

BRAKES

Front	Rear	Hand

Lighting	Headlights	Dash & Tail	Side Lamps
Single Contact	21 C.P.	In Series 2 C.P.	3 C.P.

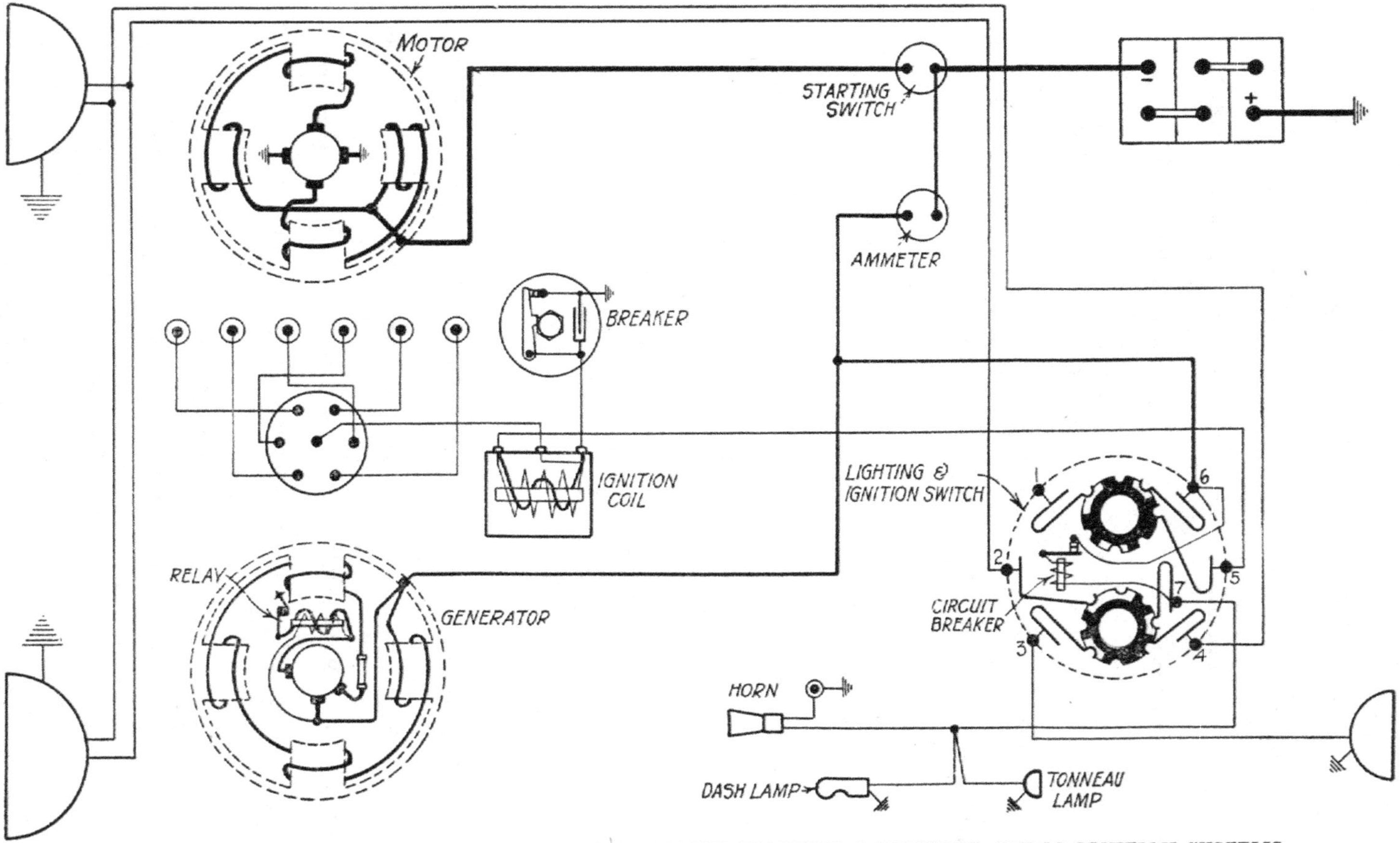

CASE WIRING DIAGRAM, 1922, MODEL V, WESTINGHOUSE STARTING & LIGHTING, DELCO IGNITION SYSTEMS

C A S E *Model V Year 1922

Westinghouse Starter & Generator Delco Ignition

GENERATOR Model	Regulation	Max. Chg. rate
No. 760	Third Brush	17 Amps.

Amps. Cut-in	Volts Warm Test	R.P.M.	Amps. Cut-in	Volts Cold Test	R.P.M.
12-14	7.5	1250	0	6.5	500
7-11	6.5-7.5	2500	17	7.5	1250

RELAY Model	Closes	Opens	Contact Gap	Air Gap
301210 - A	10 M P H	8 M P H	.013"-.018"	.031

STARTING Motor Model	Drive	Rotation
No. 710	Bendix	Anti-Clockwise

BATTERY Willard Type Capacity Bat. to Frame Con.

S 8 R G 6 Volts 120 Amps. Positive

IGNITION Coil Model	DISTRIBUTOR Model	Breaker Contact Gap
No. 2159	No. 5175	.018

FIRING ORDER	Ignition Timing	SPARK PLUG
1-5-3-6-2-4	"1-6" Sp 1/3 Adv.	Size 7/8R Gap .025

LIGHTING	Headlamps	Dash & Tail	Side Lamps
Single Contact	20 C.P.	2 C.P.	4 C.P.

ENGINE	Cylinder	Valve tappet clearance
Bore $3\frac{1}{2}$ Stroke $5\frac{1}{4}$	No. 6 L Head	Inlet .... Exhaust ....

CARBURETOR	COOLING	PISTON RING	OILING SYSTEM
Rayfield 1 1/4	Pump	3 1/2 X 3/16	Cap 6 Type F&S

CLUTCH	REAR AXLE	Gear Ratio	TRANSMISSION
Disc 10"	3/4 Flt	4.45	Grant-Lees

BRAKES	Position	Brake lining width & length
Rear wheel	Trans. or wheel W	Service $2\frac{1}{2}$ X 48 Hand $2\frac{1}{4}$ X $46\frac{1}{2}$

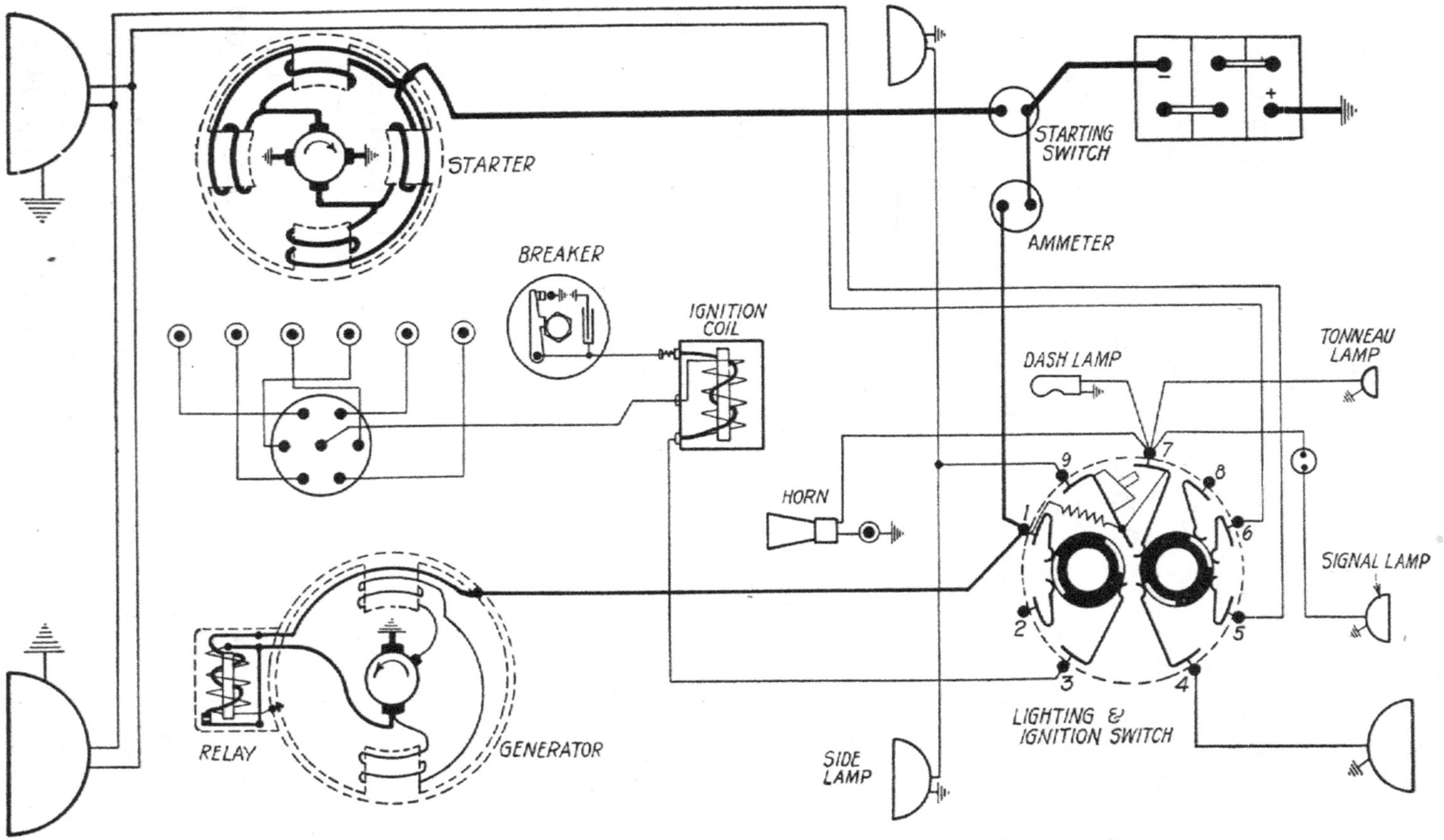

CASE WIRING DIAGRAM, 1922-1924, MODELS W, AND Y, DELCO EQUIPMENT

C.A.S.E. Model W, and Y Year 1922-24

Delco Starter & Generator Delco Ignition

GENERATOR Model			Regulation		Max. Chg. rate
No. 165			Third Brush		17 Amps.
Amps. Cut-in	Volts Warm Test	R.P.M.	Amps. Cut-in	Volts Cold Test	R.P.M.
0	7.0	650	4.5	7.0	800
13-17	7.5	1600	9-12	7.5	2300

RELAY Model	Closes	Opens	Contact Gap	Air Gap
No. 165	9 M P H	8 M P H	.046"-.062"	.030

STARTING Motor Model	Drive	Rotation
No. 208	Bendix	Clockwise

BATTERY Willard Type Capacity Bat. to Frame Con.

S J R G - 5 6 Volts 135 Amps. Positive

IGNITION Coil Model	DISTRIBUTOR Model	Breaker Contact Gap
No. 2159	No. 5251-5252	.020

FIRING ORDER	Ignition Timing	SPARK PLUG
1-5-3-6-2-4	T.D.C. at Retard	Size 7/8R Gap .030

LIGHTING	Headlamps	Dash & Tail	Side Lamps
Single Contact	21 C.P.	2 C.P.	4 C.P.

ENGINE Cylinder Valve tappet clearance

Bore $3\frac{5}{8}$ Stroke $5\frac{1}{4}$ No. 6 L Head Inlet .004 Exhaust .006

CARBURETOR	COOLING	PISTON RING	OILING SYSTEM
Rayfield 1 1/4"	Pump	3 5/8 X 3/16	Cap. 6qt. Type F & S

CLUTCH	REAR AXLE	Gear Ratio	TRANSMISSION
Disc 8"	3/4 Floating	4.45 4.70	Own 3 Speed

BRAKES Position Brake lining width & length

Rear wheel Trans. or wheel W Service $2\frac{1}{2}$ X 48 Hand $2\frac{1}{4}$ X $46\frac{1}{2}$

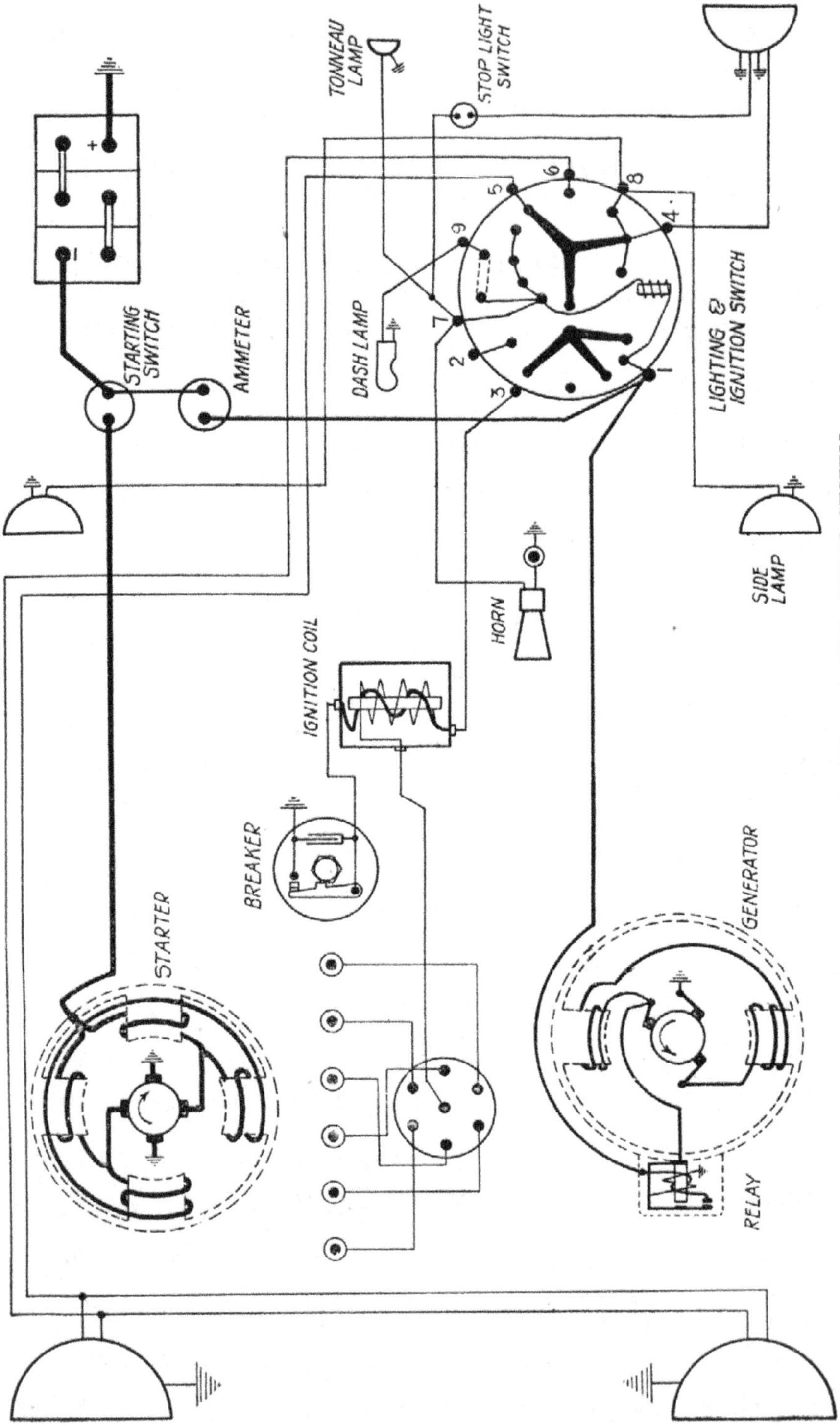

CASE WIRING DIAGRAM, 1925, J.I.C., DELCO SYSTEM

CASE Model J. I. C. Year 1925

Delco Starter & Generator Delco Ignition

Regulation	Max. Chg. rate and speed
3rd Brush	13-17 Amps. 1600

RELAY Air Gap	Contact Gap	Cut-in R.P.M.
.025"-.035"	.046"-.062"	650 or 6-8 M.P.H.

BATTERY Willard Type S J R N - 4 Volts 6 Amps 127

Bat. to Frame Con. Pos. CONTACT BREAKER Gap .020"

Firing Order 1-5-3-6-2-4 Ignition Timing T.D.C. Ret.

SPARK PLUG	ENGINE	Taxable Hp.
Size 7/8" Gap .025"	Bore 3 3/8" Stroke 4 1/2"	27

INTAKE VALVE TIMING	EXHAUST VALVE TIMING
Open 12° A.T.C. Close 40° A.B.C.	Open 40° B.B.C. Close 8° A.T.C.

VALVE CLEARANCE Hot Intake .004" Exhaust .006"

CARBURETOR	COOLING SYSTEM	OILING SYSTEM
Schebler	Type P Cap $3\frac{1}{2}$ gal.	Type Sp.Pr. Cap 6 qts.

PISTON RING: Width 3/16" Diam. 3 3/8" Gap .002"

CLUTCH Disc 8 $\frac{7"}{16}$ X $\frac{5"}{32}$ GEAR RATIO 4.9 AXLE Semi F

BRAKES

Front	Rear	Hand
35" X $2\frac{1}{2}$" X $\frac{5"}{32}$	$35\frac{1}{4}$" X $2\frac{1}{2}$" X $\frac{5"}{32}$	Trans. 20" X $2\frac{1}{2}$" X $\frac{5"}{32}$

Lighting	Headlights	Dash & Tail	Side Lamps
Single Contact	21 C.P.	2 C.P.	2 C.P.

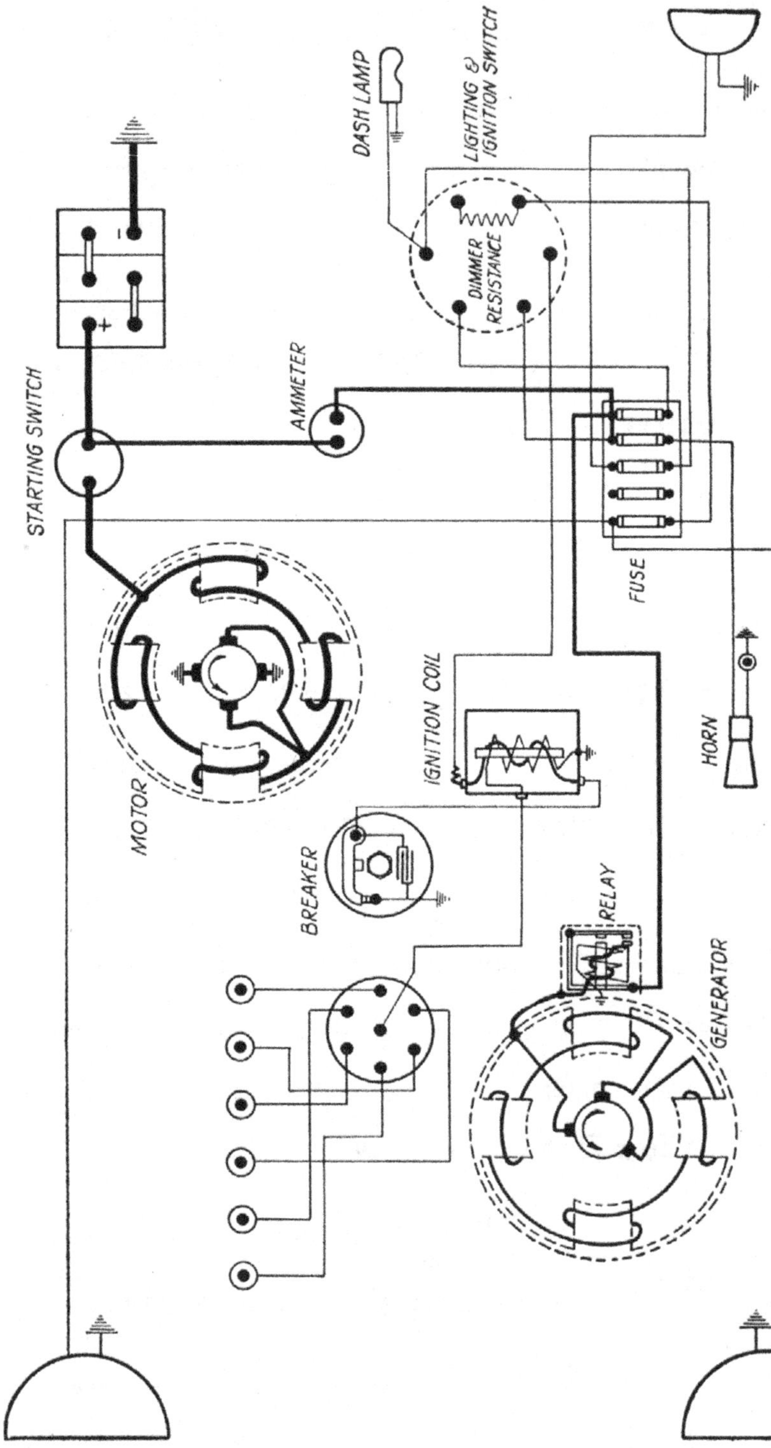

CHALMERS WIRING DIAGRAM, 1923, MODELS B & C, AUTO-LITE EQUIPMENT

CHALMERS Model 35 B & C Year 1923

Auto-Lite Starter & Generator Auto-Lite Ignition

GENERATOR Model	Regulation	Max. Chg. rate
G J - 1057	Third Brush	16 Amps.

Amps. Cut-in	Volts Warm Test	R.P.M.	Amps. Cut-in	Volts Cold Test	R.P.M.
1	6.5	600	8	7.6	800
16	8.3	1500	10	8.1	4500

RELAY Model	Closes	Opens	Contact Gap	Air Gap
No. CB - 1072	9 M P H	7 M P H	.025"-.035"	.010"

STARTING Motor Model	Drive	Rotation
No. M H - 1003	Bendix	Anti-Clockwise

BATTERY U S L Type 3 - C V X - 6 X Capacity 6 Volts 106 Amps. Bat. to Frame Con. Negative

IGNITION Coil Model	DISTRIBUTOR Model	Breaker Contact Gap
I G 4031	No. G J - 1057	.015"-.020"

FIRING ORDER	Ignition Timing	SPARK PLUG
1-4-2-6-3-5	1 7/8" past T D C	Size 7/8" Gap .025

LIGHTING	Headlamps	Dash & Tail	Side Lamps
Single Contact	21 C.P.	2 C.P.	C.P.

ENGINE — Bore 3¼ Stroke 4½ — Cylinder No. 6 L Head — Valve tappet clearance Inlet .004 Exhaust .006

CARBURETOR	COOLING	PISTON RING	OILING SYSTEM
Stromberg 1¼"	Thermo	3 1/4 X 3/16	Cap. 7 qt. Type F & S

CLUTCH	REAR AXLE	Gear Ratio	TRANSMISSION
Disc	Semi	5.18	Warner - Gear

BRAKES	Position	Brake lining width & length
Rear wheel	Trans. or wheel W	Service 2 X 43½ Hand 1½ X 41¼

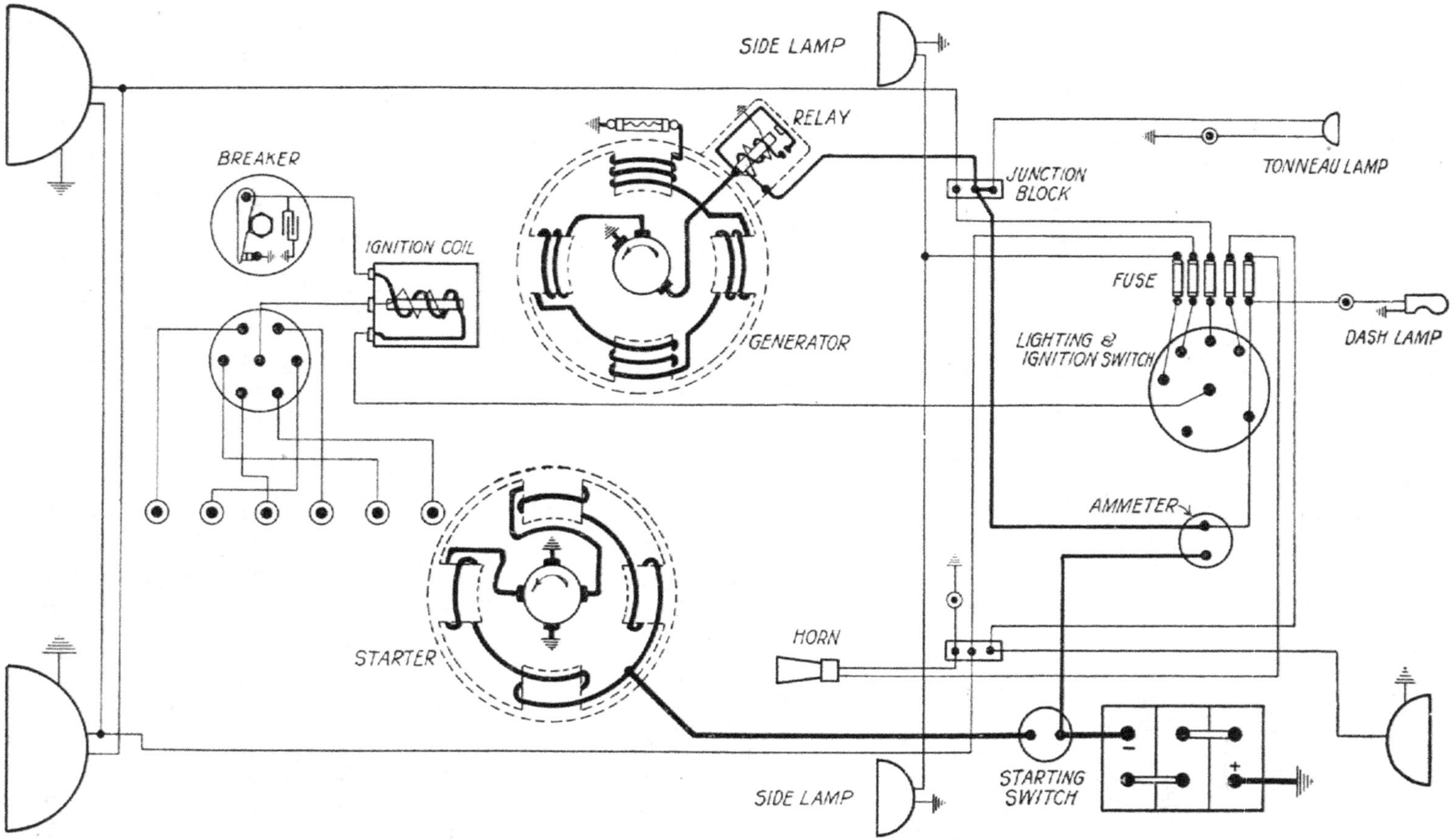

CHANDLER WIRING DIAGRAM, 1923-1924, MODEL "SIX," BOSCH EQUIPMENT

C H A N D L E R Model Six Year 1923-24

Bosch Starter & Generator Bosch Ignition

GENERATOR Model	Regulation	Max. Chg. rate
No. 1001	Third Brush	14 Amps.

Amps. Cut-in	Volts Warm Test	R.P.M.	Amps. Cut-in	Volts Cold Test	R.P.M.
0	6.0	600	7	6.9	800
9.8	7.3	1600	13.7	7.6	1600

RELAY Model	Closes	Opens	Contact Gap	Air Gap
No. CR-1000	8 M P H	6 M P H	.015"	.010"

STARTING Motor Model	Drive	Rotation
No. 935	Bendix	Anti-Clockwise

BATTERY Prest-O-L. Type Capacity Bat. to Frame Con.

613 - R H N 6 Volts 100 Amps. Positive

IGNITION Coil Model	DISTRIBUTOR Model	Breaker Contact Gap
No. T C - 20	No. T - 6158	.020

FIRING ORDER	Ignition Timing	SPARK PLUG
1-5-3-6-2-4	3/4" past T D C	Size 7/8" Gap .020

LIGHTING	Headlamps	Dash & Tail	Side Lamps
Single Contact	21 C.P.	2 C.P.	4 C.P.

ENGINE Cylinder Valve tappet clearance

Bore $3\frac{1}{2}$ Stroke 5 No. 6 L Head Inlet .004 Exhaust .004

CARBURETOR	COOLING	PISTON RING	OILING SYSTEM
Stromberg	Pump	$3\frac{1}{2}$ X 3/16	Cap. 8qt. Type F&S

CLUTCH	REAR AXLE	Gear Ratio	TRANSMISSION
Disc	Full	4.40	Own - Traffic

BRAKES Position Brake lining width & length

Rear wheel Trans. or wheel T Service 2 X $20\frac{1}{2}$ Hand $1\frac{3}{4}$ X 24

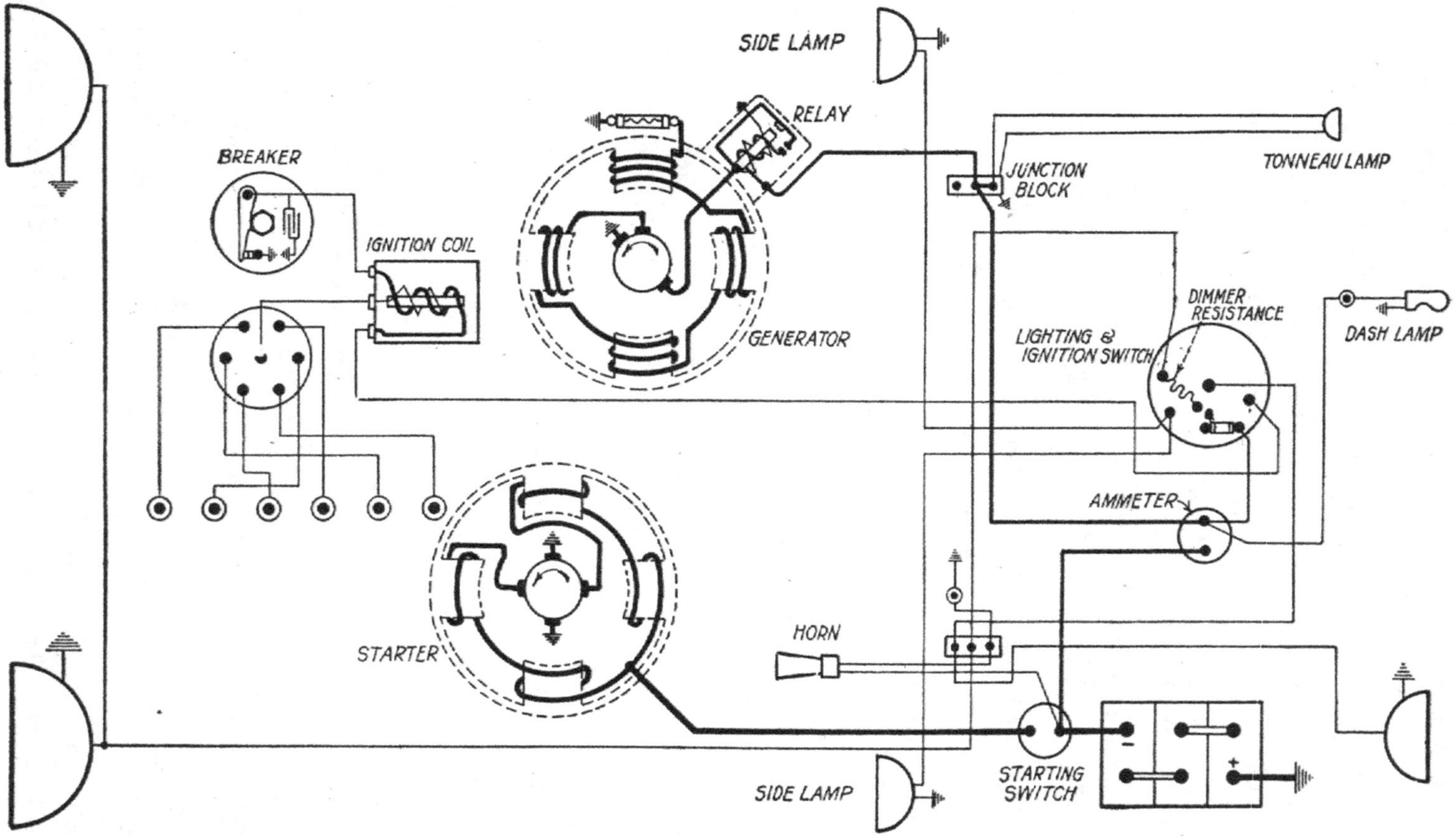

CHANDLER WIRING DIAGRAM, 1925, AMERICAN BOSCH SYSTEM

CHANDLER Model 1925 Year 1925

American Bosch Starter & Generator American Bosch Ignition

Regulation	Max. Chg. rate and speed
3rd Brush	13-17 Amps. 1600

RELAY Air Gap	Contact Gap	Cut-in R.P.M.
.010"	.015"	500 or 6-8 M.P.H.

BATTERY Prest-O-Lite Type 613 R.H.N. Volts 6 Amps. 100

Bat. to Frame Con. Pos. CONTACT BREAKER Gap .020"

Firing Order 1-5-3-6-2-4 Ignition Timing On application

SPARK PLUG		ENGINE		Taxable Hp.
Size 7/8"	Gap .020"	Bore 3 1/2"	Stroke 5"	29

INTAKE VALVE TIMING		EXHAUST VALVE TIMING	
Open 2° B.T.C.	Close 52° A.B.C.	Open 53° B.B.C.	Close 16° A.T.C.

VALVE CLEARANCE Hot Intake .004" Exhaust .004"

CARBURETOR	COOLING SYSTEM		OILING SYSTEM	
Schebler	Type P	Cap 4.6 gal.	Type Pres.	Cap 8 qts.

PISTON RING: Width 1/8" Diam. 3 1/2" Gap

CLUTCH Disc GEAR RATIO 4.4 AXLE 3/4 F

BRAKES

Front	Rear	Hand

Lighting	Headlights	Dash & Tail	Side Lamps
Single Contact	21 C.P.	2 C.P.	C.P.

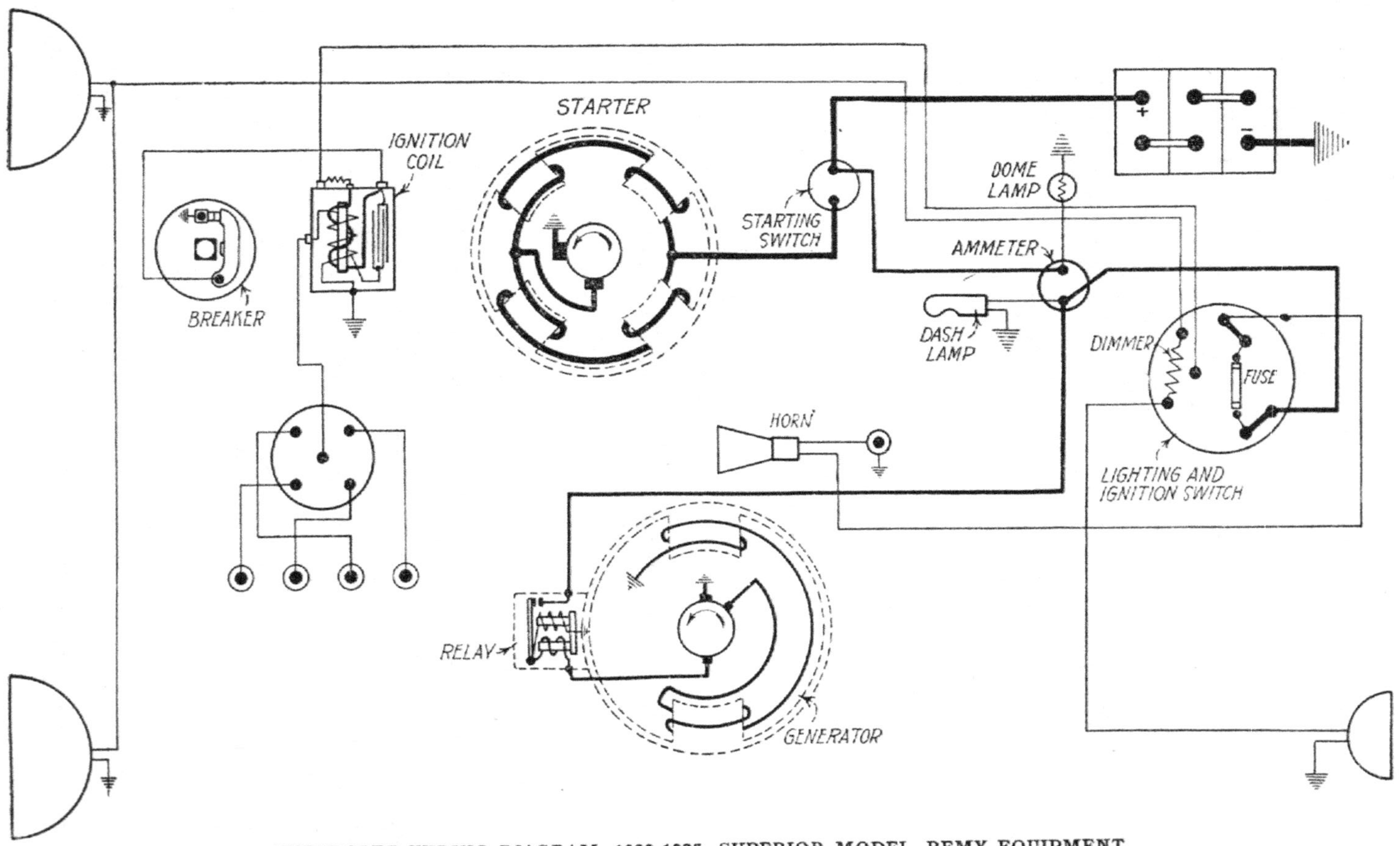

CHEVROLET WIRING DIAGRAM, 1923-1925, SUPERIOR MODEL, REMY EQUIPMENT

CHEVROLET Model Superior Year 1923-24

Remy Starter & Generator Remy Ignition

GENERATOR Model	Regulation	Max. Chg. rate
No. 950 - A	Third Brush	18 Amps.

Amps. Cut-in	Volts Warm Test	R.P.M.	Amps. Cut-in	Volts Cold Test	R.P.M.
0	7.2	750	10	7.8	1030
16	8.3	1800	16	8.3	1725

RELAY Model	Closes	Opens	Contact Gap	Air Gap
No. 265 - A	10 M P H	9 M P H	.025"-.030"	.020"

STARTING Motor Model	Drive	Rotation
No. 711 - B	Bendix	Anti-Clockwise

BATTERY Exide Type Capacity Bat. to Frame Con.

3 X 3 - 13 - 1 6 Volts 90 Amps. Negative

IGNITION Coil Model	DISTRIBUTOR Model	Breaker Contact Gap
No. 284 - L	No. 366 - P	.030"

FIRING ORDER	Ignition Timing	SPARK PLUG
1-2-4-3	T.D.C. at Retard	Size 7/8L Gap .022"

LIGHTING	Headlamps	Dash & Tail	Side Lamps
Single Contact	21 C.P.	2 C.P.	4 C.P.

ENGINE Cylinder Valve tappet clearance

Bore $3\frac{11}{16}$ Stroke 4 No. 4 I Head Inlet .004 Exhaust .004

CARBURETOR	COOLING	PISTON RING	OILING SYSTEM
Zenith 1"	Pump	3 11/16X3/16	Cap. 4qt. Type C.S.

CLUTCH	REAR AXLE	Gear Ratio	TRANSMISSION
Cone 3 11/16	Semi	3.78	Own 3 Speeds

BRAKES Position Brake lining width & length

Rear wheel Trans. or wheel W Service $1\frac{1}{4}$ X $27\frac{1}{2}$ Hand $1\frac{1}{4}$ X $26\frac{1}{2}$

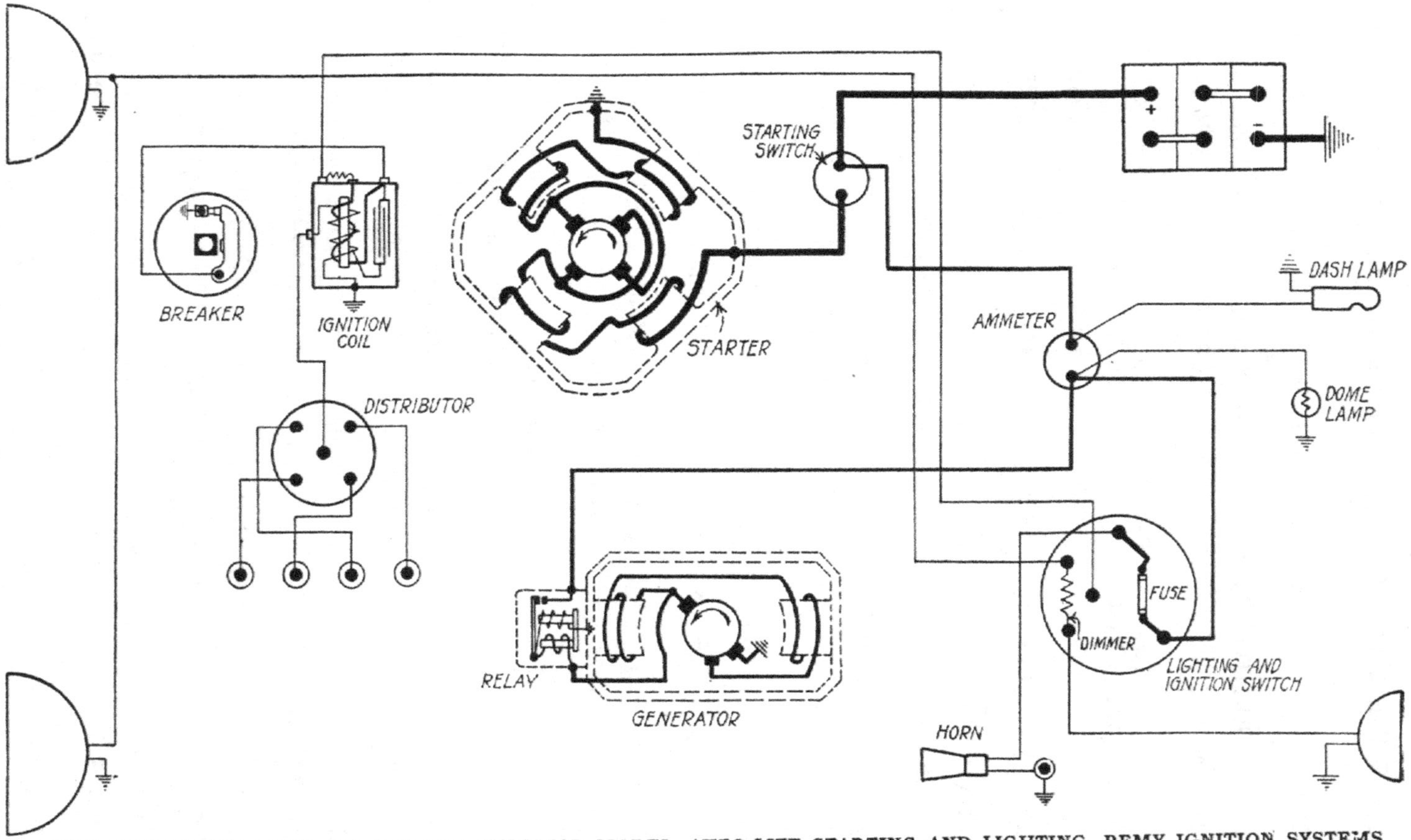

CHEVROLET WIRING DIAGRAM, 1923-1924, SUPERIOR MODEL, AUTO-LITE STARTING AND LIGHTING, REMY IGNITION SYSTEMS

CHEVROLET Model Superior Year 1923-24

Auto-Lite Starter & Generator Remy Ignition

GENERATOR Model	Regulation	Max. Chg. rate
G H 1012	Third Brush	16 Amperes

Amps. Cut-in	Volts Warm Test	R.P.M.	Amps. Cut-in	Volts Cold Test	R.P.M.
6	7.0-7.3	1200	6	8.0-8.5	1000
13	8.3-8.8	2400	10	8.0-8.5	2400

RELAY Model	Closes	Opens	Contact Gap	Air Gap
G H	10 MPH	9 MPH	.025-.030	.020

STARTING Motor Model	Drive	Rotation
M F	Bendix	Anti-Clockwise

BATTERY Willard Type SJXR-3 Capacity 6 Volts 90 Amps. Bat. to Frame Con. Negative

IGNITION Coil Model	DISTRIBUTOR Model	Breaker Contact Gap
No. 284-L	No. 366-P	.030

FIRING ORDER	Ignition Timing	SPARK PLUG
1-2-4-3	TDC at Retard	Size 7/8 Gap .022

LIGHTING Single Contact; Headlamps 21 C.P.; Dash & Tail 2 C.P.; Side Lamps 4 C.P.

ENGINE Bore $3\frac{11}{16}$ Stroke 4; Cylinder No. 4 L Head; Valve tappet clearance Inlet .004 Exhaust .004

CARBURETOR	COOLING	PISTON RING	OILING SYSTEM
Zenith 1"	Pump	$3\frac{11}{16}$ X $\frac{3}{16}$	Cap. 4 qt. Type Cir. Sp.

CLUTCH	REAR AXLE	Gear Ratio	TRANSMISSION
Cone $12\frac{3}{16}$	Semi	3.78	Own 3 Speeds

BRAKES Rear wheel; Position Trans. or wheel W; Brake lining width & length Service $1\frac{1}{4}$ X $27\frac{1}{2}$ Hand $1\frac{1}{4}$ X $26\frac{1}{2}$

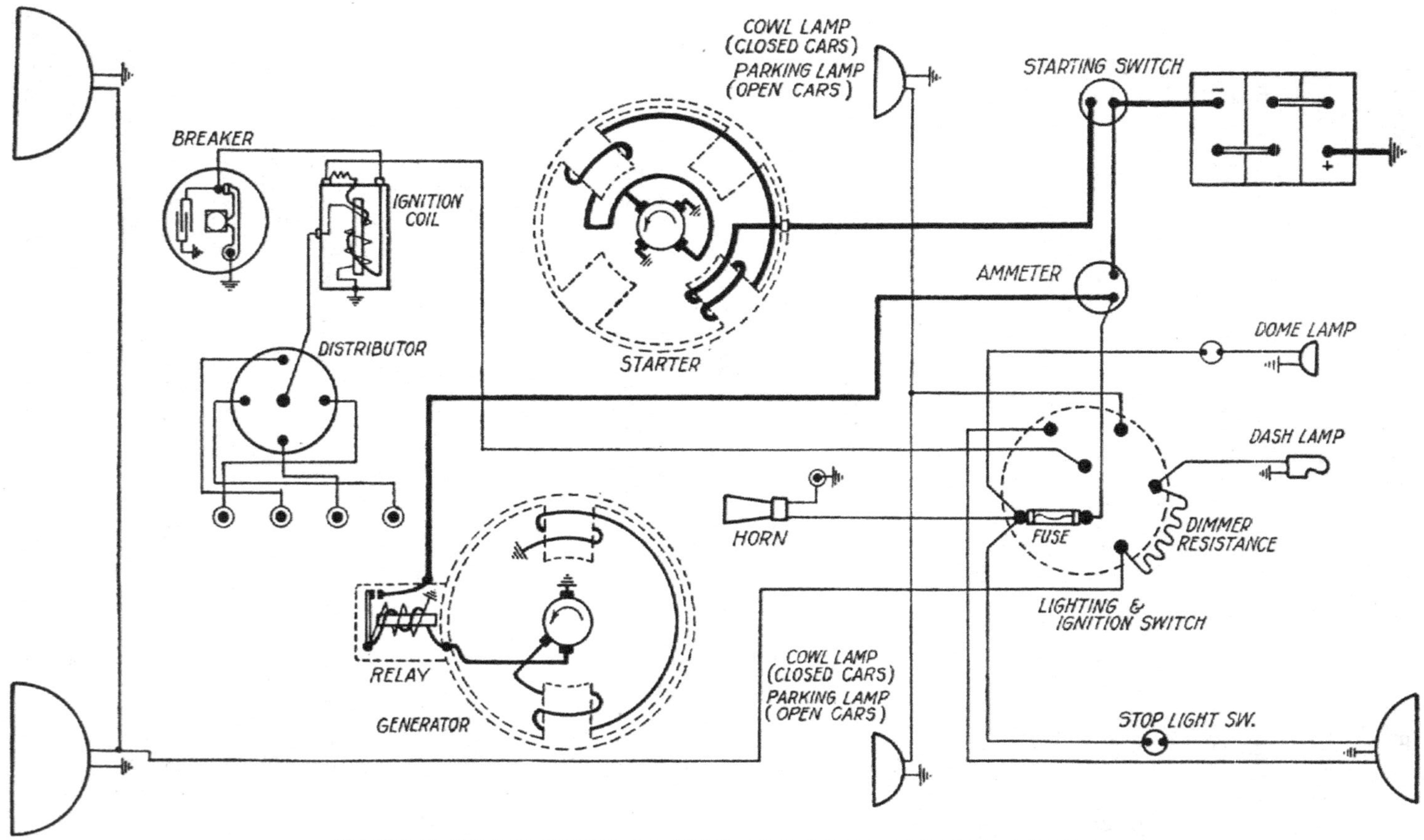

CHRYSLER "FOUR" WIRING DIAGRAM, 1925, MODEL 25F, REMY SYSTEM

CHRYSLER 4 Model 22 F Year 1924-1925

Remy Starter & Generator Remy Ignition

Regulation	Max. Chg. rate and speed
3rd Brush	16 Amps. 1800

RELAY Air Gap	Contact Gap	Cut-in R.P.M.
.016"	.020"	750

BATTERY Willard Type LWR-13 Volts 6 Amps. 93

Bat. to Frame Con. Pos. CONTACT BREAKER Gap .022"

Firing Order 1-3-4-2 Ignition Timing .060" P.A.T.C. Adv.

SPARK PLUG	ENGINE	Taxable Hp.
Size 7/8" Gap .027"	Bore 3 5/8" Stroke 4½"	21

INTAKE VALVE TIMING	EXHAUST VALVE TIMING
Open Close	Open Close

VALVE CLEARANCE Hot Intake .004" Exhaust .006"

CARBURETOR	COOLING SYSTEM	OILING SYSTEM
Stewart	Type Ther. Cap 3½ gal.	Type Pres. Cap 4 qts.

PISTON RING: Width .125" Diam. 3 5/8" Gap .005"-.015"

CLUTCH Plate 5 7/8" X 8 5/8" GEAR RATIO AXLE Semi F.

BRAKES

Front	Rear	Hand

Lighting	Headlights	Dash & Tail	Side Lamps
Single Contact	21 C.P.	3 C.P.	3 C.P.

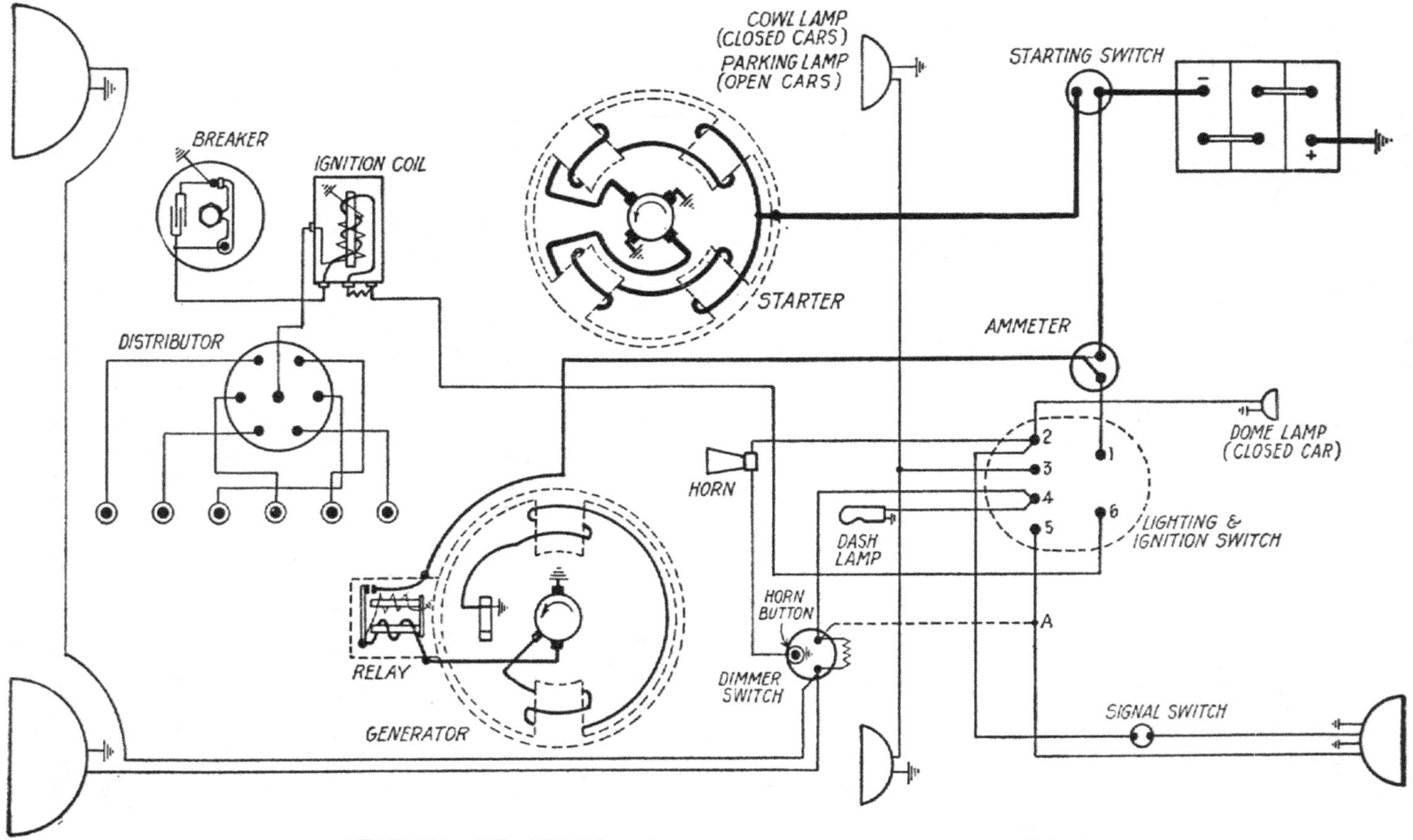

CHRYSLER "SIX" WIRING DIAGRAM, 1925, MODEL B, REMY SYSTEM

CHRYSLER 6 Model B Year 1925

Remy Starter & Generator Remy Ignition

Regulation: 3rd Brush

Max. Chg. rate and speed: 9-12 Amps. 1900

RELAY Air Gap: .014" Contact Gap: .014" Cut-in R.P.M.: 675 or 10.5 M.P.H.

BATTERY Willard Type CRR-15 Volts 6 Amps 100

Bat. to Frame Con. Pos. CONTACT BREAKER Gap .022"

Firing Order 1-5-3-6-2-4 Ignition Timing 14°B.T.C. "AD-SP" on Fl. Adv.

SPARK PLUG: Size 7/8" Gap .025"

ENGINE: Bore 3" Stroke $4\frac{1}{2}$"

Taxable Hp.: 21

INTAKE VALVE TIMING: Open T.D.C. Close 40°A.B.C.

EXHAUST VALVE TIMING: Open 48°B.B.C. Close 2°A.T.C.

VALVE CLEARANCE Hot Intake .004" Exhaust .006"

CARBURETOR: Ball - Ball

COOLING SYSTEM: Type P Cap 4 gal.

OILING SYSTEM: Type Pres. Cap 6 qts.

PISTON RING: Width 1/8" Diam. 3" Gap .005"-.016"

CLUTCH Disc $6\frac{5"}{8}$ X $7\frac{5"}{8}$ GEAR RATIO 4.6 AXLE Semi F.

BRAKES

Front: $33\frac{13"}{16}$ X $1\frac{1}{2}$" X $\frac{1"}{32}$

Rear: $33\frac{13"}{16}$ X $1\frac{1}{2}$" X $\frac{1"}{32}$

Hand: $18\frac{1"}{8}$ X $2\frac{1}{4}$" X $\frac{3"}{16}$

Lighting: Single Contact

Headlights: 21 C.P.

Dash & Tail: 3 C.P.

Side Lamps: 2 C.P.

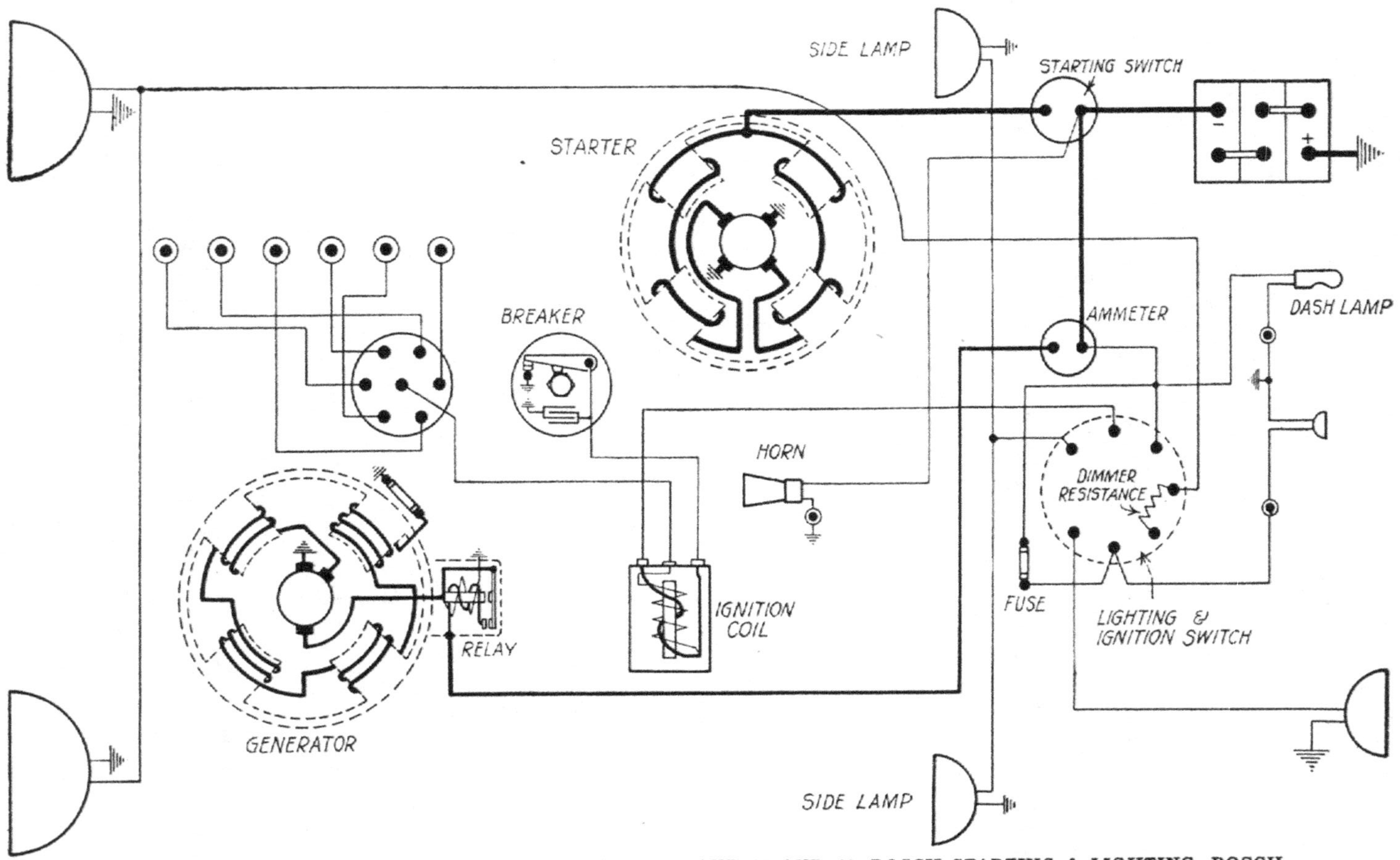

CLEVELAND WIRING DIAGRAM, 1923-1925, MODELS 6-42 AND 31 AND 43, BOSCH STARTING & LIGHTING, BOSCH BATTERY IGNITION SYSTEMS

CLEVELAND Model 42 Year 1923-24

Bosch Starter & Generator Bosch Ignition

GENERATOR Model	Regulation	Max. Chg. rate
No. 1021 - 1040	Third Brush	15 Amps.

Amps. Cut-in	Volts Warm Test	R.P.M.	Amps. Cut-in	Volts Cold Test	R.P.M.
0	6.8	550	4	7.5	640
10	8.5	1100	15.5	8.5	1600

RELAY Model	Closes	Opens	Contact Gap	Air Gap
No. C R - 1000	7 Volts	2 Amps.	.012"-.014"	.018

STARTING Motor Model	Drive	Rotation
No. 900	Bendix	Anti-Clockwise

BATTERY Prest-O-L. Type Capacity Bat. to Frame Con.

611 - R.H.N. 6 Volts 90 Amps. Positive

IGNITION Coil Model	DISTRIBUTOR Model	Breaker Contact Gap
No. T C - 30	No. T M - 684	.014 - .016

FIRING ORDER Ignition Timing SPARK PLUG

1-5-3-6-2-4 1 1/8" before T.D.C Size 7/8" L Gap .021

LIGHTING Headlamps Dash & Tail Side Lamps

Single Contact 21 C.P. 2 C.P. 4 C.P.

ENGINE Cylinder Valve tappet clearance

Bore $3\frac{1}{16}$ Stroke $4\frac{1}{2}$ No. 6 I Head Inlet .006 Exhaust .006

CARBURETOR	COOLING	PISTON RING	OILING SYSTEM
Stromberg 1"	Pump	3 1/6 X 3/16	Cap. 6 qt. Type C.S.

CLUTCH	REAR AXLE	Gear Ratio	TRANSMISSION
Plate 10"	Full	4.90	Own 3 Sp.

BRAKES Position Brake lining width & length

Rear wheel Trans. or wheel T. Service 2 X $20\frac{1}{2}$ Hand $1\frac{1}{2}$ X $21\frac{1}{2}$

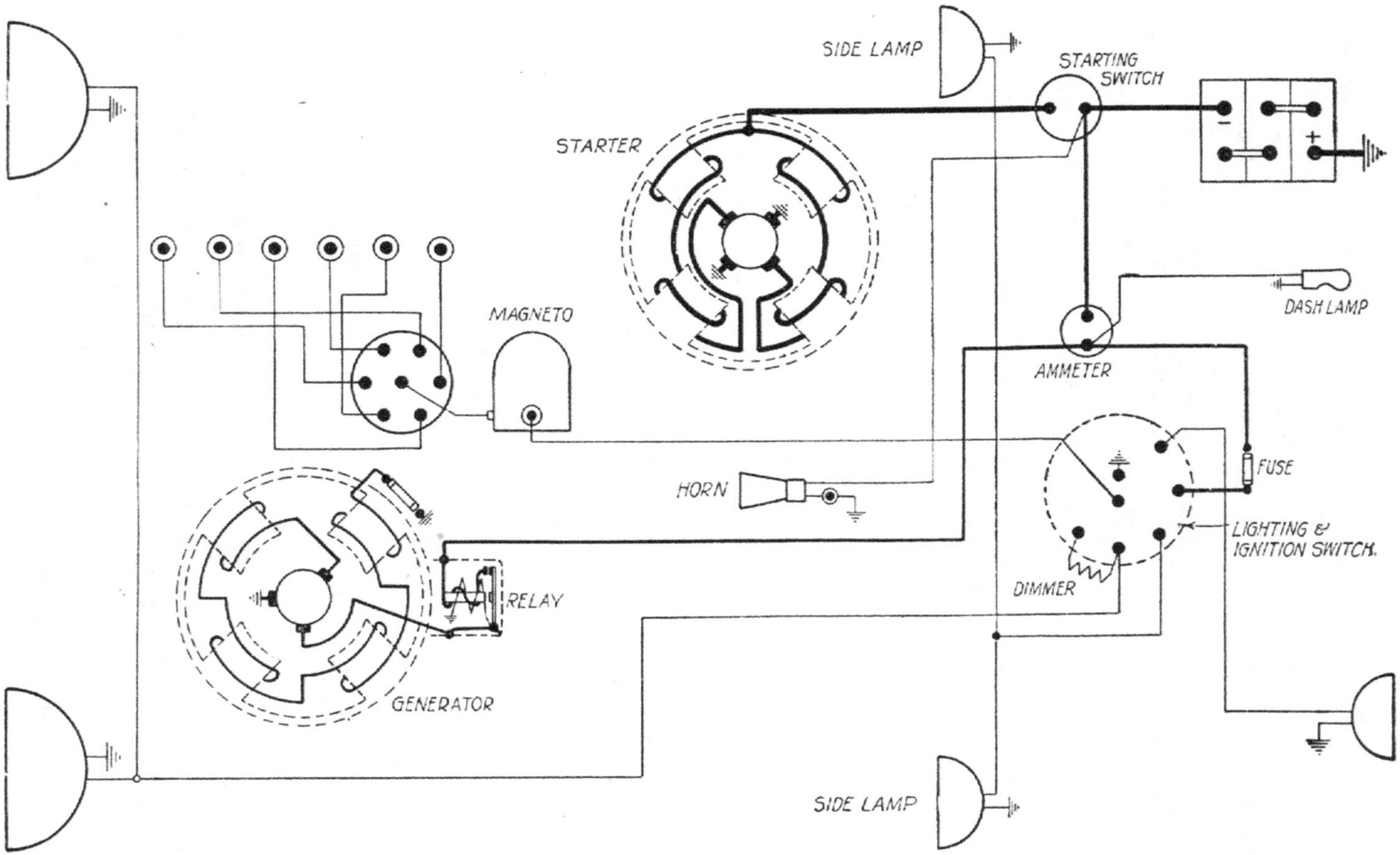

CLEVELAND WIRING DIAGRAM. 1923-1924. MODEL 6-42. BOSCH STARTING AND LIGHTING. BOSCH MAGNETO IGNITION SYSTEMS

CLEVELAND Model 6 - 42 Year 1923-24

Bosch Starter & Generator Bosch Magneto Ignition

GENERATOR Model	Regulation	Max. Chg. rate
No. 1021	Third Brush	14 Amps.

Amps. Cut-in	Volts Warm Test	R.P.M.	Amps. Cut-in	Volts Cold Test	R.P.M.
0.0	6.0	600	7.	7.0	800
9.8	7.3	1600	12.8	7.6	2000

RELAY Model	Closes	Opens	Contact Gap	Air Gap
No. C R - 1000	7 Volts	2 Amps.	.012-.014	.018

STARTING Motor Model	Drive	Rotation
No. 900	Bendix	Anti-Clockwise

BATTERY Prest-O-L Type	Capacity	Bat. to Frame Con.
611 - R H N	6 Volts 94 Amps.	Positive

IGNITION Coil Model	DISTRIBUTOR Model	Breaker Contact Gap
None used	A T - 6	.020

FIRING ORDER	Ignition Timing	SPARK PLUG
1-5-3-6-2-4	1 1/8" before TDC	Size 7/8" L Gap .021

LIGHTING	Headlamps	Dash & Tail	Side Lamps
Single Contact	21 C.P.	2 C.P.	4 C.P.

ENGINE	Cylinder	Valve tappet clearance
Bore $3\frac{1}{16}$ Stroke $4\frac{1}{2}$	No. 6 I Head	Inlet .006 Exhaust .006

CARBURETOR	COOLING	PISTON RING	OILING SYSTEM
Stromberg 1"	Pump	3 1/16 X 3/16	Cap. 6qt. Type C.S.

CLUTCH	REAR AXLE	Gear Ratio	TRANSMISSION
Plate 10"	Full	4.90	Own 3 Sp.

BRAKES	Position	Brake lining width & length
Rear wheel	Trans. or wheel T	Service 2 X $20\frac{1}{2}$ Hand $1\frac{1}{2}$ X $21\frac{1}{2}$

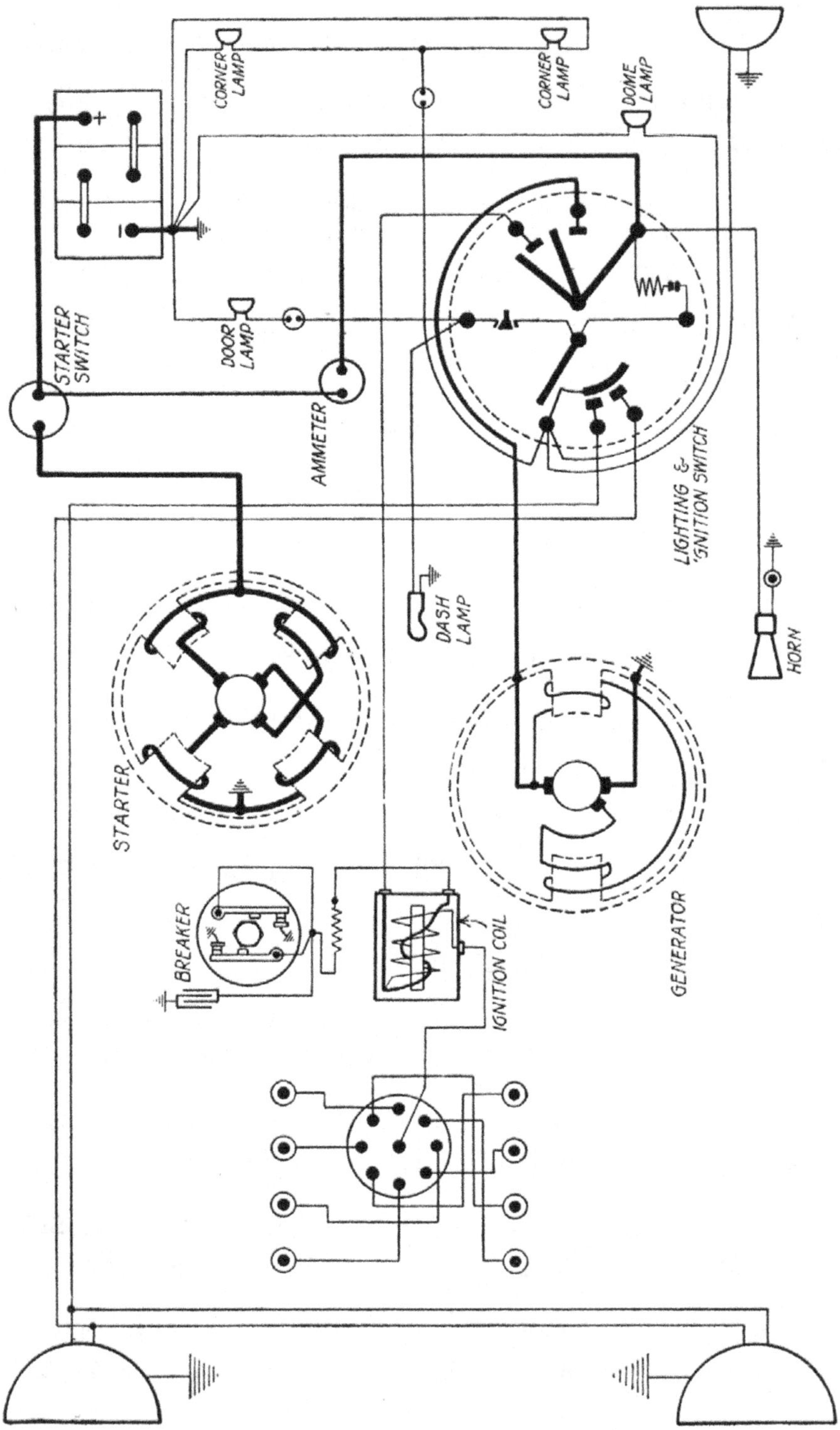

COLE WIRING DIAGRAM, 1923, MODEL 8-90, DELCO EQUIPMENT

C O L E Model 8 - 90 Year 1923

Delco Starter & Generator Delco Ignition

GENERATOR Model | Regulation | Max. Chg. rate
No. 153 | Third Brush | 16 Amps.

Amps. Cut-in	Volts Warm Test	R.P.M.	Amps. Cut-in	Volts Cold Test	R.P.M.
0	- - -	500	10-12	- - -	900
14-16	- - -	1400	11-13	- - -	2000

RELAY NONE USED. CONTACT MADE BY SWITCH.

STARTING Motor Model | Drive | Rotation
No. 72 - 259 | Bendix | Clockwise

BATTERY Willard Type | Capacity | Bat. to Frame Con.
S S R N - 5 | 6 Volts 125 Amps. | Negative

IGNITION Coil Model | DISTRIBUTOR Model | Breaker Contact Gap
No. 2127 | No. 153 | .018

FIRING ORDER | Ignition Timing | SPARK PLUG
1R-4L-3R-2L-4R-1L-2R-3L | T. D. C. at Retard | Size 7/8" Gap .030

LIGHTING | Headlamps | Dash & Tail | Side Lamps
Single Contact | 21 C.P. | 4 C.P. | 4 C.P.

ENGINE | Cylinder | Valve tappet clearance
Bore 3½ Stroke 4½ | No. 8 L Head | Inlet .005 Exhaust .005

CARBURETOR | COOLING | PISTON RING | OILING SYSTEM
Johnson 1½" | Pump | 3 1/2 X 3/16 | Cap. 6qt. Type H S

CLUTCH | REAR AXLE | Gear Ratio | TRANSMISSION
Cone | Full | 4.46 | Northway

BRAKES | Position | Brake lining width & length
Rear wheel | Trans. or wheel W | Service 2½ X 48 Hand 2¼X46½

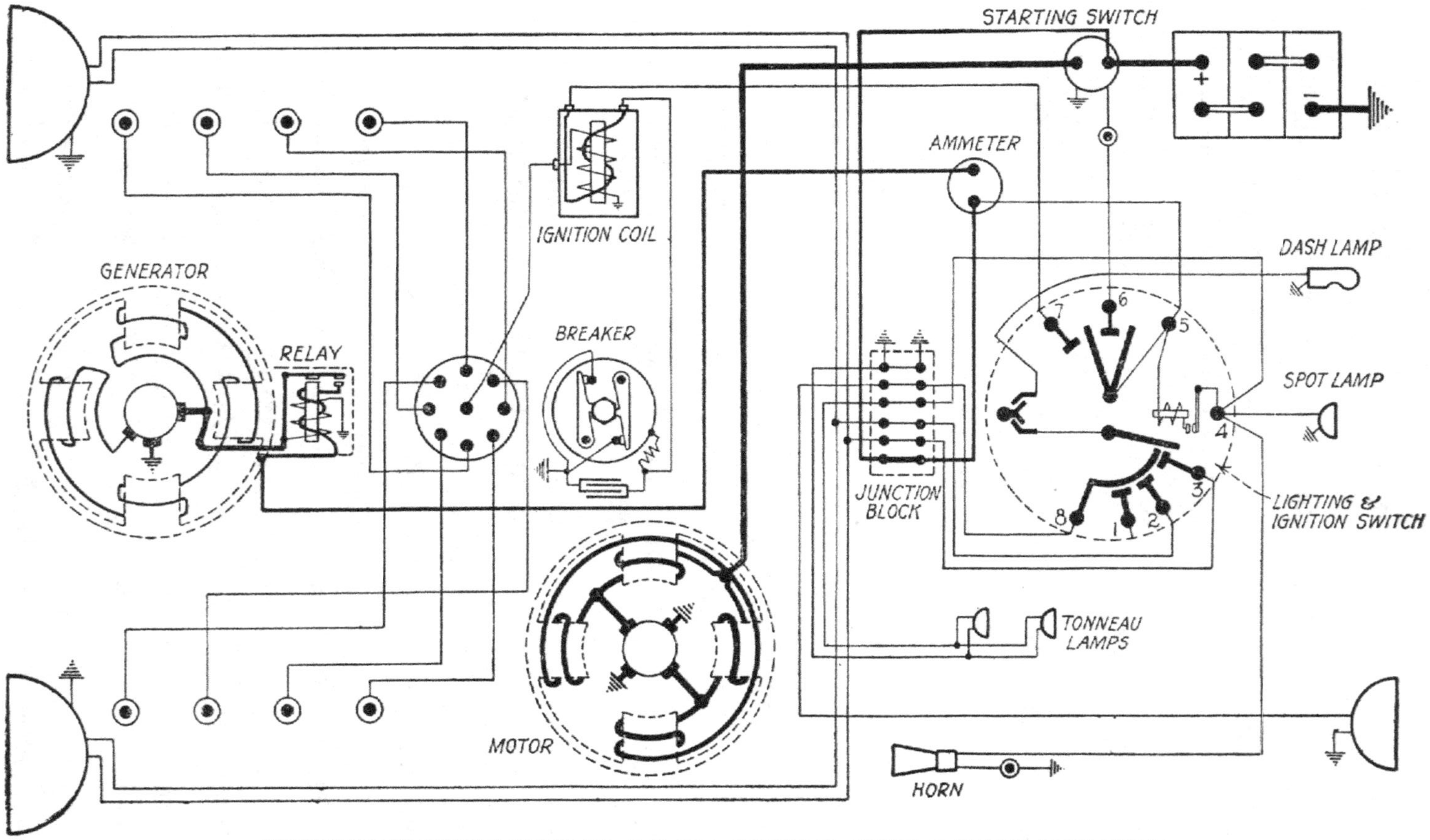

CUNNINGHAM WIRING DIAGRAM, 1923-1924, MODEL V, SERIES 5, DELCO EQUIPMENT

CUNNINGHAM Model **V Series 5** Year **1923-24**

Delco Starter & Generator **Delco** Ignition

GENERATOR Model	Regulation	Max. Chg. rate
No. 182	Third Brush	18 Amps.

Amps. Cut-in	Volts Warm Test	R.P.M.	Amps. Cut-in	Volts Cold Test	R.P.M.
0	- - -	450	10	- - -	800
13-17	- - -	1000	14	- - -	1450

RELAY Model	Closes	Opens	Contact Gap	Air Gap
No. 5745	10MPH	8MPH	.046-.062	.030

STARTING Motor Model	Drive	Rotation
No. 183	Bendix	Anti-Clockwise

BATTERY **Willard** Type S J R N - 5 — Capacity **6** Volts **110** Amps. — Bat. to Frame Con. **Negative**

IGNITION Coil Model	DISTRIBUTOR Model	Breaker Contact Gap
No. 2135	No. 5218	.018

FIRING ORDER	Ignition Timing	SPARK PLUG
1R-4L-3R-2L-4R-1L-2R-3L	TDC Sp 1/3 Adv.	Size 7/8"R Gap .030

LIGHTING	Headlamps	Dash & Tail	Side Lamps
Single Contact	21 C.P.	2 C.P.	4 C.P.

ENGINE — Bore $3\frac{3}{4}$ Stroke 5 — Cylinder No. 8 L Head — Valve tappet clearance Inlet .001 Exhaust .001

CARBURETOR	COOLING	PISTON RING	OILING SYSTEM
Stromberg $1\frac{1}{2}$"	Pump	3 3/4 X 3/16	Cap. 10qt. Type Force

CLUTCH	REAR AXLE	Gear Ratio	TRANSMISSION
Disc $8\frac{7}{16}$	Full	4.23	Own

BRAKES **Rear** wheel — Position Trans. or wheel **W** — Brake lining width & length Service 3 X $22\frac{1}{4}$ Hand 3 X $45\frac{1}{2}$

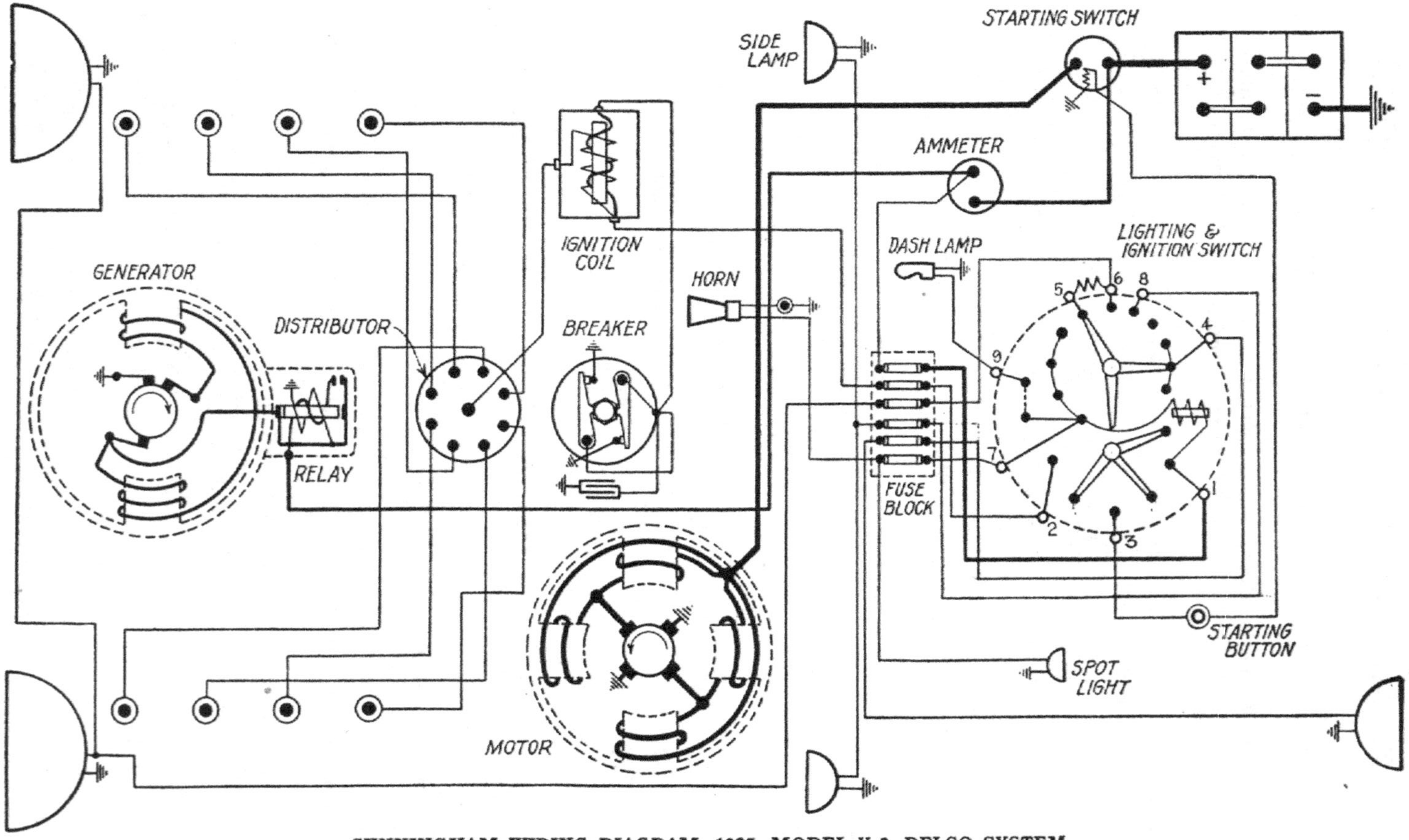

CUNNINGHAM WIRING DIAGRAM, 1925, MODEL V-6, DELCO SYSTEM

C U N N I N G H A M Model V 6 Year 1925

Delco Starter & Generator Delco Ignition

Regulation	Max. Chg. rate and speed
3rd Brush	18-20 Amps. 1600

RELAY Air Gap	Contact Gap	Cut-in R.P.M.
.020"-.035"	.015"-.020"	

BATTERY Willard Type SJRN-5 Volts 6 Amps. 125

Bat. to Frame Con. Neg. CONTACT BREAKER Gap .015"

Firing Order 1R-1L-4R-4L-2L-3R-3L-2R Ignition Timing 5°A.T.C. Ret.

SPARK PLUG	ENGINE	Taxable Hp.
Size 7/8" Gap .030"	Bore 3 3/4" Stroke 5"	45

INTAKE VALVE TIMING	EXHAUST VALVE TIMING
Open 5°A.T.C. Close 51°A.B.C.	Open 41°B.B.C. Close 5°A.T.C.

VALVE CLEARANCE Cold Intake .001" Exhaust .0015"

CARBURETOR	COOLING SYSTEM	OILING SYSTEM
Stromberg	Type P Cap. 10 gal.	Type Pres. Cap. 8 qts.

PISTON RING: Width 3/16" Diam. 3 3/4" Gap .008"

CLUTCH Disc 8 7"/16 X 5"/32 GEAR RATIO 4.23 AXLE F.F.

BRAKES

Front	Rear	Hand
	46" X 3" X 3"/16	46" X 2½" X 3"/16

Lighting	Headlights	Dash & Tail	Side Lamps
Single Contact	21 C.P.	2 C.P.	2 C.P.

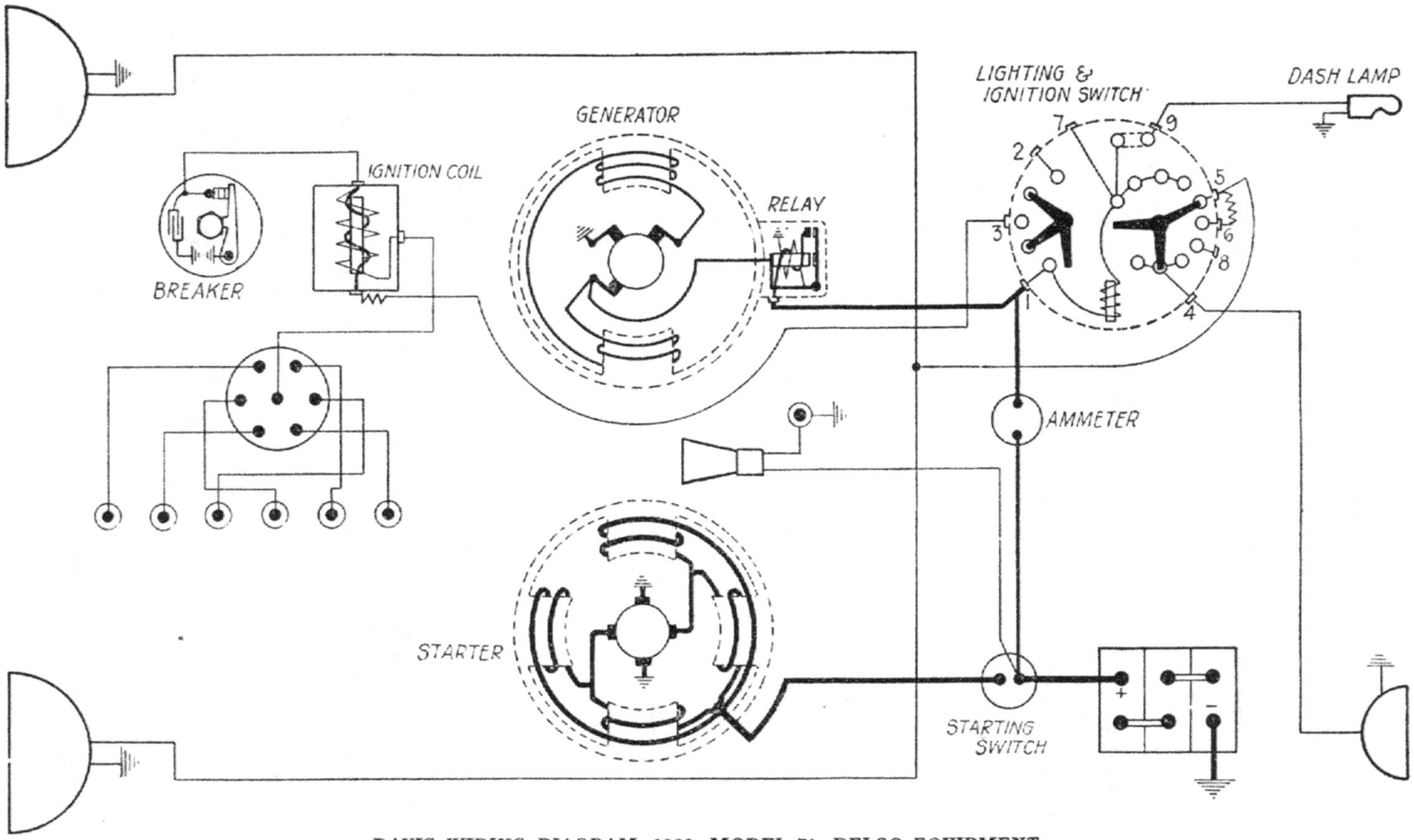

DAVIS WIRING DIAGRAM, 1923, MODEL 71, DELCO EQUIPMENT

D A V I S Model 71 Year 1923

Delco Starter & Generator Delco Ignition

GENERATOR Model			Regulation	Max. Chg. rate	
No. 257			Third Brush	16 Amps.	

Amps. Cut-in	Volts Warm Test	R.P.M.	Amps. Cut-in	Volts Cold Test	R.P.M.
4	- - -	800	6	- - -	800
15	- - -	1600	16-18	- - -	1600

RELAY Model	Closes	Opens	Contact Gap	Air Gap
No. 257	10 MPH	7 MPH	.025 - .035	.030

STARTING Motor Model	Drive	Rotation
No. 240	Bendix	Anti-Clockwise

BATTERY Willard Type Capacity Bat. to Frame Con.

S S R - 3 6 Volts 80 Amps. Negative

IGNITION Coil Model	DISTRIBUTOR Model	Breaker Contact Gap
No. 2178	No. 257	.020 - .027

FIRING ORDER	Ignition Timing	SPARK PLUG
1-5-3-6-2-4	T.D.C.at Retard	Size 7/8" Gap .025

LIGHTING	Headlamps	Dash & Tail	Side Lamps
Single Contact	21 C.P.	2 C.P.	2 C.P.

ENGINE Cylinder Valve tappet clearance

Bore $3\frac{3}{8}$ Stroke $4\frac{1}{2}$ No. 6 L Head Inlet .005 Exhaust .005

CARBURETOR	COOLING	PISTON RING	OILING SYSTEM
Stromberg $1\frac{1}{4}$"	Pump	3 1/4 X 3/16	Cap. 6 qt Type H S

CLUTCH	REAR AXLE	Gear Ratio	TRANSMISSION
Disc	Semi	5.09	Warner-Gear

BRAKES Position Brake lining width & length

Rear wheel Trans. or wheel T Service $2\frac{1}{4}$ X 38 Hand $2\frac{1}{4}$ X 24

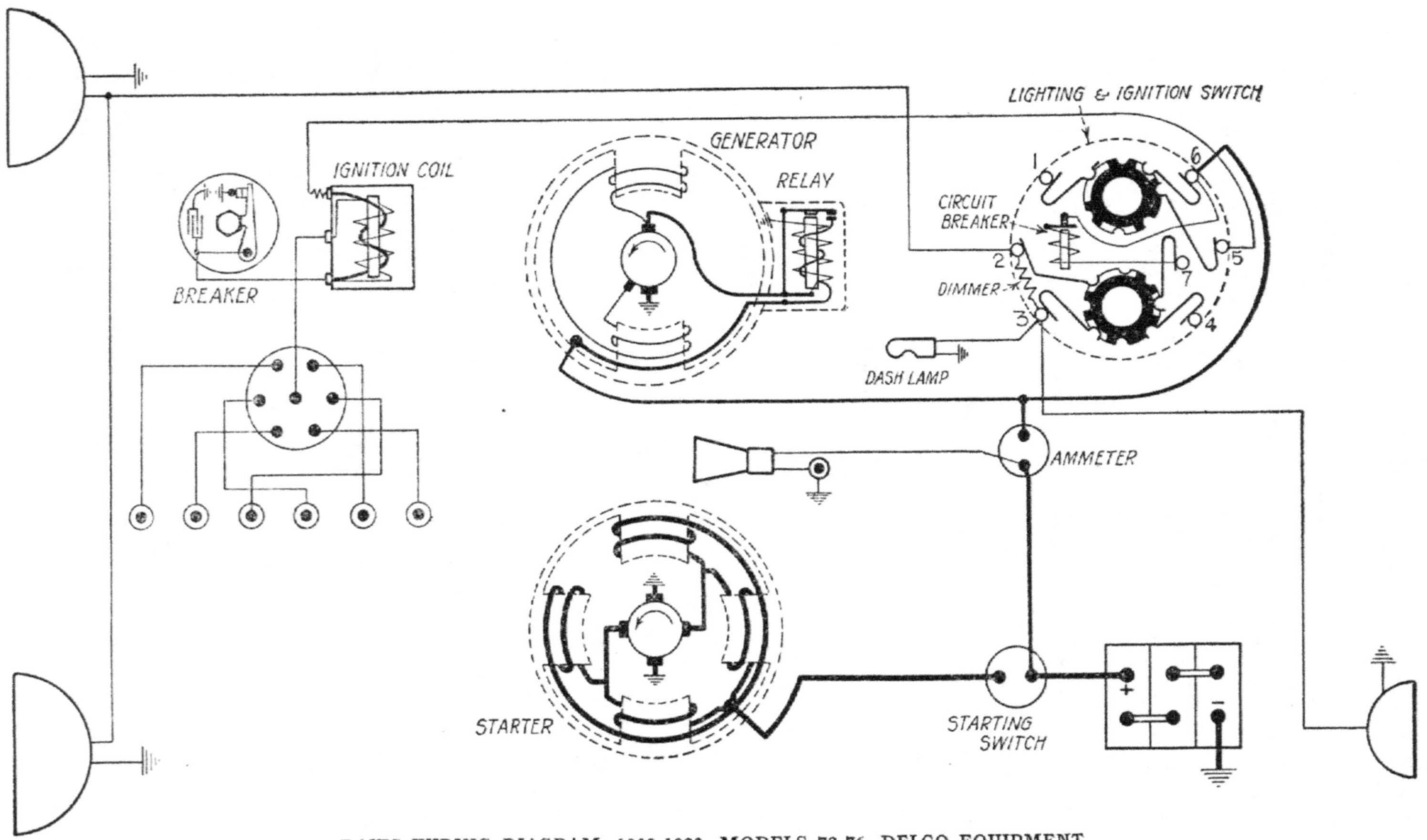

DAVIS WIRING DIAGRAM, 1922-1923, MODELS 72-76, DELCO EQUIPMENT

D A V I S Model 72 - 76 Year 1922-23

Delco Starter & Generator Delco Ignition

GENERATOR Model	Regulation	Max. Chg. rate
No. 241	Third Brush	16 Amps.

Amps. Cut-in	Volts Warm Test	R.P.M.	Amps. Cut-in	Volts Cold Test	R.P.M.
5	- - -	700	11	- - -	1000
16	- - -	1600	15	- - -	1800

RELAY Model	Closes	Opens	Contact Gap	Air Gap
No. 241	10 MPH	7 MPH	.025-.035	.030

STARTING Motor Model	Drive	Rotation
No. 240	Bendix	Anti-Clockwise

BATTERY Willard Type Capacity Bat. to Frame Con.

S J R - 3 6 Volts 80 Amps. Negative

IGNITION Coil Model	DISTRIBUTOR Model	Breaker Contact Gap
No. 2159	No. 241	.020 - .027

FIRING ORDER	Ignition Timing	SPARK PLUG
1-5-3-6-2-4	TDC at Retard	Size 7/8" Gap .025

LIGHTING	Headlamps	Dash & Tail	Side Lamps
Single Contact	21 C.P.	2 C.P.	2 C.P.

ENGINE Cylinder Valve tappet clearance

Bore $3\frac{3}{8}$ Stroke $4\frac{1}{2}$ No. 6 L Head Inlet .004 Exhaust .005

CARBURETOR	COOLING	PISTON RING	OILING SYSTEM
Stromberg $1\frac{1}{4}$"	Pump	3 1/4 X 3/16	Cap. 6qt Type H. S.

CLUTCH	REAR AXLE	Gear Ratio	TRANSMISSION
Disc	Semi	5.09	Warner-Gear

BRAKES Position Brake lining width & length

Rear wheel Trans. or wheel T Service $2\frac{1}{4}$ X 34 Hand $2\frac{1}{4}$ X 24

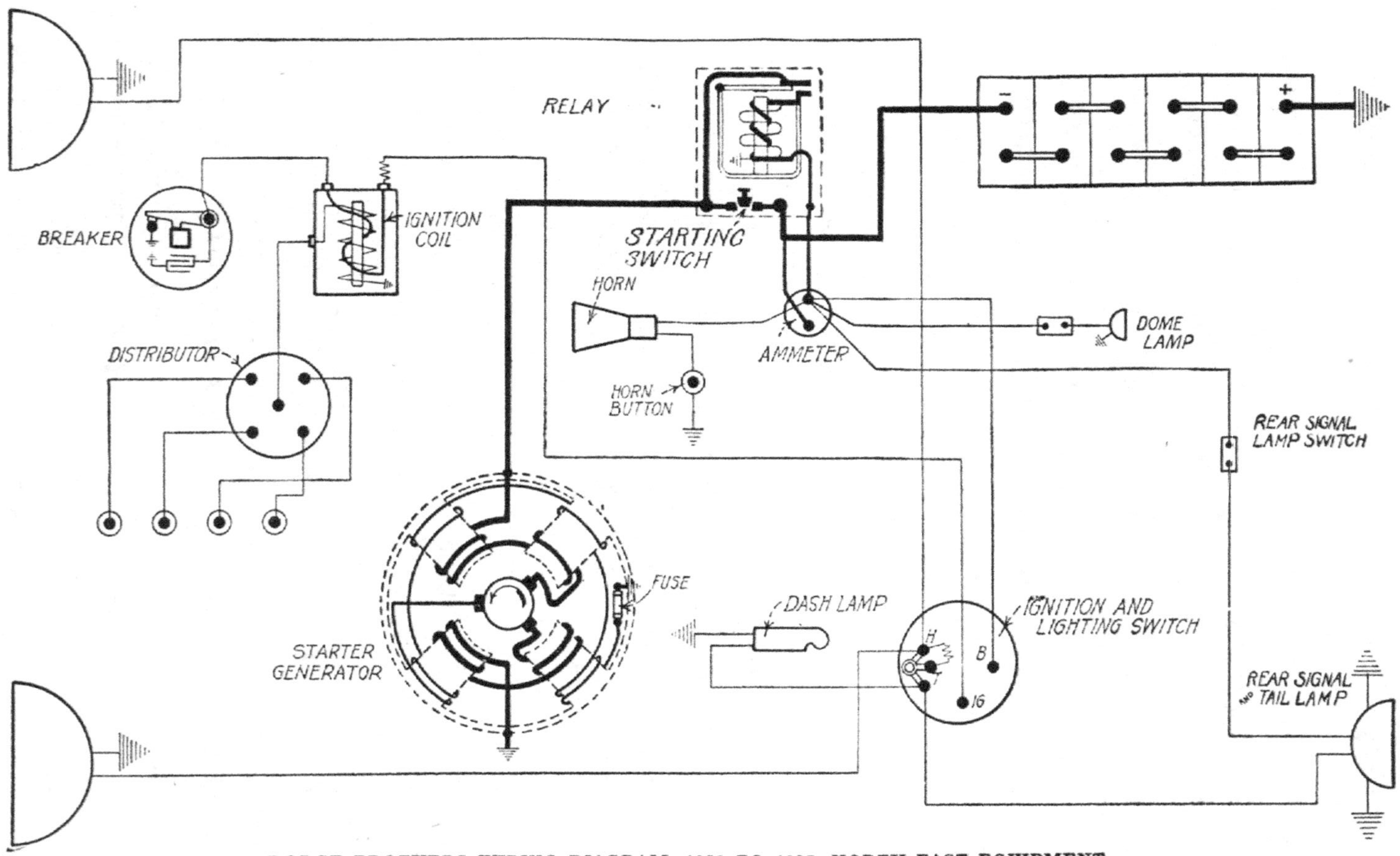

DODGE BROTHERS WIRING DIAGRAM, 1923 TO 1925, NORTH EAST EQUIPMENT

DODGE BROS. Model All Year 1923-24

North East Starter & Generator Dodge N.E. Ignition

GENERATOR Model	Regulation	Max. Chg. rate
G Type 3804	Reverse series	9 Amps.

Amps. Cut-in	Volts Warm Test	R.P.M.	Amps. Cut-in	Volts Cold Test	R.P.M.
0	12	1000	0	12	900
8.0	16	1800	9.0	16	1800

RELAY Model	Closes	Opens	Contact Gap	Air Gap
No. 8100	10 M P H	8½ M P H	.020	.003

STARTING Motor Model	Drive	Rotation
G Type 3804	Chain	Anti-clockwise

BATTERY Willard Type Capacity Bat. to Frame Con.

S J R - 26 12 Volts 50 Amps. Positive

IGNITION Coil Model	DISTRIBUTOR Model	Breaker Contact Gap
No. 10205	O Type 10004	.020"

FIRING ORDER	Ignition Timing	SPARK PLUG
1 - 3 - 4 - 2	#4 Ext. closed	Size 7/8 R Gap .030"

LIGHTING	Headlamps	Dash & Tail	Side Lamps
Single Contact	12-16V-21 C.P	2 C.P	4 C.P

ENGINE Cylinder Valve tappet clearance

Bore $3\frac{7}{8}$ Stroke $4\frac{1}{2}$ No. 4 L Head Inlet .004 Exhaust .004

CARBURETOR	COOLING	PISTON RING	OILING SYSTEM
Stewart 1 1/8"	Pump	3 7/8 X 1/8	Cap 5 Type F & S

CLUTCH	REAR AXLE	Gear Ratio	TRANSMISSION
Disc.	Semi	4.55	Own

BRAKES Position Brake lining width & length

Rear wheel Trans. or wheel W. Service 2¼ X 19¼ Hand 1¾ X 14½

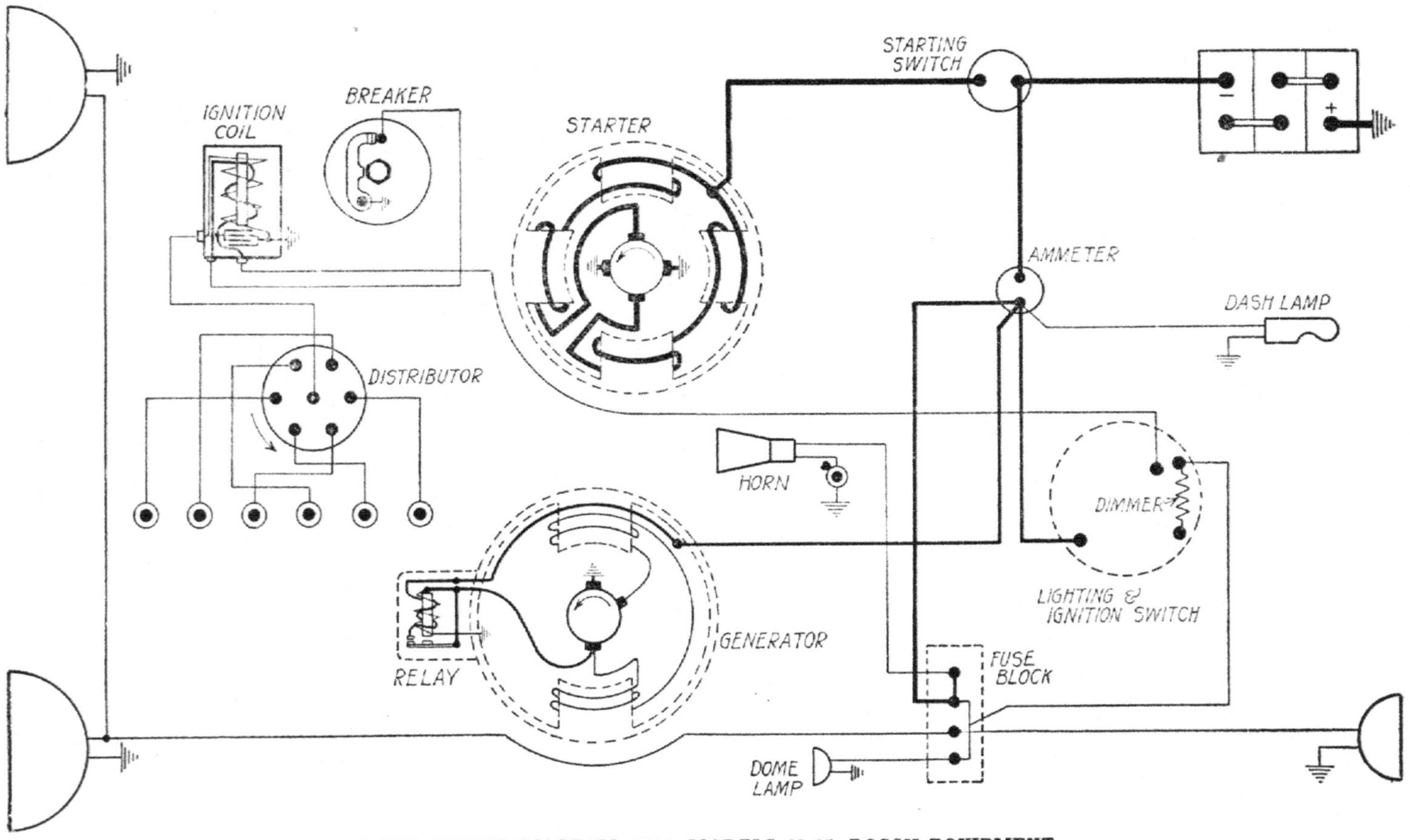

DORT WIRING DIAGRAM, 1923, MODELS 20-25, BOSCH EQUIPMENT

D O R T Model 20-25 Year 1923

Bosch Starter & Generator Bosch Ignition

GENERATOR Model	Regulation	Max. Chg. rate
No. G D - 636	Third Brush	14 Amps.

Amps. Cut-in	Volts Warm Test	R.P.M.	Amps. Cut-in	Volts Cold Test	R.P.M.
0	6.9	500	7.5	7.5	660
12	8.8	1100	16	8.8	1200

RELAY Model	Closes	Opens	Contact Gap	Air Gap
No. C R - 600	7 M P H	7 M P H	.013"-.018"	.030

STARTING Motor Model	Drive	Rotation
No. G - D 525	Bendix	Anti-Clockwise

BATTERY U S L Type Capacity Bat. to Frame Con.

H D N - 311 - X 6 Volts 90 Amps. Positive

IGNITION Coil Model	DISTRIBUTOR Model	Breaker Contact Gap
No. T - C - 20	No. T C - 20	.020

FIRING ORDER	Ignition Timing	SPARK PLUG
1-4-2-6-3-5	TDC at Retard	Size 7/8R Gap .025

LIGHTING	Headlamps	Dash & Tail	Side Lamps
Single Contact	21 C.P.	2 C.P.	4 C.P.

ENGINE Cylinder Valve tappet clearance

Bore $3\frac{1}{8}$ Stroke $4\frac{1}{4}$ No. 6 I Head Inlet .010 Exhaust .010

CARBURETOR	COOLING	PISTON RING	OILING SYSTEM
Carter 1"	Thermo	3 1/8 X 1/8	Cap. 4qt. Type Force

CLUTCH	REAR AXLE	Gear Ratio	TRANSMISSION
Disc 7 7/8	3/4 Flt	4.66	Own

BRAKES Position Brake lining width & length

Rear wheel Trans. or wheel W Service $1\frac{3}{4}$ X $35\frac{1}{2}$ Hand $1\frac{3}{4}$ X 36

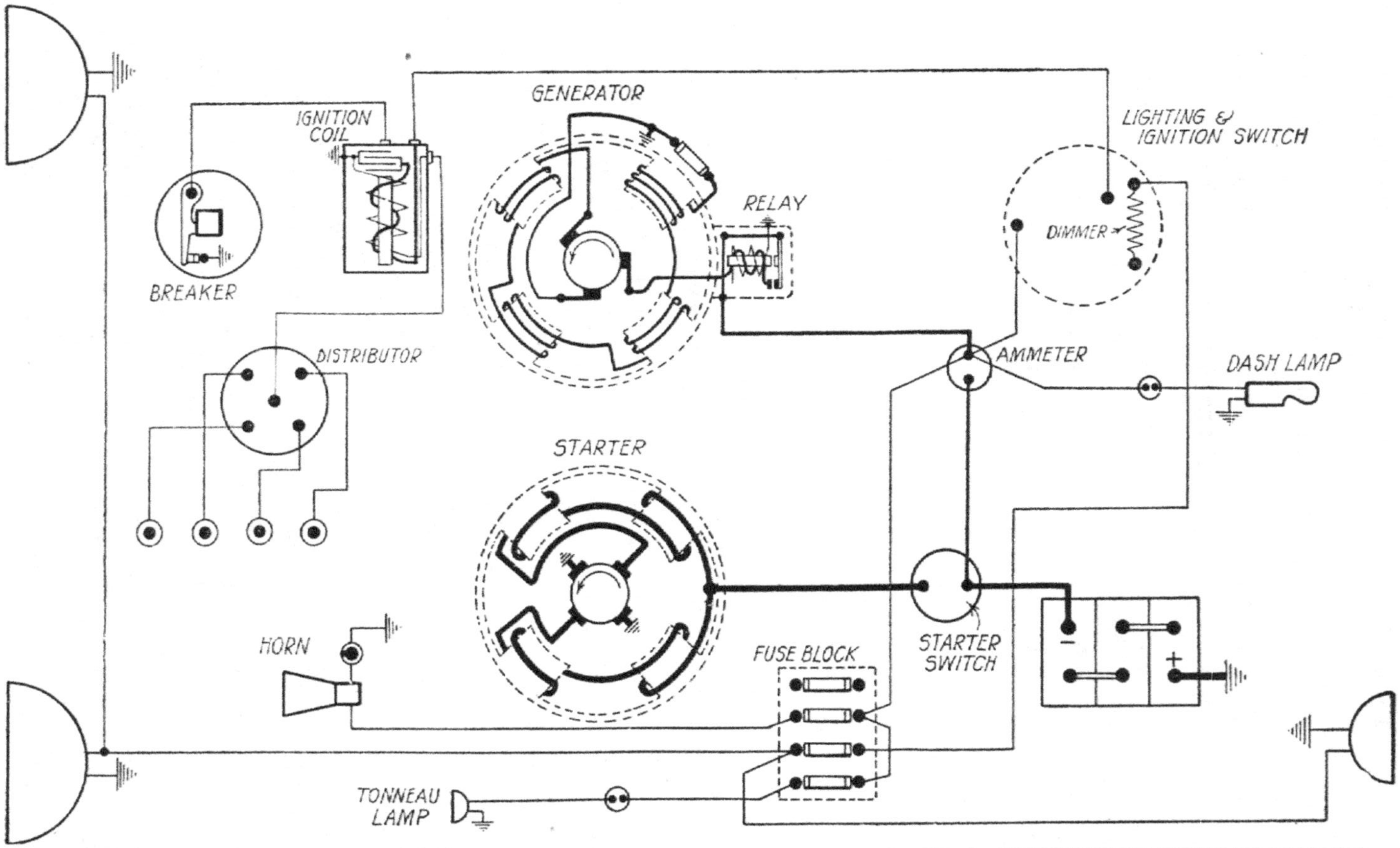

DORT WIRING DIAGRAM, 1923, MODELS 4-18 AND 4-23, BOSCH STARTING AND LIGHTING, CONNECTICUT IGNITION SYSTEMS

D O R T Model 18 and 23 Year 1923

Bosch Starter & Generator Connecticut Ignition

GENERATOR Model	Regulation	Max. Chg. rate
No. 1223	Third Brush	10.5 Amps.

Amps. Cut-in	Volts Warm Test	R.P.M.	Amps. Cut-in	Volts Cold Test	R.P.M.
0	6.9	450	6	7.2	700
10.2	8.8	1300	9.5	8.0	1700

RELAY Model	Closes	Opens	Contact Gap	Air Gap
No. CR - 1200	8 M P H	6 M P H	.013"-.018"	.012"

STARTING Motor Model	Drive	Rotation
No. 525	Bendix	Anti-Clockwise

BATTERY U S L Type Capacity Bat. to Frame Con.

H D N - 311 - X 6 Volts 90 Amps. Positive

IGNITION Coil Model	DISTRIBUTOR Model	Breaker Contact Gap
No. 225 - Y	225 - Y	.025

FIRING ORDER	Ignition Timing	SPARK PLUG
1 - 3 - 4 - 2	TDC at Retard	Size 7/8"R Gap .030

LIGHTING	Headlamps	Dash & Tail	Side Lamps
Single Contact	21 C.P.	2 C.P.	4 C.P.

ENGINE Cylinder Valve tappet clearance

Bore $3\frac{1}{2}$ Stroke 5 No. 4 L Head Inlet .005 Exhaust .005

CARBURETOR	COOLING	PISTON RING	OILING SYSTEM
Carter 1"	Thermo	3 1/2 X 3/16	Cap 5 Type C.S.

CLUTCH	REAR AXLE	Gear Ratio	TRANSMISSION
Disc 7 7/8	3/4 Flt	4.45	Own

BRAKES	Position	Brake lining width & length
Rear wheel	Trans. or wheel W	Service $1\frac{3}{4}$ X $35\frac{1}{2}$ Hand $1\frac{3}{4}$ X 36

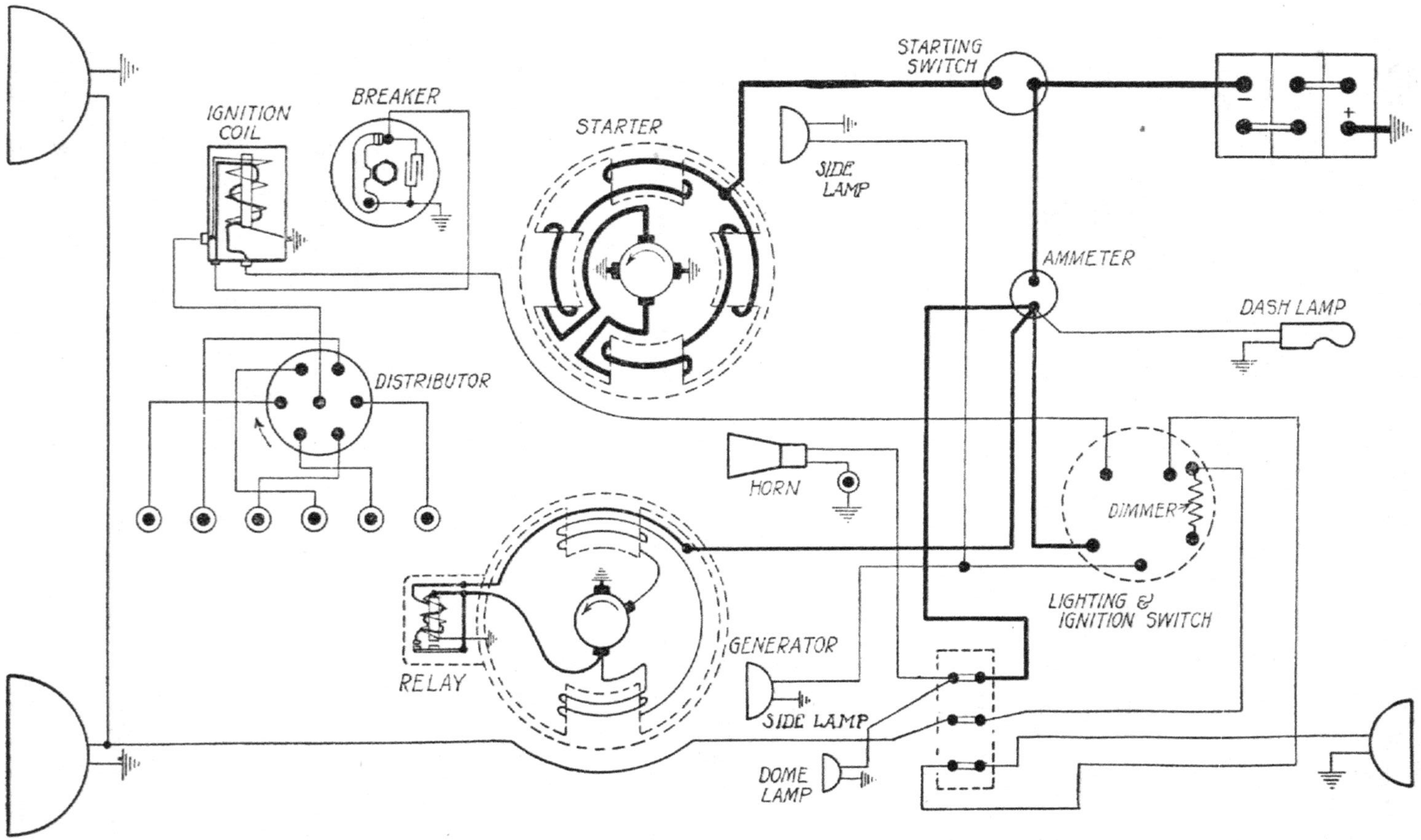

DORT WIRING DIAGRAM, 1924, MODELS 6-27 B, C, K, & S, BOSCH EQUIPMENT

DORT Model 27,B,C,K,&S Year 1924

Bosch Starter & Generator Bosch Ignition

GENERATOR Model	Regulation	Max. Chg rate
No. G. D. - 636	Third Brush	14 Amps.

Amps. Cut-in	Volts Warm Test	R.P.M.	Amps. Cut-in	Volts Cold Test	R.P.M.
0	6.9	500	7.5	7.2	660
14.0	8.8	1200	13	8.0	1100

RELAY Model	Closes	Opens	Contact Gap	Air Gap
No. C R - 600	9 M P H	7 M P H	.013 - .018	.030

STARTING Motor Model	Drive	Rotation
No. G - D 525	Bendix	Anti-Clockwise

BATTERY U. S. L. Type Capacity Bat. to Frame Con.

H D N - 311 - Y 6 Volts 90 Amps. Positive

IGNITION Coil Model	DISTRIBUTOR Model	Breaker Contact Gap
No. T - C - 20	No. T - C - 20	.020

FIRING ORDER	Ignition Timing	SPARK PLUG
1-4-2-6-3-5	TDC at Retard	Size 7/8"R Gap .025

LIGHTING	Headlamps	Dash & Tail	Side Lamps
Single Contact	21 C.P.	2 C.P.	4 C.P.

ENGINE Cylinder Valve tappet clearance

Bore $3\frac{1}{8}$ Stroke $4\frac{1}{4}$ No. 6 I Head Inlet .010 Exhaust .010

CARBURETOR	COOLING	PISTON RING	OILING SYSTEM
Carter 1"	Thermo S	3 1/8 X 1/8	Cap 4qt. Type Force

CLUTCH	REAR AXLE	Gear Ratio	TRANSMISSION
Disc 9"	3/4 Float	4.66	Own 3 Sp.

BRAKES	Position	Brake lining width & length
Rear wheel	Trans. or wheel W	Service $1\frac{3}{4}$ X $35\frac{1}{2}$ Hand $1\frac{3}{4}$ X 36

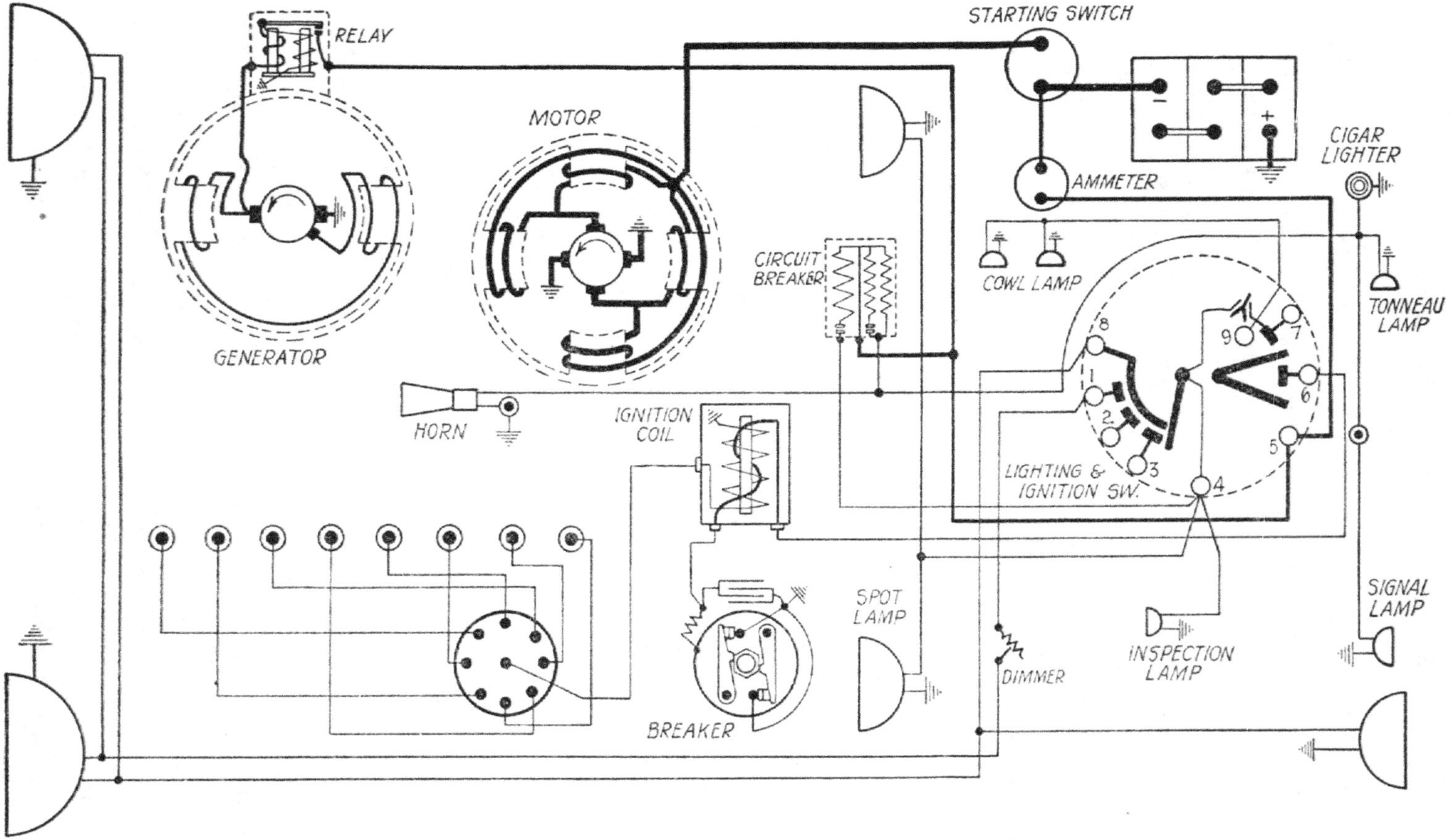

DUESENBERG WIRING DIAGRAM, 1922-1924, MODEL STRAIGHT 8, DELCO EQUIPMENT

D U E S E N B E R G Model Straight 8 Year 1922-24

Delco Starter & Generator Delco Ignition

GENERATOR Model	Regulation	Max. Chg. rate
No. 237-242	Third Brush	19 Amps.

Amps. Cut-in	Volts Warm Test	R.P.M.	Amps. Cut-in	Volts Cold Test	R.P.M.
3	- - -	600	13.5	- - -	1000
17	- - -	1600	19.	- - -	1600

RELAY Model	Closes	Opens	Contact Gap	Air Gap
No. 5745	7 M P H	4 M P H	.025"-.035"	.030"

STARTING Motor Model	Drive	Rotation
No. 200	Bendix	Anti-Clockwise

BATTERY Willard Type S R - 5 Capacity 6 Volts 120 Amps. Bat. to Frame Con. Positive

IGNITION Coil Model	DISTRIBUTOR Model	Breaker Contact Gap
No. 2135	No. 15735	.015" - .020"

FIRING ORDER	Ignition Timing	SPARK PLUG
1-5-3-7-4-8-2-6	½" past T.D.C.	Size 18 MM Gap .030"

LIGHTING	Headlamps	Dash & Tail	Side Lamps
Single Contact	21 C.P.	2 C.P.	4 C.P.

ENGINE — Cylinder — Valve tappet clearance

Bore $2\frac{7}{8}$ Stroke 5 No. 8 I Head Inlet .002 Exhaust .002

CARBURETOR	COOLING	PISTON RING	OILING SYSTEM
Stromberg 1½"	Pump	2 7/8 X 1/8"	Cap. 6 qt. Type F&S

CLUTCH	REAR AXLE	Gear Ratio	TRANSMISSION
Plate 13"	Semi	4.90 5.30	Own

BRAKES	Position	Brake lining width & length
Four wheel	Trans. or wheel W	Service 2¼ X 14½ Hand 2¼ X 7½

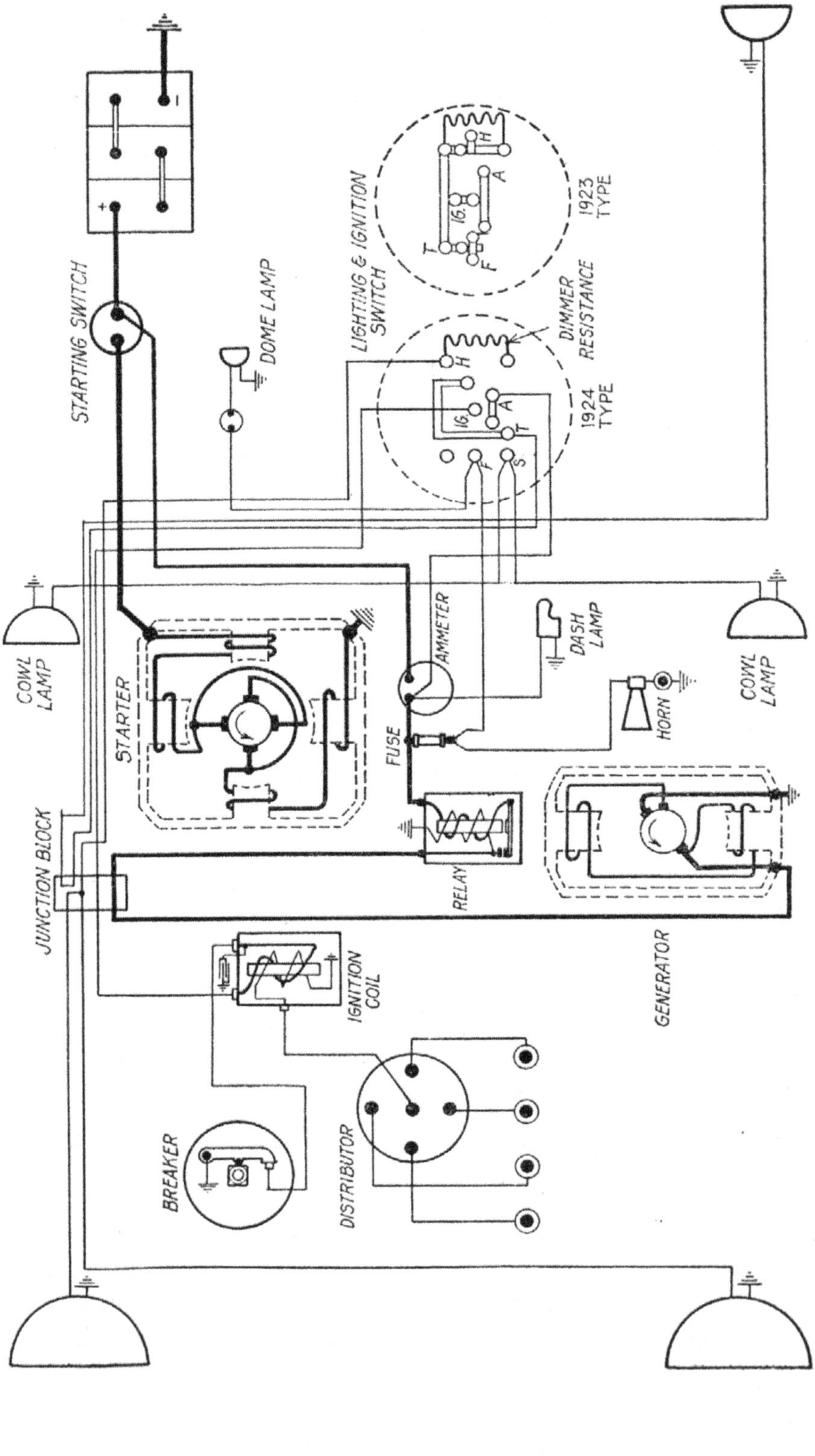

DURANT WIRING DIAGRAM, 1923-1925, MODEL A-22, AUTO-LITE SYSTEM

DURANT Model A 22 Year 1925

Auto Lite Starter & Generator Auto Lite Ignition

Regulation	Max. Chg. rate and speed
3rd Brush	16 Amps. 2400

RELAY Air Gap	Contact Gap	Cut-in R.P.M.
.010"	.020"	675

BATTERY U.S.L. Type 3-HVX-5X Volts 6 Amps 92

Bat. to Frame Con. Neg. CONTACT BREAKER Gap .020"

Firing Order 1-2-4-3 Ignition Timing T.D.C. Ret.

SPARK PLUG	ENGINE	Taxable Hp.
Size 7/8" Gap .027"	Bore 3 7/8" Stroke 4 1/4"	24

INTAKE VALVE TIMING	EXHAUST VALVE TIMING
Open 16° A.T.C. Close 52° A.B.C.	Open Close 16° A.T.C.

VALVE CLEARANCE Hot Intake .004" Exhaust .004"

CARBURETOR	COOLING SYSTEM	OILING SYSTEM
Tillotson	Type Cap	Type Pres. Cap

PISTON RING: Width .125" Diam 3 7/8" Gap

CLUTCH Plate GEAR RATIO AXLE Semi F

BRAKES

Front	Rear	Hand

Lighting	Headlights	Dash & Tail	Side Lamps
Single Contact	21 C.P.	2 C.P.	C.P.

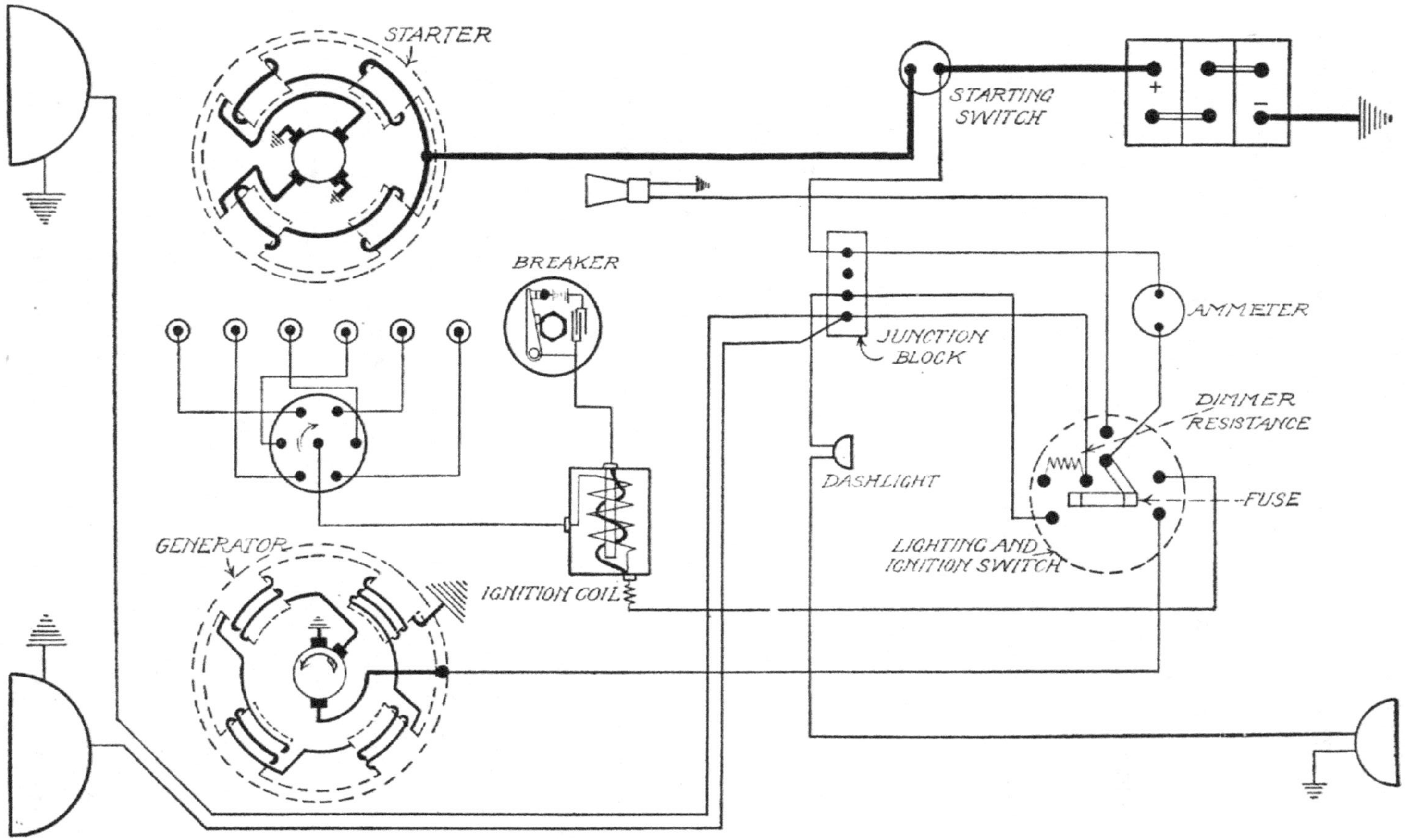

ESSEX "SIX" WIRING DIAGRAM, 1924-1925, BOSCH EQUIPMENT

E S S E X Model Six Year 1925

American Bosch Starter & Generator American Bosch Ignition

Regulation: 3rd Brush

Max. Chg. rate and speed: 8 Amps. 800

RELAY Air Gap: None | Contact Gap: | Cut-in R.P.M.:

BATTERY Prest-O-Lite Type 613 J F K Volts 6 Amps 105

Bat. to Frame Con. Neg. CONTACT BREAKER Gap .018"

Firing Order 1-5-3-6-2-4 Ignition Timing T.D.C. Ret.

SPARK PLUG: Size Met. Gap .025"

ENGINE: Bore 2 11/16" Stroke 4 1/4"

Taxable Hp.: 17

INTAKE VALVE TIMING: Open 7°A.T.C. Close 50°A.B.C.

EXHAUST VALVE TIMING: Open 50°B.B.C. Close 8°A.T.C.

VALVE CLEARANCE Hot Intake .004" Exhaust .005"

CARBURETOR: Stewart

COOLING SYSTEM: Type P Cap 4 gal.

OILING SYSTEM: Type Spl. Cap 4½ qts.

PISTON RING: Width 1/8" Diam 2 11/16" Gap .006"-.008"

CLUTCH Disc GEAR RATIO 5.6 AXLE 3/4 F

BRAKES

Front	Rear	Hand
	38" X 1 ¾" X 3"/16	35" X 1½" X 3"/16

Lighting	Headlights	Dash & Tail	Side Lamps
Single Contact	21 C.P.	3 Volt 2 C.P.	3 Volt 2 C.P.

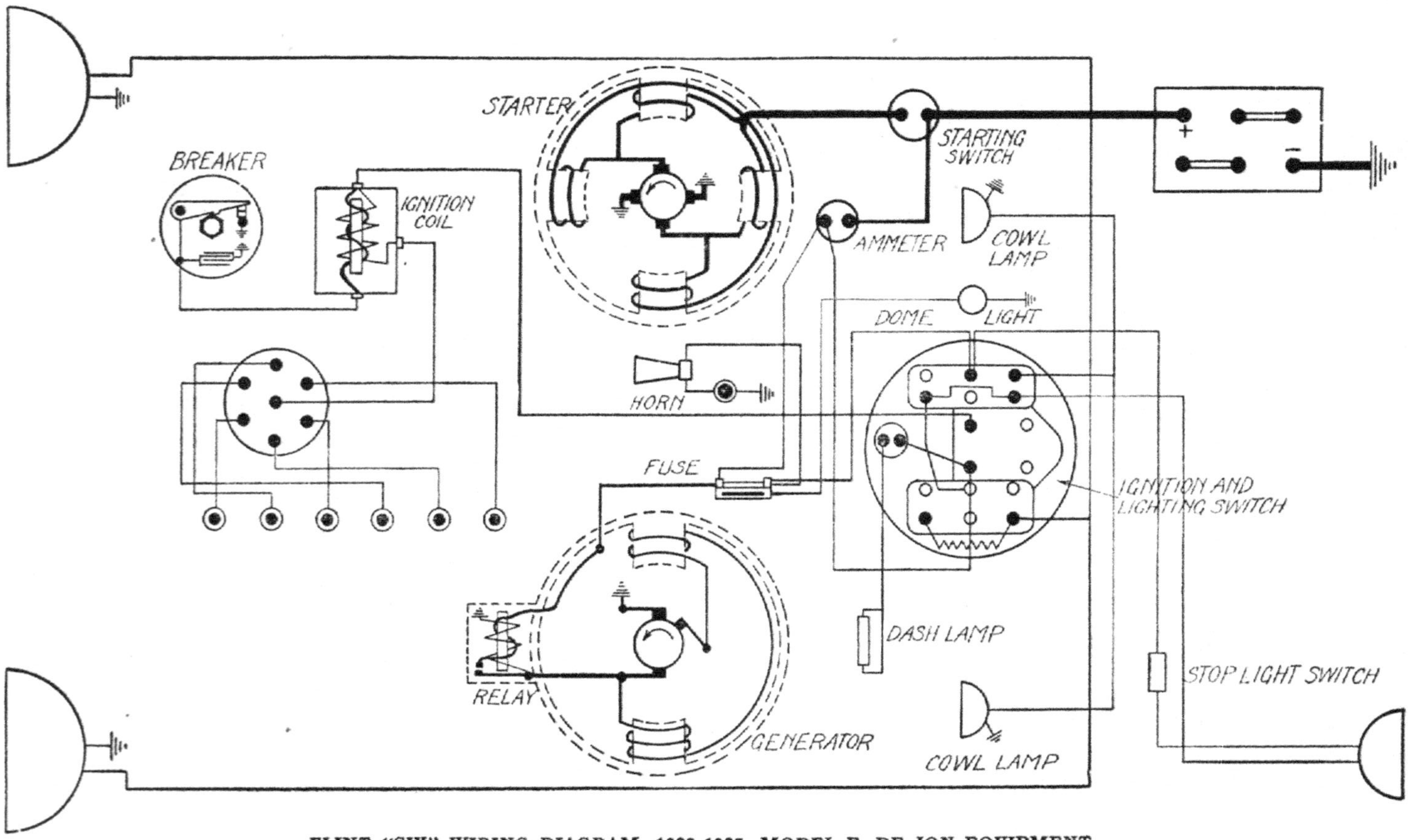

FLINT "SIX" WIRING DIAGRAM, 1923-1925, MODEL E, DE JON EQUIPMENT

FLINT Model E Year 1923-24

DeJon Starter & Generator DeJon Ignition

GENERATOR Model	Regulation	Max. Chg. rate
DA 4001	Third Brush	15.5

Amps. Cut-in	Volts Warm Test	R.P.M.	Amps. Cut-in	Volts Cold Test	R.P.M.
0	6.5	400	3.75	7.0	500
12	8.0	1200	15.5	8.0	1200

RELAY Model	Closes	Opens	Contact Gap	Air Gap
RA4001	8 volts	1 Amp.	.025-.035	.015

STARTING Motor Model	Drive	Rotation
SA4001	Bendix	Anti-Clockwise

BATTERY U.S.L. Type Capacity Bat. to Frame Con.
6 Volts 117 Amps. Negative

IGNITION Coil Model	DISTRIBUTOR Model	Breaker Contact Gap
CA4001	1 A 4001	.018-.020

FIRING ORDER 1-5-3-6-2-4 Ignition Timing TDC at Retard SPARK PLUG Size 7/8 Gap .025

LIGHTING Single Contact Headlamps 21 C.P. Dash & Tail 2 C.P. Side Lamps 2 C.P.

ENGINE Bore $3\frac{3}{8}$ Stroke 5 Cylinder No. 6 L Head Valve tappet clearance Inlet .003 Exhaust .003

CARBURETOR	COOLING	PISTON RING	OILING SYSTEM
Stromberg $1\frac{1}{4}$	Pump	$3\frac{3}{8}$ x $\frac{3}{16}$	Cap 7 qt. Type F&S

CLUTCH	REAR AXLE	Gear Ratio	TRANSMISSION
Plate 10"	Semi	4.88	Own

BRAKES Rear wheel Position Trans. or wheel W Brake lining width & length Service 2 X $42\frac{3}{8}$ Hand $1\frac{3}{4}$ X 40

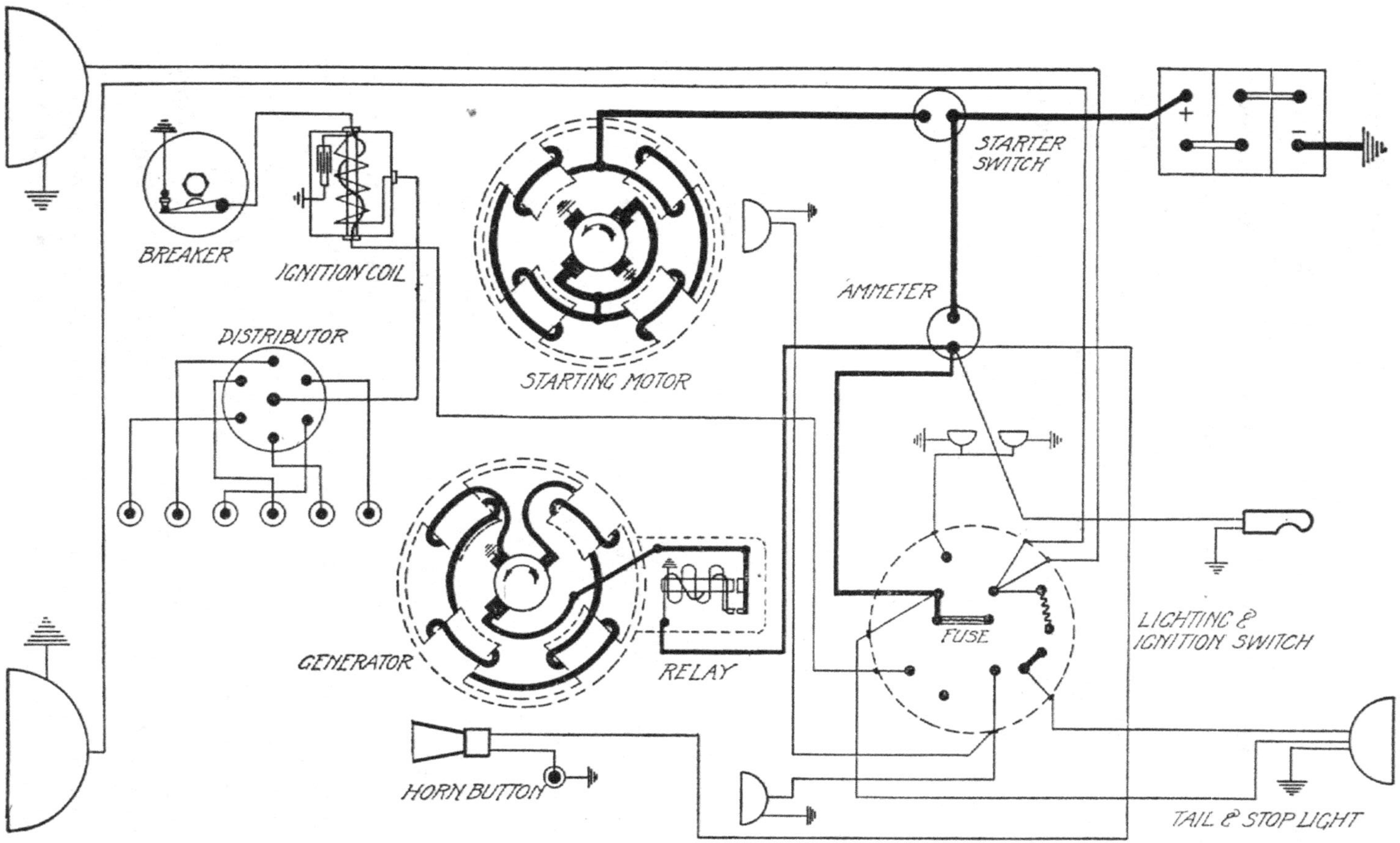

FLINT WIRING DIAGRAM, 1925, MODEL 6-40, AUTO-LITE SYSTEM

F L I N T Model 6-40 Year 1925

Auto-Lite Starter & Generator Auto-Lite Ignition

Regulation	Max. Chg. rate and speed
3rd Brush	15.5 Amps. 1700

RELAY Air Gap	Contact Gap	Cut-in R.P.M.
.015"-.030"	.025"-.035"	600 or 8 M.P.H.

BATTERY U.S.L. Type 3-HVX-5X Volts 6 Amps. 92

Bat. to Frame Con. Neg. CONTACT BREAKER Gap .025"

Firing Order 1-5-3-6-2-4 Ignition Timing T.D.C. Ret.

SPARK PLUG	ENGINE	Taxable Hp.
Size 7/8" Gap .022"	Bore 3 1/8" Stroke 4 1/4"	23

INTAKE VALVE TIMING	EXHAUST VALVE TIMING
Open 4°A.T.C. Close 46°A.B.C.	Open 41°B.B.C. Close 1°A.T.C.

VALVE CLEARANCE Hot Intake .006" Exhaust .006"

CARBURETOR	COOLING SYSTEM	OILING SYSTEM
Tillotson	Type P Cap. 3 6/10 gal.	Type Pres. Cap 6 qts.

PISTON RING: Width Diam Gap

CLUTCH Plate 9 1"/2 X 5"/32 GEAR RATIO 4.77 AXLE Semi F

BRAKES

Front	Rear	Hand
8 7"/16 X 1 3"/4 X 5"/32	8 7"/16 X 1 5"/16 X 5"/32	

Lighting	Headlights	Dash & Tail	Side Lamps
Single Contact	21 C.P.	2 C.P.	2 C.P.

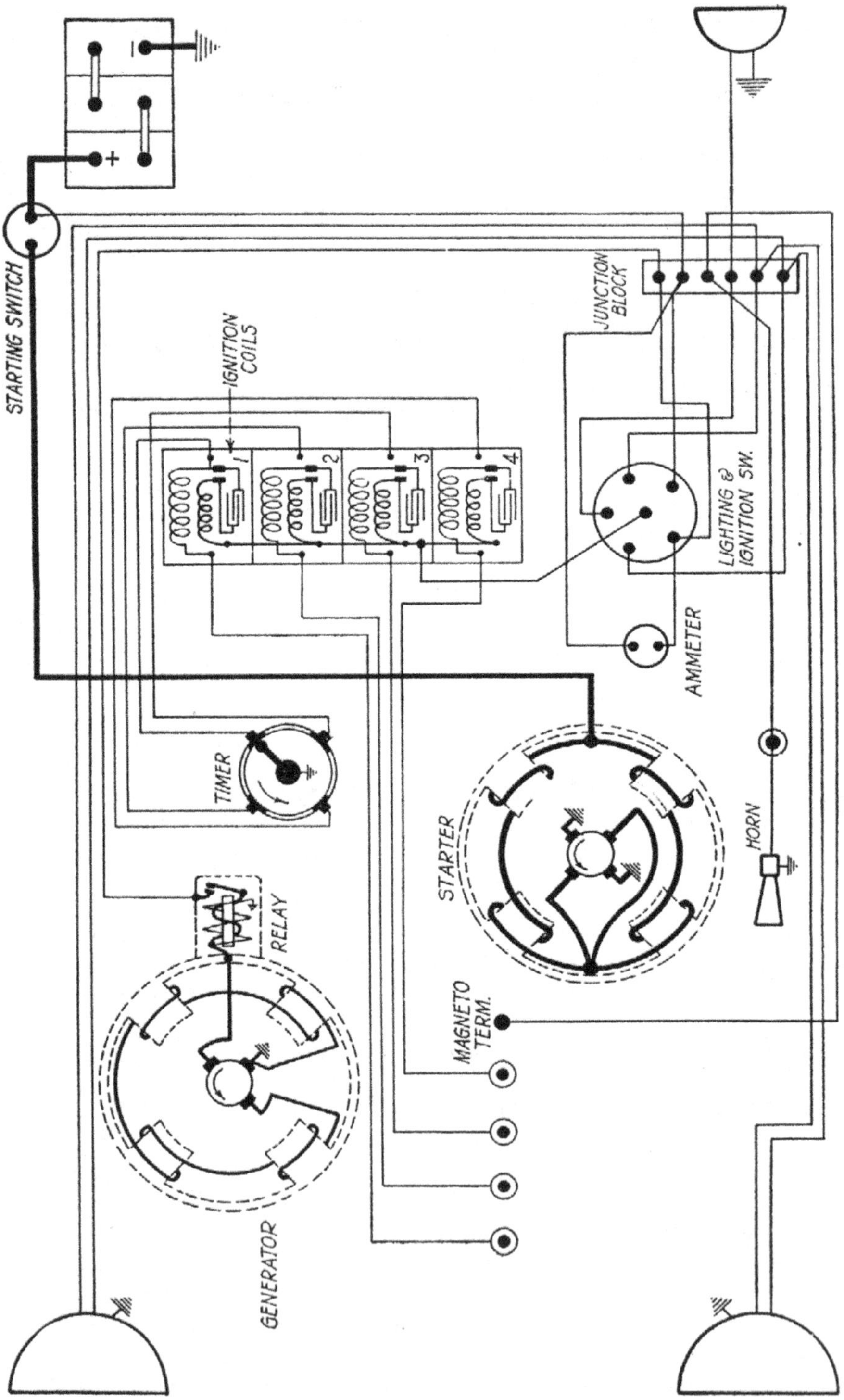

FORD WIRING DIAGRAM, 1924-1925, ALL MODELS

FORD Model T Year 1925

Own Starter & Generator Own Ignition

Regulation	Max. Chg. rate and speed
3rd Brush	12 Amps. 1800

RELAY Air Gap	Contact Gap	Cut-in R.P.M.
	.032"	750 or 10 M.P.H.

BATTERY Various Type Volts 6 Amps 80

Bat. to Frame Con. Neg. CONTACT BREAKER Gap .031 Vib

Firing Order 1-2-4-3 Ignition Timing T.D.C. Ret.

SPARK PLUG		ENGINE		Taxable Hp.
Size Pipe	Gap .030"	Bore $3\frac{3}{4}$"	Stroke 4"	23

INTAKE VALVE TIMING		EXHAUST VALVE TIMING	
Open $\frac{1"}{16}$ A.T.C. Piston	Close $\frac{9"}{16}$ A.B.C. Piston	Open $\frac{5"}{16}$ B.B.C. Piston	Close T.D.C. Piston

VALVE CLEARANCE Intake 1/32" Exhaust 1/32"

CARBURETOR	COOLING SYSTEM	OILING SYSTEM
Ford	Type Ther. Cap 3 gal.	Type Spl. Cap 4 qts.

PISTON RING: Width 1/4" Diam 3.75" Gap .004"-.015"

CLUTCH Disc GEAR RATIO 3.63 AXLE Semi F

BRAKES

Front	Rear	Hand
Trans $23\frac{1}{2}$"X1 $\frac{3"}{16}$ X $\frac{5"}{32}$		Shoe

Lighting	Headlights	Dash & Tail	Side Lamps
Single Contact	21 & 2 C.P.	3 C.P.	C.P.

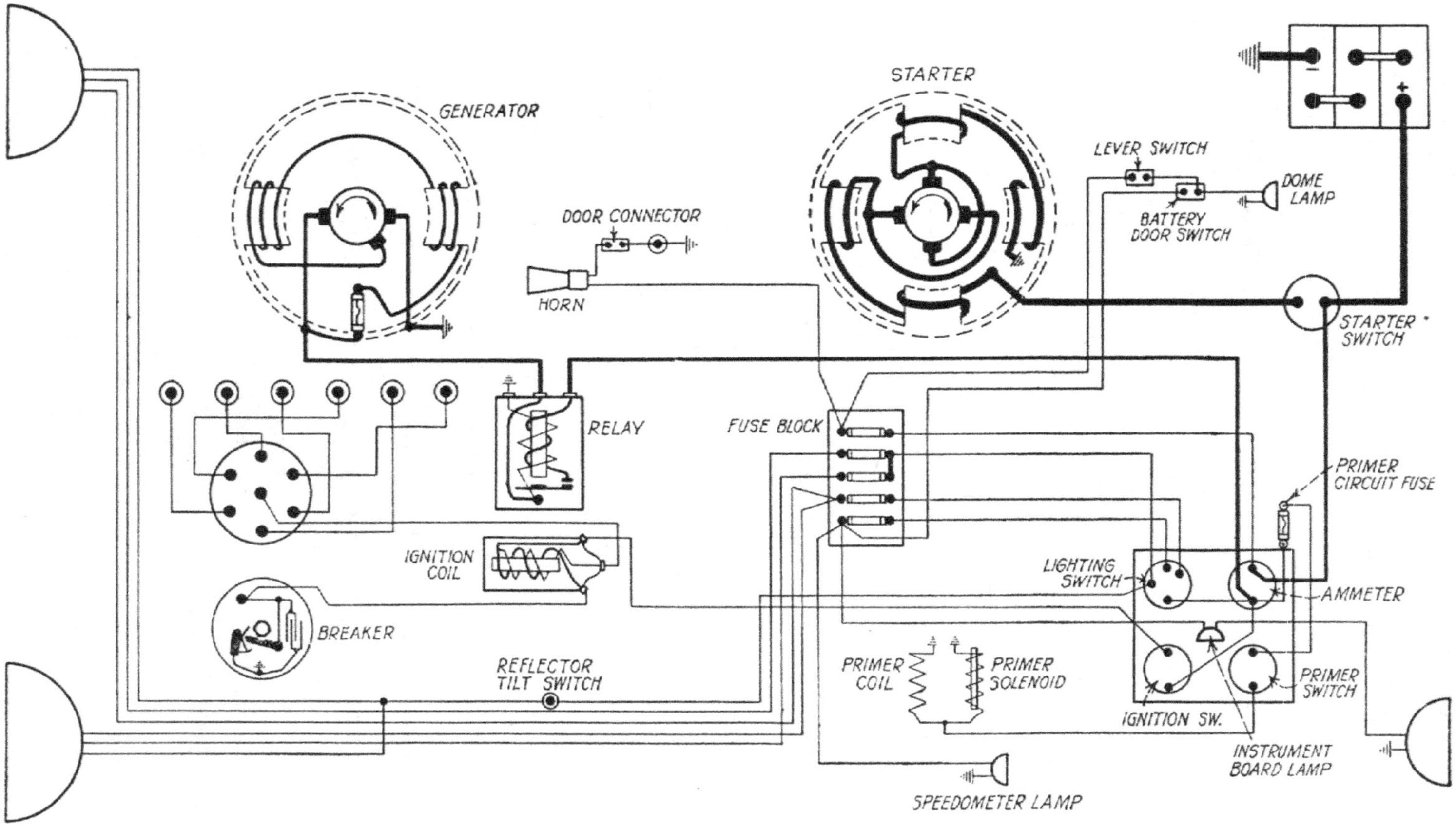

FRANKLIN WIRING DIAGRAMS, 1922-1923, MODEL 10-A, NORTH EAST STARTING AND LIGHTING, ATWATER-KENT IGNITION SYSTEMS

FRANKLIN Model 10 - A Year 1922-23.

North East Starter & Generator Atwater-Kent Ignition

GENERATOR Model	Regulation	Max. Chg. rate
No. L A - 3938	Third Brush	15 - 20 Amps.

Amps. Cut-in	Volts Warm Test	R.P.M.	Amps. Cut-in	Volts Cold Test	R.P.M.
0	6.5	600	18	7.0	1320
18	8.0	1700	19	7.5	1440

RELAY Model	Closes	Opens	Contact Gap	Air Gap
No. 8100	10 M P H	8 M P H	.025 - .040	.010

STARTING Motor Model	Drive	Rotation
K	Chain	Anti-Clockwise

BATTERY	Type	Capacity	Bat. to Frame Con.
Willard	S J R - 5	6 Volts 132 Amps.	Negative

IGNITION Coil Model	DISTRIBUTOR Model	Breaker Contact Gap
R A	R A	.015

FIRING ORDER	Ignition Timing	SPARK PLUG
1-4-2-6-3-5	1" B H C on F.W.	Size 7/8L Gap .030

LIGHTING	Headlamps	Dash & Tail	Side Lamps
Single Contact	6V - 21 C.P.	3V - 2 C.P.	6V - 4 C.P.

ENGINE		Cylinder	Valve tappet clearance
Bore $3\frac{1}{4}$	Stroke 4	No. 6 I Head	Inlet .01 Exhaust .01

CARBURETOR	COOLING	PISTON RING	OILING SYSTEM
Own 1"	Air	3 1/4 X 1/8	Cap. 5 qt. Type F & S

CLUTCH	REAR AXLE	Gear Ratio	TRANSMISSION
Plate 10"	Semi	4.72	Own

BRAKES	Position	Brake lining width & length
Rear wheel	Trans. or wheel T	Service $3\frac{1}{2}$ X 24 Hand $2\frac{5}{8}$ X 28

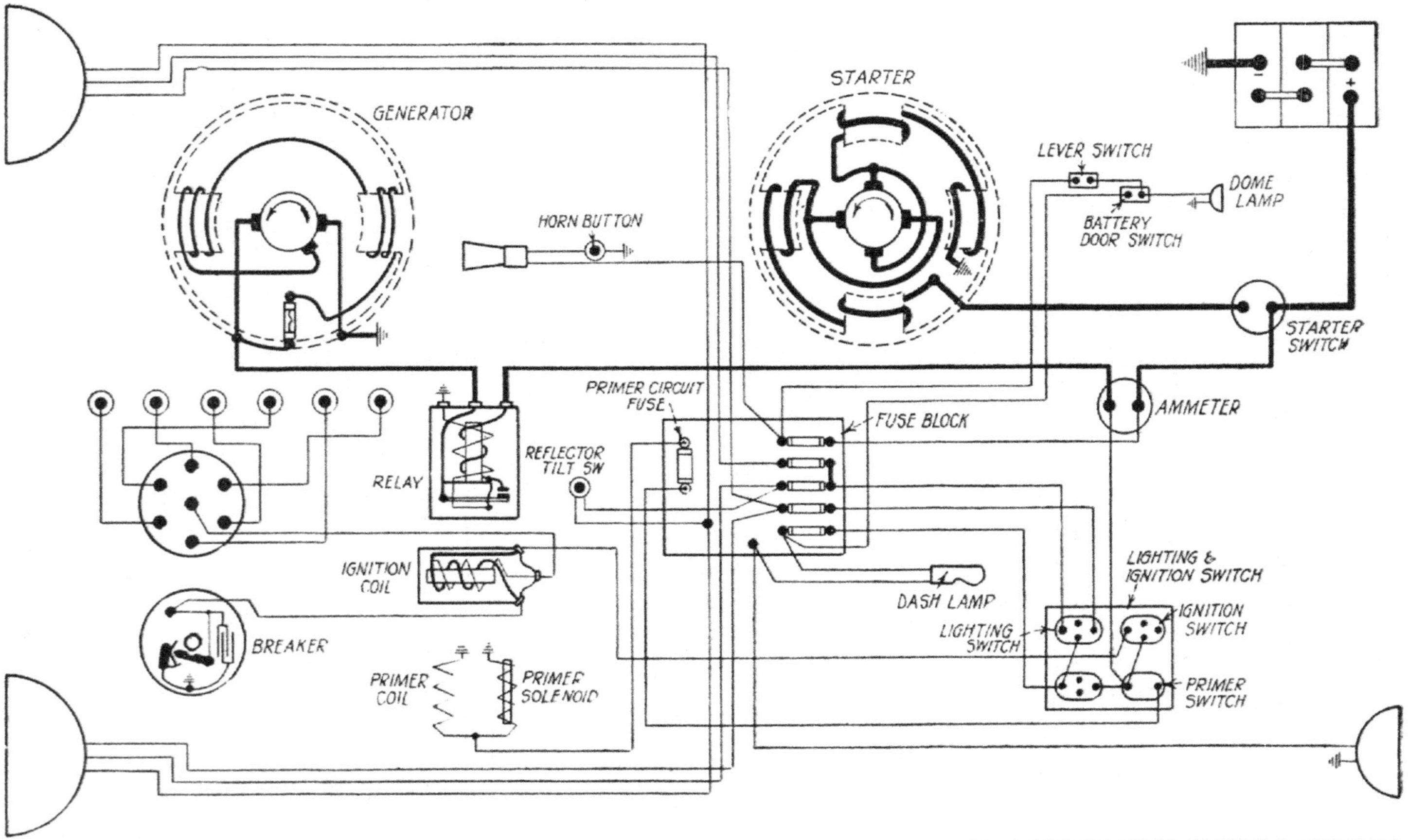

FRANKLIN WIRING DIAGRAM, 1923-1925, MODEL 10-B, NORTH EAST STARTING & LIGHTING, ATWATER KENT IGNITION SYSTEMS

Same for **10-C, 1925** Cars, except relay is on generator

FRANKLIN Model 10 - B Year 1923-24

North East Starter & Generator Atwater-Kent Ignition

GENERATOR Model	Regulation	Max. Chg. rate
No. L A - 3938	Third Brush	20 Amps.

Amps. Cut-in	Volts Warm Test	R.P.M.	Amps. Cut-in	Volts Cold Test	R.P.M.
0	6.5	600	18	7.0	1320
18	8.0	1700	19	7.5	1440

RELAY Model	Closes	Opens	Contact Gap	Air Gap
No. 8100	10 M P H	8 M P H	.025 - .040	.010

STARTING Motor Model	Drive	Rotation
K	Chain	Anti-Clockwise

BATTERY Willard Type Capacity Bat. to Frame Con.

S J R - 5 6 Volts 132 Amps. Negative

IGNITION Coil Model	DISTRIBUTOR Model	Breaker Contact Gap
R A	R A	.015

FIRING ORDER	Ignition Timing	SPARK PLUG
1-4-2-6-3-5	1" before T D C	Size 7/8" L Gap .030

LIGHTING	Headlamps	Dash & Tail	Side Lamps
Single Contact	6-8V-21 C.P.	3-4V-2 C.P.	6-8V-4 C.P.

ENGINE Cylinder Valve tappet clearance

Bore 3 1/4 Stroke 4 No. 6 I Head Inlet .010 Exhaust .010

CARBURETOR	COOLING	PISTON RING	OILING SYSTEM
Own 1"	Air	3 1/4 X 1/8	Cap 5 qt. Type F & S

CLUTCH	REAR AXLE	Gear Ratio	TRANSMISSION
Plate 10"	Semi	4.72	Own

BRAKES Position Brake lining width & length

Rear wheel Trans. or wheel T Service 3½ X 24 Hand 2⅝ X 28

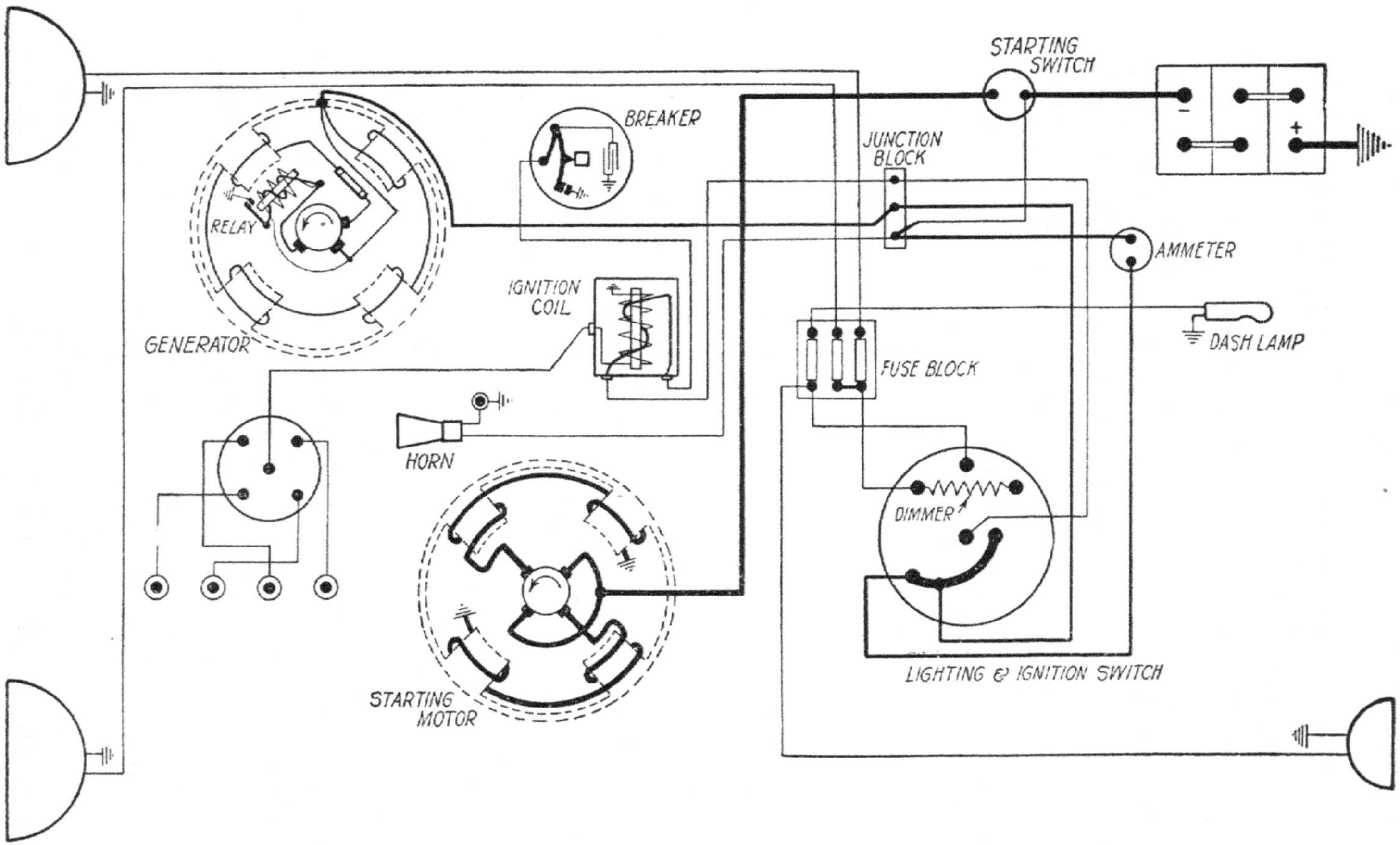

GARDNER WIRING DIAGRAM, 1923-1924, MODEL 55, WESTINGHOUSE EOUIPMENT

GARDNER Model 55 Year 1923-24

Westinghouse Starter & Generator Westinghouse Ignition

GENERATOR Model	Regulation	Max. Chg. rate
No. 331328	Third Brush	17 Amps.

Amps. Cut-in	Volts Warm Test	R.P.M.	Amps. Cut-in	Volts Cold Test	R.P.M.
0	6.5	500	8	8.0	750
10	8.0	1200	16	8.0	1600

RELAY Model	Closes	Opens	Contact Gap	Air Gap
No. 301210-A	9 M P H	7 M P H	.030 - .035	.005

STARTING Motor Model	Drive	Rotation
No. 297903	Bendix	Anti-Clockwise

BATTERY Willard Type	Capacity	Bat. to Frame Con.
S J R N - 3 or S W 4	6 Volts 132 Amps.	Positive

IGNITION Coil Model	DISTRIBUTOR Model	Breaker Contact Gap
No. 288761 A	No. 290356 B	.008 - .012

FIRING ORDER	Ignition Timing	SPARK PLUG
1 - 3 - 4 - 2	20° past T D C	Size 7/8 Gap .025

LIGHTING	Headlamps	Dash & Tail	Side Lamps
Single Contact	21 C.P.	2 C.P.	4 C.P.

ENGINE	Cylinder	Valve tappet clearance
Bore 3 11/16 Stroke 5	No. 4 L Head	Inlet .004 Exhaust .004

CARBURETOR	COOLING	PISTON RING	OILING SYSTEM
Zenith 1"	Thermo	3 11/16 X 1/8	Cap. 5qt. Type Force

CLUTCH	REAR AXLE	Gear Ratio	TRANSMISSION
Plate 10"	3/4 Flt	4.80	Mechanics

BRAKES	Position	Brake lining width & length
Rear wheel	Trans. or wheel T	Service 2 X 17 Hand 1¾ X 20¾

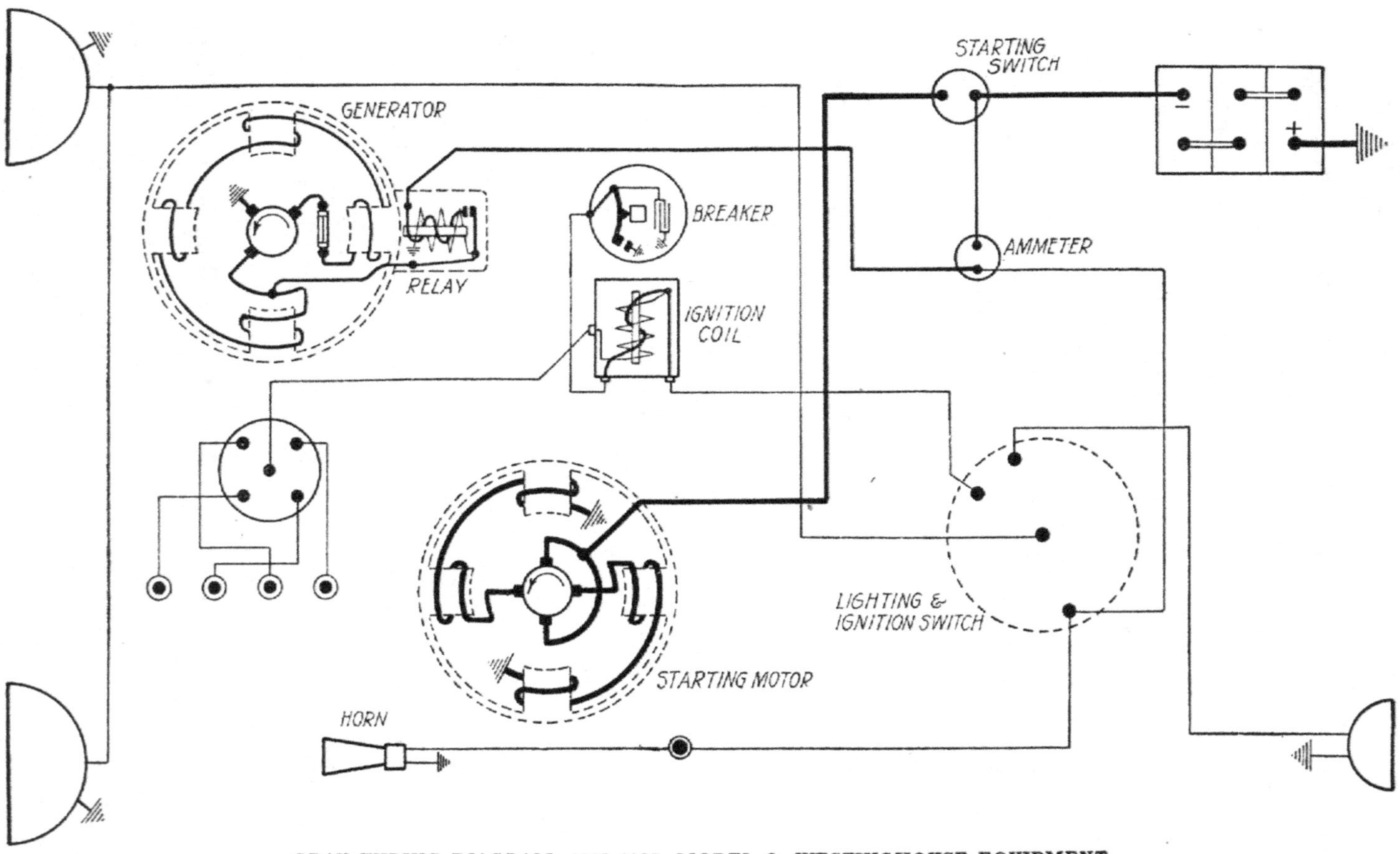

GRAY WIRING DIAGRAM, 1922-1925, MODEL O, WESTINGHOUSE EQUIPMENT

G R A Y Model All Year 1922-24

Westinghouse Starter & Generator Westinghouse Ignition

GENERATOR Model	Regulation	Max. Chg. rate
No. 34 - A T 1	Third Brush	14 Amps.

Amps. Cut-in	Volts Warm Test	R.P.M.	Amps. Cut-in	Volts Cold Test	R.P.M.
0	6.5	650	0	6.5	600
8-11	7.5	1500	12-14	7.5	1600

RELAY Model	Closes	Opens	Contact Gap	Air Gap
No. 301210 A	9 M P H	7 M P H	.030	.012

STARTING Motor Model	Drive	Rotation
No. 33 - A B	Bendix	Anti-Clockwise

BATTERY U S L Type Capacity Bat. to Frame Con.

C D N - 311 - X 6 Volts 80 Amps. Positive

IGNITION Coil Model	DISTRIBUTOR Model	Breaker Contact Gap
S C - 288761	S C - 303015	.012 - .016

FIRING ORDER	Ignition Timing	SPARK PLUG
1 - 3 - 4 - 2	T.D.C. at Retard	Size 7/8" Gap .025

LIGHTING	Headlamps	Dash & Tail	Side Lamps
Single Contact	15 C.P.	2 C.P.	C.P

ENGINE Cylinder Valve tappet clearance

Bore $3\frac{5}{8}$ Stroke 4 No. 4 L Head Inlet .004 Exhaust .006

CARBURETOR	COOLING	PISTON RING	OILING SYSTEM
Scoe 1	Thermo	3 5/8 X 3/16	Cap. 4 qt. Type C. S.

CLUTCH	REAR AXLE	Gear Ratio	TRANSMISSION
Disc. $9\frac{3}{4}$"	Semi	3.70	D. G. & M.

BRAKES Position Brake lining width & length

Rear wheel Trans. or wheel T ; Service $1\frac{1}{4}$ X 28 Hand $1\frac{1}{4}$ X $19\frac{1}{2}$

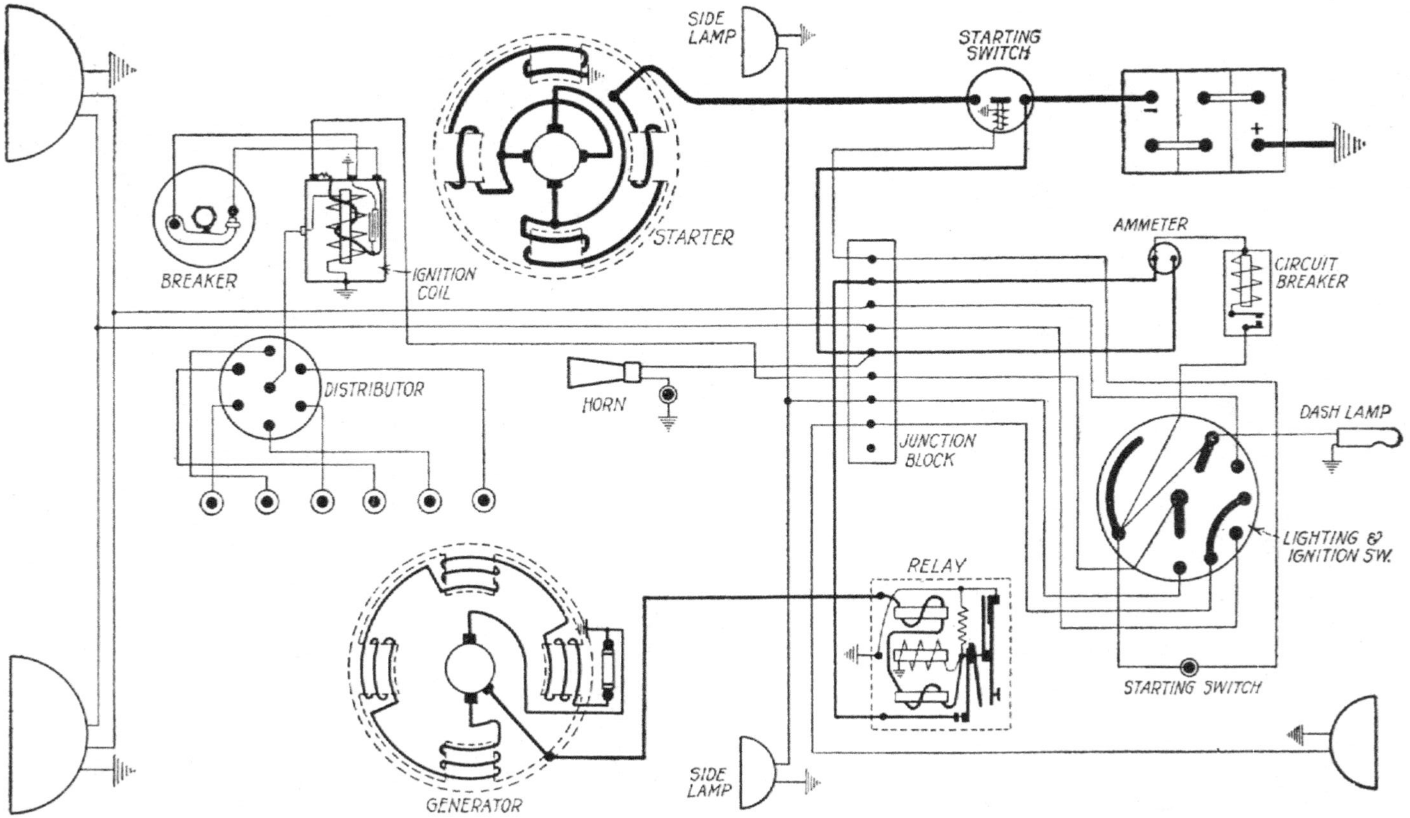

HAYNES WIRING DIAGRAM, 1923-1924, MODELS 50-55 & 60, LEECE-NEVILLE STARTING AND LIGHTING, KINGSTON IGNITION SYSTEMS

HAYNES Models 50-55 & 60 Year 1923-24

Leece-Neville Starter & Generator Kingston Ignition

GENERATOR Model	Regulation	Max. Chg. rate
No. 454 - G	Third Brush	18 Amps.

Amps. Cut-in	Volts Warm Test	R.P.M.	Amps. Cut-in	Volts Cold Test	R.P.M.
0	- - -	430	14.5	- - -	800
15	- - -	1200	18	- - -	120

RELAY Model	Closes	Opens	Contact Gap	Air Gap
No. 13-C C - 6	10 M P H	7 M P H	.047"	.063"

STARTING Motor Model	Drive	Rotation
No. 388 - M	Bendix	Anti-Clockwise

BATTERY Prest-O-L. Type	Capacity	Bat. to Frame Con.
615 S.H.C.	6 Volts 132 Amps.	Positive

IGNITION Coil Model	DISTRIBUTOR Model	Breaker Contact Gap
No. 68	No. 30 A	.018"

FIRING ORDER	Ignition Timing	SPARK PLUG
1-4-2-6-3-5	T. D. C. at Retard	Size 7/8" Gap .025"

LIGHTING	Headlamps	Dash & Tail	Side Lamps
Single Contact	21 C.P.	2 C.P.	6 C.P.

ENGINE	Cylinder	Valve tappet clearance
Bore 3½ Stroke 5	No. 6 L Head	Inlet .004" Exhaust .008"

CARBURETOR	COOLING	PISTON RING	OILING SYSTEM
Rayfield 1¼"	Pump	3½ X 1/4	Cap. 7 qt. Type F&S

CLUTCH	REAR AXLE	Gear Ratio	TRANSMISSION
Disc 10"	3/4 Flt	4.70	Own

BRAKES	Position	Brake lining width & length
Rear wheel	Trans. or wheel W	Service 1¾X46½ Hand 1¾X46½

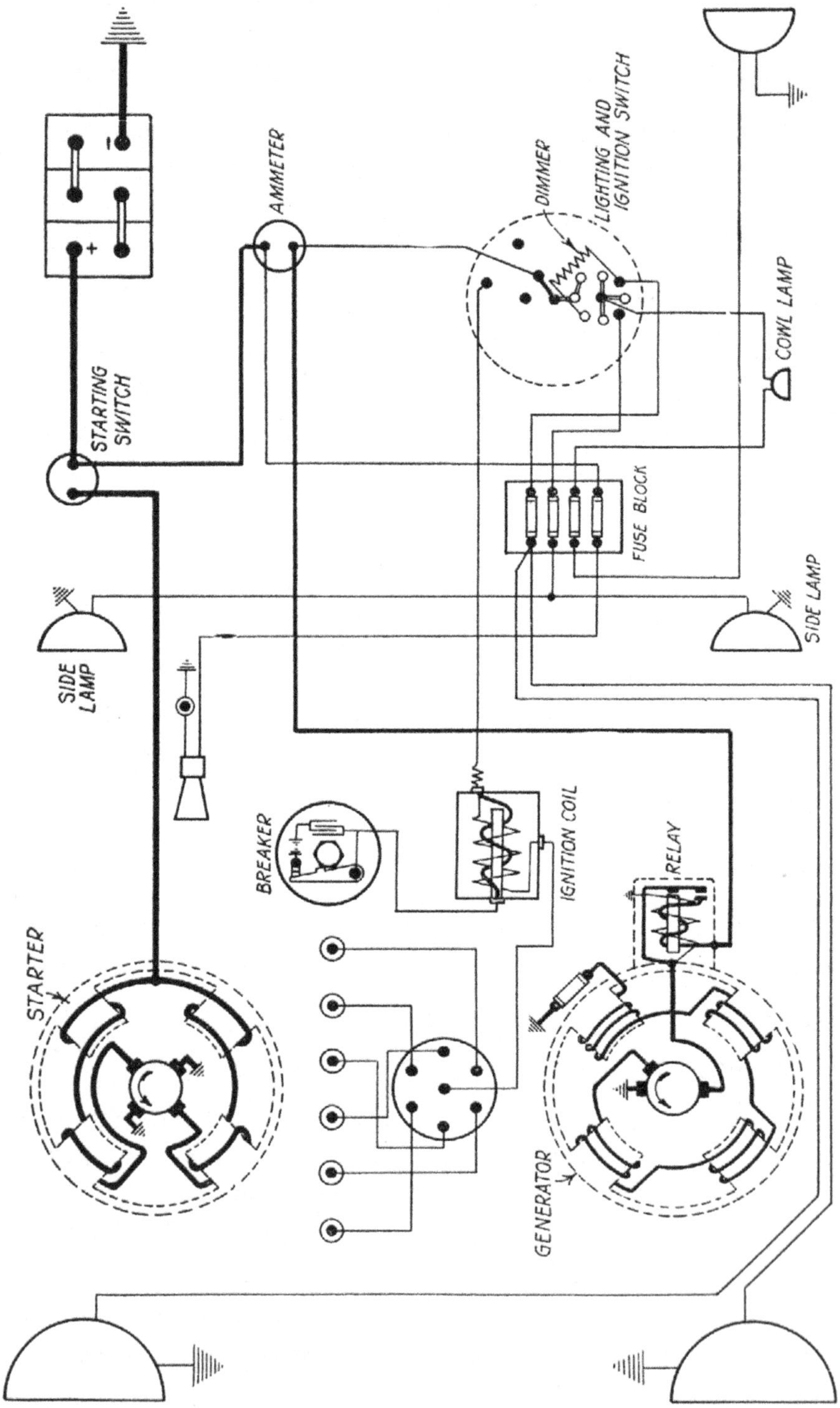

HUDSON WIRING DIAGRAM, 1923-1924, MODEL SUPER-SIX, AMERICAN-BOSCH EQUIPMENT

HUDSON Model Super-Six Year 1923-24

American-Bosch Starter & Generator American-Bosch Ignition

GENERATOR Model	Regulation	Max. Chg. rate
No. 1238	Third Brush	18 Amps.

Amps. Cut-in	Volts Warm Test	R.P.M.	Amps. Cut-in	Volts Cold Test	R.P.M.
10-12	8.5	1100	15-18	8.8	1100
12-14	8.7	1500	14-17	8.7	1500

RELAY Model	Closes	Opens	Contact Gap	Air Gap
No. CR - 1200	7 Volts	2 Amps.	.015	.010

STARTING Motor Model	Drive	Rotation
No. 926	O-R Clutch	Anti-Clockwise

BATTERY Exide Type Capacity Bat. to Frame Con.

3- X C - 15 - 1 6 Volts 100 Amps. Negative

IGNITION Coil Model	DISTRIBUTOR Model	Breaker Contact Gap
No. T C - 30	No. T C - 6100	.020

FIRING ORDER	Ignition Timing	SPARK PLUG
1-5-3-6-2-4	TDC at Retard	Size 7/8"R Gap .030

LIGHTING	Headlamps	Dash & Tail	Side Lamps
Single Contact	6 V 21 C.P.	3 V 2 C.P.	6 V 2 C.P.

ENGINE Cylinder Valve tappet clearance

Bore 3 1/2 Stroke 5 No. 6 L Head Inlet .003 Exhaust .005

CARBURETOR	COOLING	PISTON RING	OILING SYSTEM
Own 1 1/2	Pump	3 1/2 X 1/8	Cap. 7qt. Type C.S.

CLUTCH	REAR AXLE	Gear Ratio	TRANSMISSION
Disc 9 7/8"	Semi	4.45	Own

BRAKES	Position	Brake lining width & length	
Rear wheel	Trans. or wheel W	Service 2½ X 40	Hand 2½ X 40

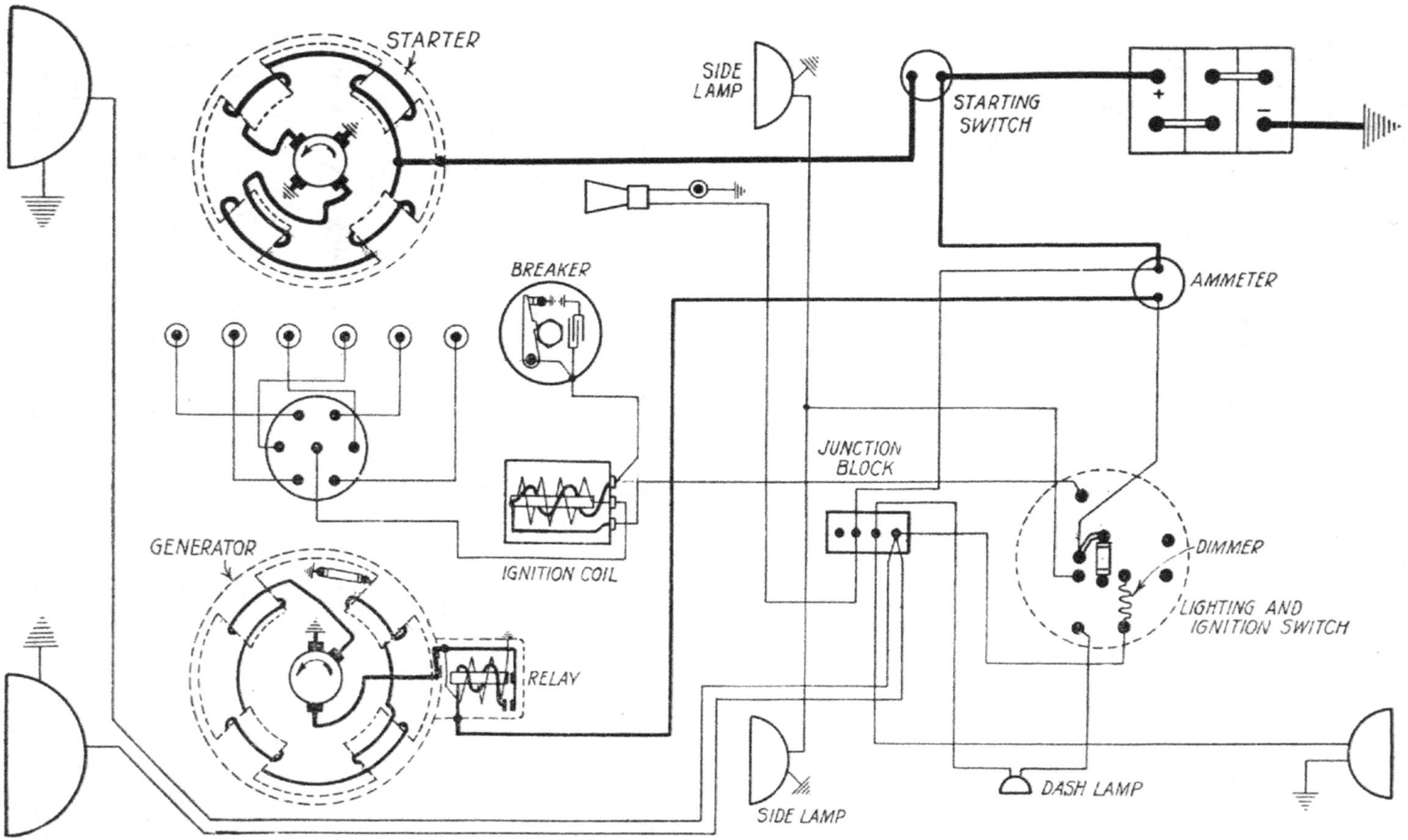

HUDSON WIRING DIAGRAM, 1925, SUPER-SIX MODEL, AMERICAN BOSCH SYSTEM

H U D S O N Model Super Six Year 1925

American Bosch Starter & Generator American Bosch Ignition

Regulation 3rd Brush

Max. Chg. rate and speed 17 Amps.

RELAY Air Gap .010" Contact Gap .015" Cut-in R.P.M.

BATTERY Prest-O-Lite Type 615 J F K Volts 6 Amps 100

Bat. to Frame Con. Neg. CONTACT BREAKER Gap .020"

Firing Order: 1-5-3-6-2-4 Ignition Timing "A" on FL. Adv.

SPARK PLUG Size Met Gap .025"

ENGINE Bore 3½" Stroke 5"

Taxable Hp. 29

INTAKE VALVE TIMING Open 7° A.T.C. Close 42° A.B.C.

EXHAUST VALVE TIMING Open 55° B.B.C. Close 8° A.T.C.

VALVE CLEARANCE Hot Intake .002" Exhaust .004"

CARBURETOR Stewart

COOLING SYSTEM Type P Cap 4 gal.

OILING SYSTEM Type Spl. Cap 7½ qts.

PISTON RING: Width Diam 3½" Gap .006"-.008"

CLUTCH Disc GEAR RATIO 4.55 AXLE ¾ F

BRAKES

Front

Rear 44½" X 2¼" X 3"/16

Hand 40" X 2¼" X 3"/16

Lighting Single Contact

Headlights 21 C.P.

Dash & Tail 3 V Series 2 C.P.

Side Lamps 2 C.P.

Dash D.C.

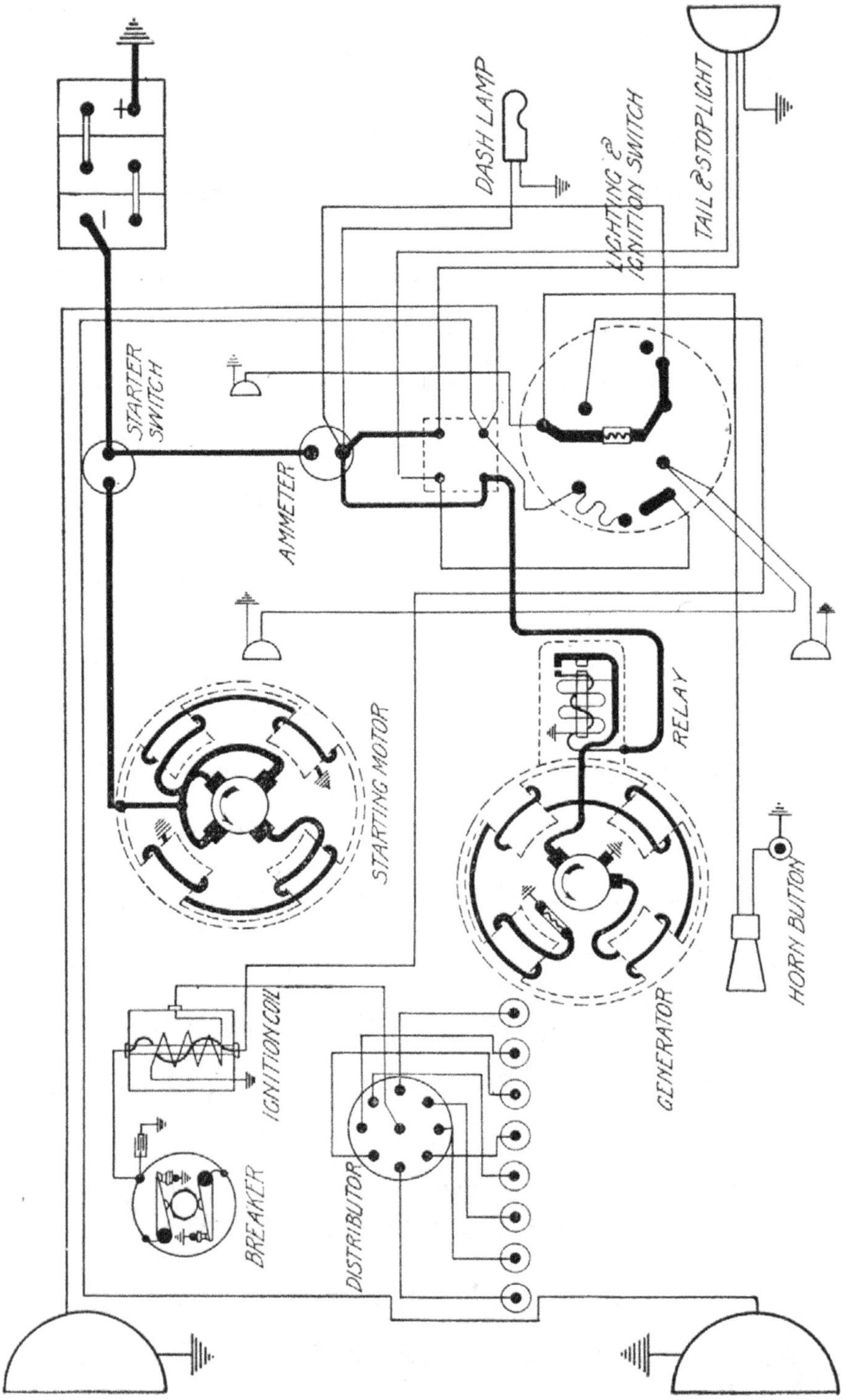

HUPMOBILE WIRING DIAGRAM, 1925, MODEL E-1, SERIAL NO. E-1 UP, WESTINGHOUSE LIGHTING AND STARTING, ATWATER KENT IGNITION

HUPMOBILE Model E 1 Year 1925

Westinghouse Starter & Generator Atwater Kent Ignition

Regulation	Max. Chg. rate and speed
3rd Brush	16-17 Amps. 1400

RELAY Air Gap	Contact Gap	Cut-in R.P.M.
.015"	.030"-.035"	

BATTERY Willard Type S J R - 6 Volts 6 Amps 155

Bat. to Frame Con. Pos. CONTACT BREAKER Gap .010"

Firing Order 1-5-2-3-8-4-7-6 Ignition Timing 8° A.T.C. Ret.

SPARK PLUG	ENGINE	Taxable Hp.
Size 7/8" Gap .022"	Bore 2 7/8" Stroke 4 3/4"	27

INTAKE VALVE TIMING	EXHAUST VALVE TIMING
Open 1°-5' B.T.C. Close 50° A.B.C.	Open 47°-5' B.B.C. Close 4° A.T.C.

VALVE CLEARANCE Hot Intake .007" Exhaust .007"

CARBURETOR	COOLING SYSTEM	OILING SYSTEM
Stromberg	Type P Cap 4¾ gal.	Type Pres. Cap 8 qts.

PISTON RING: Width 1/8" Diam 2 7/8" Gap

CLUTCH Plate 8¼" X 9"/64 GEAR RATIO 4.9 AXLE Semi F

BRAKES.

Front	Rear	Hand
35 5"/16 X 2" X 3"/16	35 5"/16 X 2" X 3"/16	22 3"/8 X 2" X 3"/16

Lighting	Headlights	Dash & Tail	Side Lamps
Single Contact	21 C.P.	3 C.P.	3 C.P.

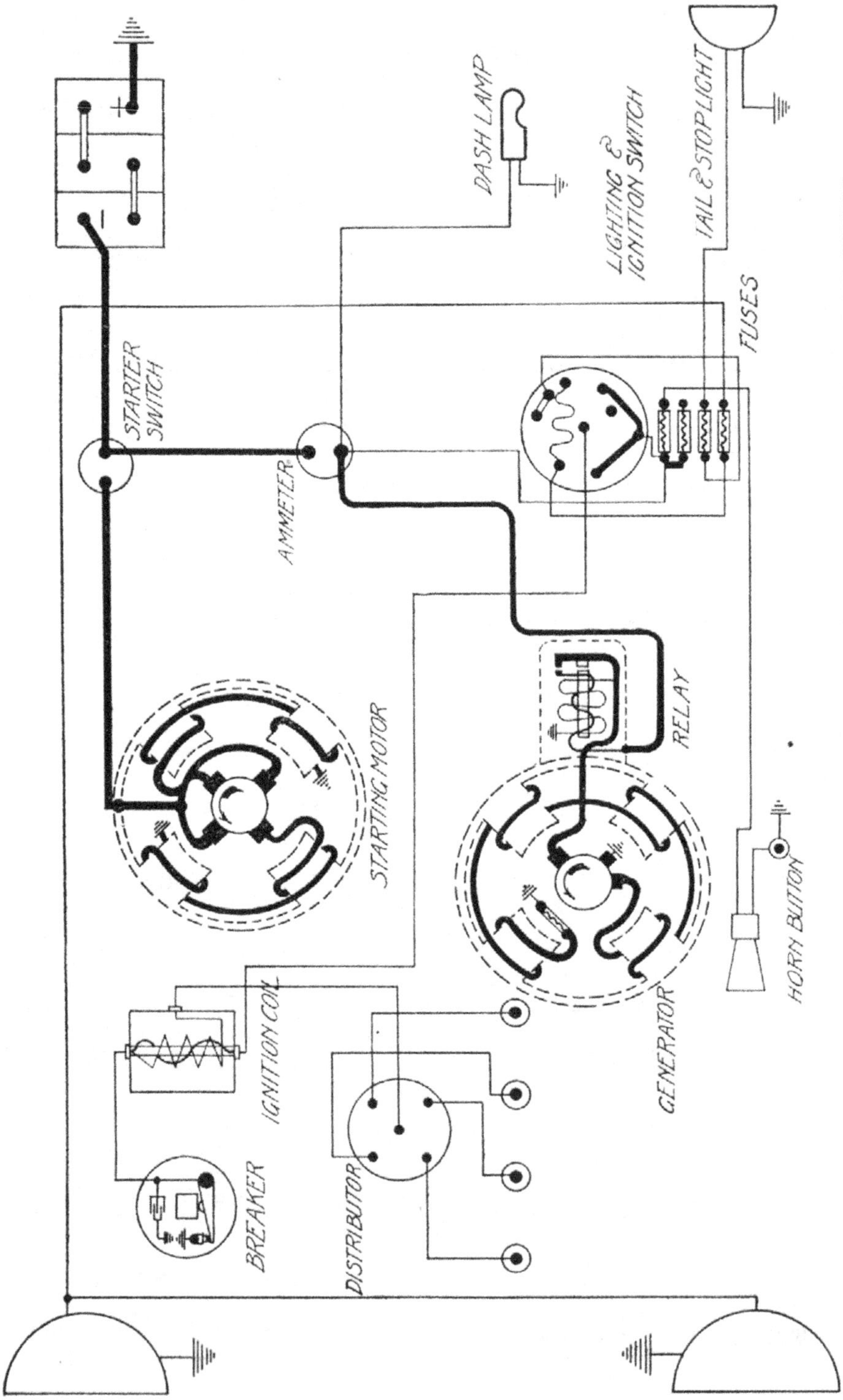

HUPMOBILE WIRING DIAGRAM, 1925, MODEL R-14, SERIAL NO. R-160397 UP, WESTINGHOUSE SYSTEM

H U P M O B I L E Model R 14 Year 1925

Westinghouse Starter & Generator Westinghouse Ignition

Regulation	Max. Chg. rate and speed
3rd Brush	15-17 Amps. 1250

RELAY Air Gap	Contact Gap	Cut-in R.P.M.
.005"	.030"-.035"	500 or 10 M.P.H.

BATTERY Willard Type CRR-15 Volts 6 Amps 105

Bat. to Frame Con. Pos. CONTACT BREAKER Gap .013"

Firing Order 1-2-4-3 Ignition Timing T.D.C. FL. 2" Past Ind. 1/2 Adv.

SPARK PLUG	ENGINE	Taxable Hp.
Size 7/8" Gap .020"	Bore 3 1/4" Stroke 5 1/2"	17

INTAKE VALVE TIMING	EXHAUST VALVE TIMING
Open 6° A.T.C. Close 46° A.B.C.	Open 48° B.B.C. Close 8° A.T.C.

VALVE CLEARANCE Hot Intake .007" Exhaust .007"

CARBURETOR	COOLING SYSTEM	OILING SYSTEM
Stromberg	Type Ther. Cap 6 gal.	Type Pres. Cap 6 qts.

PISTON RING: Width 1/8" Diam 3 1/4" Gap —

CLUTCH Plate 8 3"/4 X 9"/64 GEAR RATIO 4.9 AXLE Semi F

BRAKES

Front	Rear	Hand
	39 3"/8 X 2" X 3"/16	40 1"/8 X 1 3"/4 X 5"/32

Lighting	Headlights	Dash & Tail	Side Lamps
Single Contact	21 C.P.	2 C.P.	2 C.P.

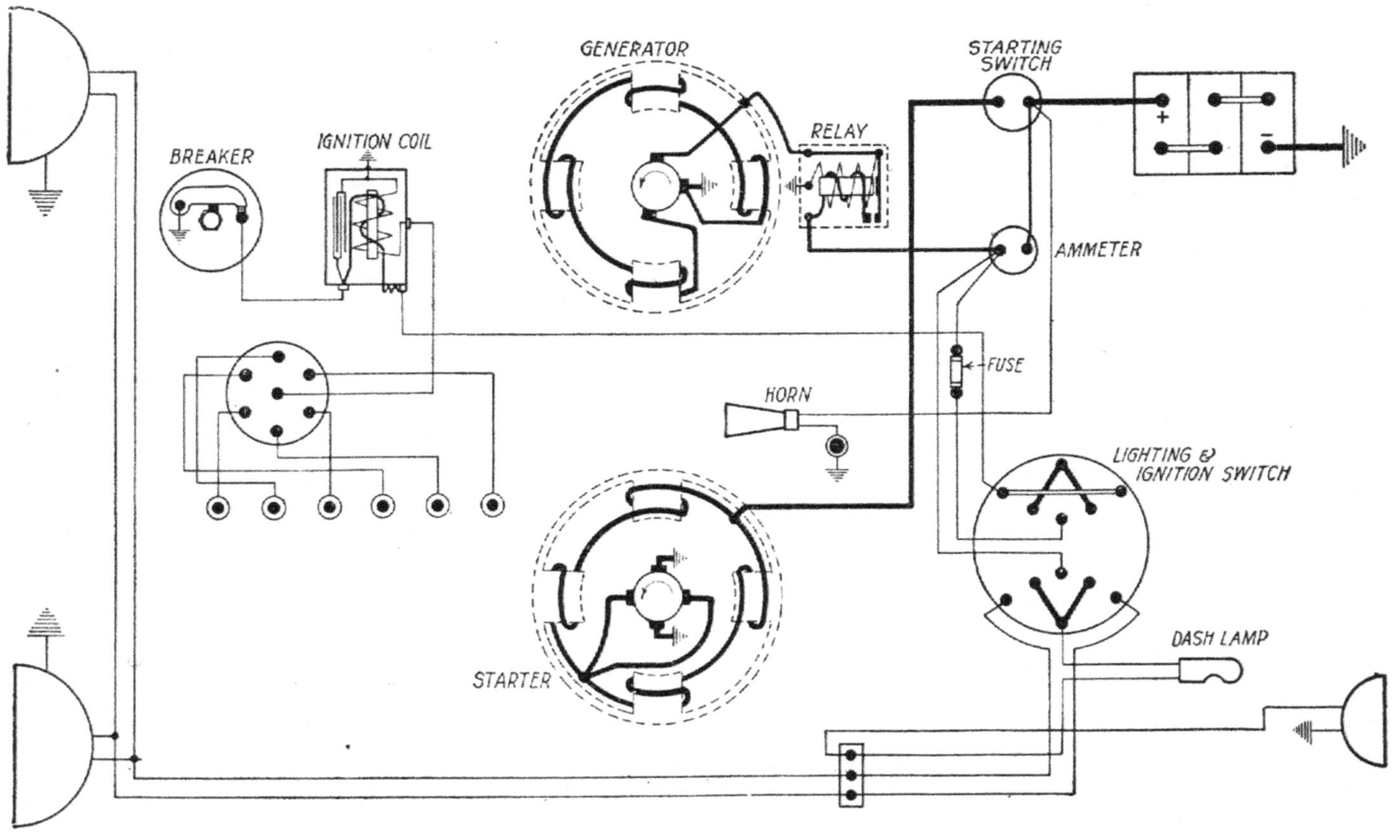

JACKSON WIRING DIAGRAM, 1923, MODEL 6-38, AUTO-LITE STARTING AND LIGHTING, REMY IGNITION SYSTEMS

JACKSON Model 6 - 38 Year 1923

Auto-Lite Starter & Generator Remy Ignition

GENERATOR Model	Regulation	Max. Chg. rate
G J - 1025 - E	Third Brush	16.5 Amps.

Amps. Cut-in	Volts Warm Test	R.P.M.	Amps. Cut-in	Volts Cold Test	R.P.M.
0	6.5	600	10	8.0	1000
8	7.5	800	16	8.0	1400

RELAY Model	Closes	Opens	Contact Gap	Air Gap
G J	10 MPH	8 MPH	.030 - .045	Fixed

STARTING Motor Model	Drive	Rotation
M H - 1025	Bendix	Anti-Clockwise

BATTERY Exide Type 3-XC-13-1 Capacity 6 Volts 90 Amps. Bat. to Frame Con. Negative

IGNITION Coil Model	DISTRIBUTOR Model	Breaker Contact Gap
No. 284 - B	No. 350 B-C	.020

FIRING ORDER	Ignition Timing	SPARK PLUG
1-5-3-6-2-4	TDC at Retard	Size 7/8 Gap .025

LIGHTING	Headlamps	Dash & Tail	Side Lamps
Single Contact	6 V 21 C.P.	3 V 2 C.P.	6 V 2 C.P.

ENGINE — Bore 3¼ Stroke 4½ — Cylinder No. 6 L Head — Valve tappet clearance Inlet .005 Exhaust .005

CARBURETOR	COOLING	PISTON RING	OILING SYSTEM
Stromberg 1¼"	Pump	3¼ x ¼	Cap 6 qt. Type Force

CLUTCH	REAR AXLE	Gear Ratio	TRANSMISSION
Disc 10"	3/4 Flt.	4.50	Covert

BRAKES — Rear wheel — Position: Trans. or wheel T — Brake lining width & length: Service 2 X 37 Hand 2 X 15¼

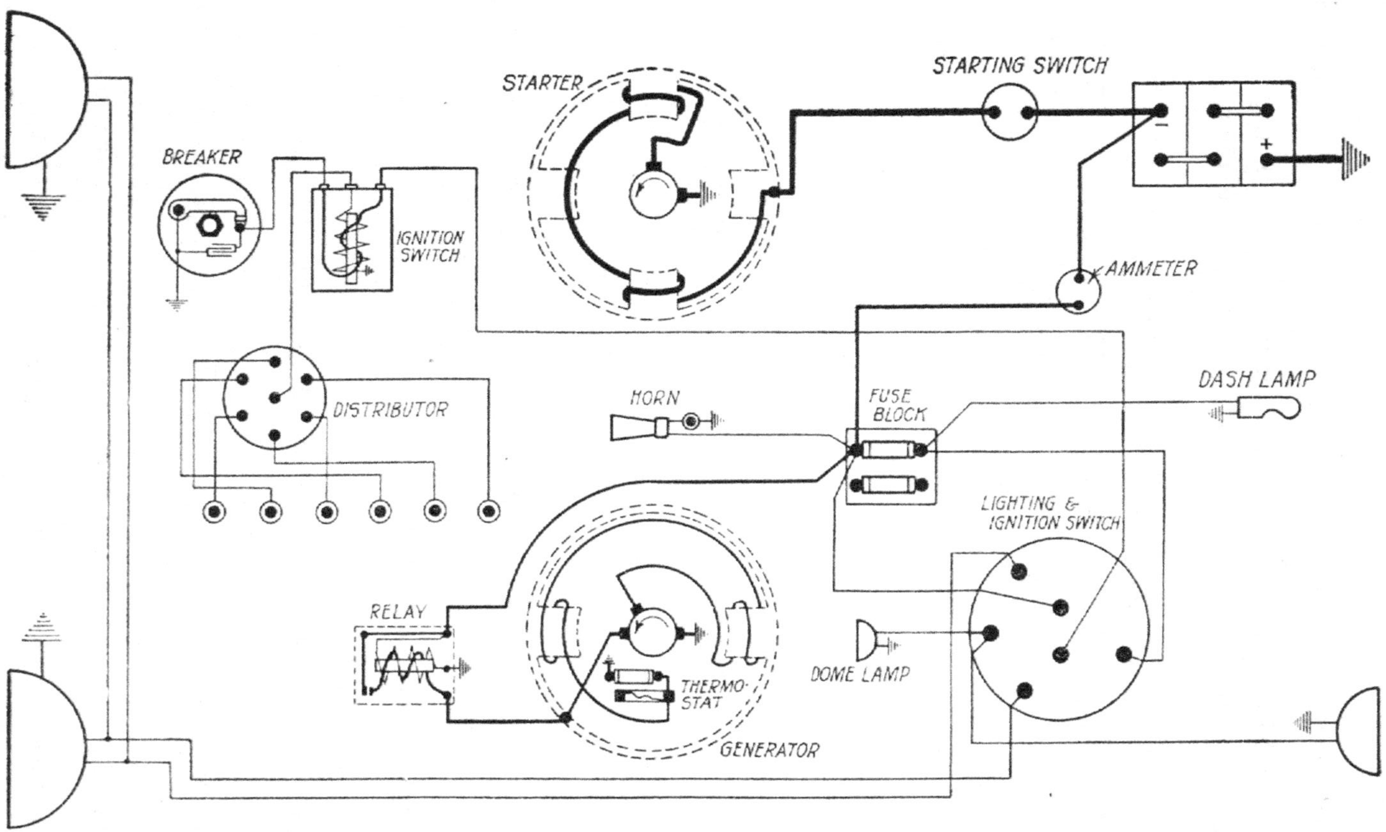

JEWETT WIRING DIAGRAM, 1923-1925, MODELS 6-50 AND 23-25, REMY STARTING & LIGHTING, ATWATER KENT IGNITION SYSTEMS

J E W E T T Model 6 - 50 Year 1923-24

Remy Starter & Generator Atwater-Kent Ignition

GENERATOR Model	Regulation	Max. Chg. rate
No. 913 E	Third Brush	18 Amps.

Amps. Cut-in	Volts Warm Test	R.P.M.	Amps. Cut-in	Volts Cold Test	R.P.M.
0	7.0	675	7	7.2	825
11	8.2	1900	20	8.5	1800

RELAY Model	Closes	Opens	Contact Gap	Air Gap
No. 266-A-B	6.9 V	6. V	.014"	.014"

STARTING Motor Model	Drive	Rotation
No. 713 - A - C	Bendix	Clockwise

BATTERY Westinghouse Type	Capacity	Bat. to Frame Con.
6 W B - 15	6 Volts 115 Amps.	Positive

IGNITION Coil Model	DISTRIBUTOR Model	Breaker Contact Gap
No. N - 3	No. 2 - A	.015"

FIRING ORDER	Ignition Timing	SPARK PLUG
1-5-3-6-2-4	"IN-OP-6"at Retard	Size 7/8" Gap .025

LIGHTING	Headlamps	Dash & Tail	Side Lamps
Single Contact	21 C.P.	4 C.P.	4 C.P.

ENGINE — Cylinder — Valve tappet clearance

Bore $3\frac{1}{4}$ Stroke 5 No. 6 L Head Inlet .006 Exhaust .010

CARBURETOR	COOLING	PISTON RING	OILING SYSTEM
Rayfield 1 1/4"	Pump	3 1/4 X 3/16	Cap. 6qt. Type F & S

CLUTCH	REAR AXLE	Gear Ratio	TRANSMISSION
Plate 7 3/4"	3/4 Flt	4.54	Warner - Gear

BRAKES	Position	Brake lining width & length
Rear wheel	Trans. or wheel T	Service 2 X 20 Hand 2 X $24\frac{1}{2}$

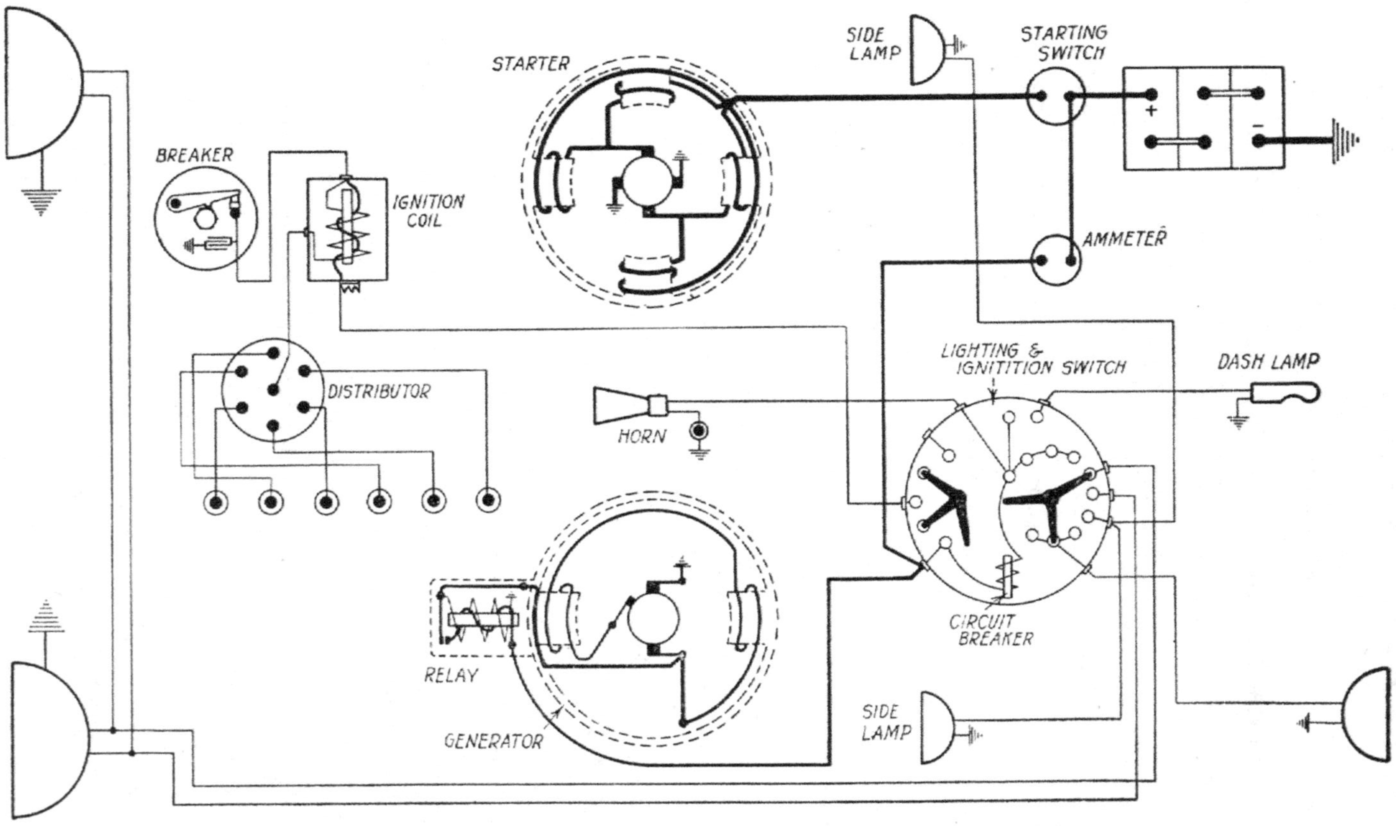

JORDAN WIRING DIAGRAM, 1923-1925, MODELS MX, H, K, AND L, DELCO EQUIPMENT

JORDAN Model M X - H Year 1923-24

Delco Starter & Generator Delco Ignition

GENERATOR Model			Regulation	Max. Chg. rate	
No. 256			Third Brush	15 Amps.	
Amps. Cut-in	Volts Warm Test	R.P.M.	Amps. Cut-in	Volts Cold Test	R.P.M.
5	- - -	700	6	- - -	700
15	- - -	1800	15	- - -	1600

RELAY Model	Closes	Opens	Contact Gap	Air Gap
No. 5754	8 MPH	5 MPH	.025 - .035	.025

STARTING Motor Model	Drive	Rotation
No. 208	Bendix	Clockwise

BATTERY Willard Type Capacity Bat. to Frame Con.

X R -15 6 Volts 110 Amps. Negative

IGNITION Coil Model	DISTRIBUTOR Model	Breaker Contact Gap
No. 2178	No. 5255	.022

FIRING ORDER	Ignition Timing	SPARK PLUG
1-5-3-6-2-4	3½" past TDC	Size 7/8" Gap .025

LIGHTING	Headlamps	Dash & Tail	Side Lamps
Single Contact	21 C.P.	2 C.P.	6 C.P.

ENGINE Cylinder Valve tappet clearance

Bore $3\frac{5}{16}$ Stroke $4\frac{3}{4}$ No. 6 L Head Inlet .004 Exhaust .006

CARBURETOR	COOLING	PISTON RING	OILING SYSTEM
Stromberg $1\frac{1}{4}$"	Pump	$3\frac{5}{16}$ X $\frac{3}{16}$	Cap 7 qt. Type H S

CLUTCH	REAR AXLE	Gear Ratio	TRANSMISSION
Plate 9 7/8	Semi	4.42 4.67	D. G. & M.

BRAKES	Position	Brake lining width & length
Rear wheel	Trans. or wheel W	Service 2 X $43\frac{1}{4}$ Hand $1\frac{3}{4}$ X 44

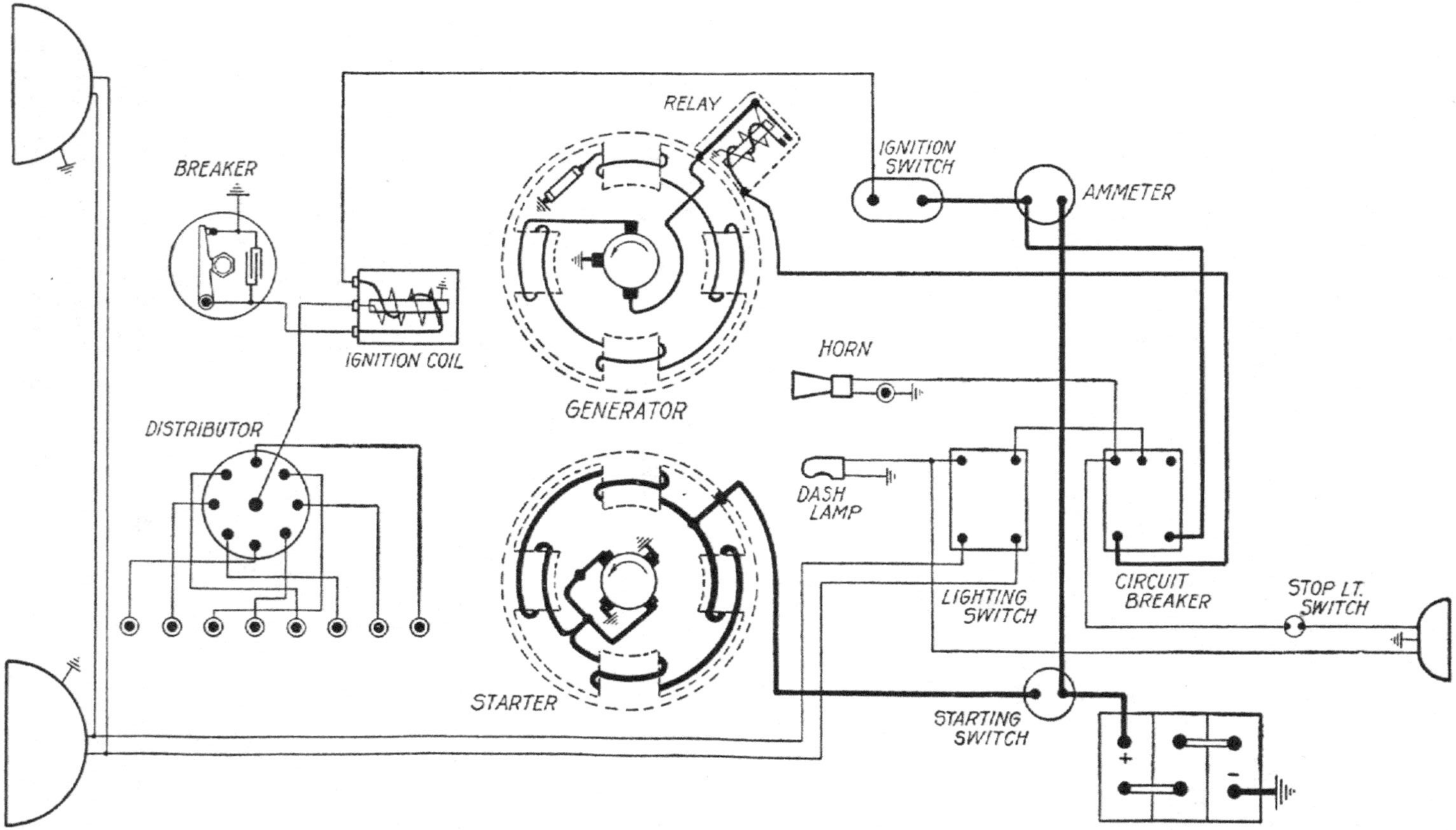

JORDAN WIRING DIAGRAM, GREAT LINE EIGHT, MODEL A, 1924-25, AMERICAN BOSCH SYSTEM

J O R D A N Model Line 8 A Year 1925

American Bosch Starter & Generator American Bosch Ignition

Regulation	Max. Chg. rate and speed
3rd Brush	17 Amps. 1300

RELAY Air Gap	Contact Gap	Cut-in R.P.M.
.010"	.015"	500

BATTERY Willard Type C R - 15 Volts 6 Amps 115

Bat. to Frame Con. Neg. CONTACT BREAKER Gap .014"-.016"

Firing Order 1-6-2-5-8-3-7-4 Ignition Timing T.D.C. ½ Adv.

SPARK PLUG	ENGINE	Taxable Hp.
Size 7/8" Gap .025"	Bore 3 5/16" Stroke 4 3/4"	26

INTAKE VALVE TIMING	EXHAUST VALVE TIMING
Open 12°A.T.C. Close 40°A.B.C.	Open 40°B.B.C. Close 8°A.T.C.

VALVE CLEARANCE Hot Intake .004" Exhaust .006"

CARBURETOR	COOLING SYSTEM	OILING SYSTEM
Stromberg	Type P Cap 5 gal.	Type Pres. Cap 8 qts.

PISTON RING: Width 3/16" Diam 3 5/16" Gap

CLUTCH Plate GEAR RATIO 4.45 AXLE Semi F

BRAKES

Front	Rear	Hand

Lighting	Headlights	Dash & Tail	Side Lamps
Single Contact	21 & 4 C.P.	2 C.P.	2 C.P.

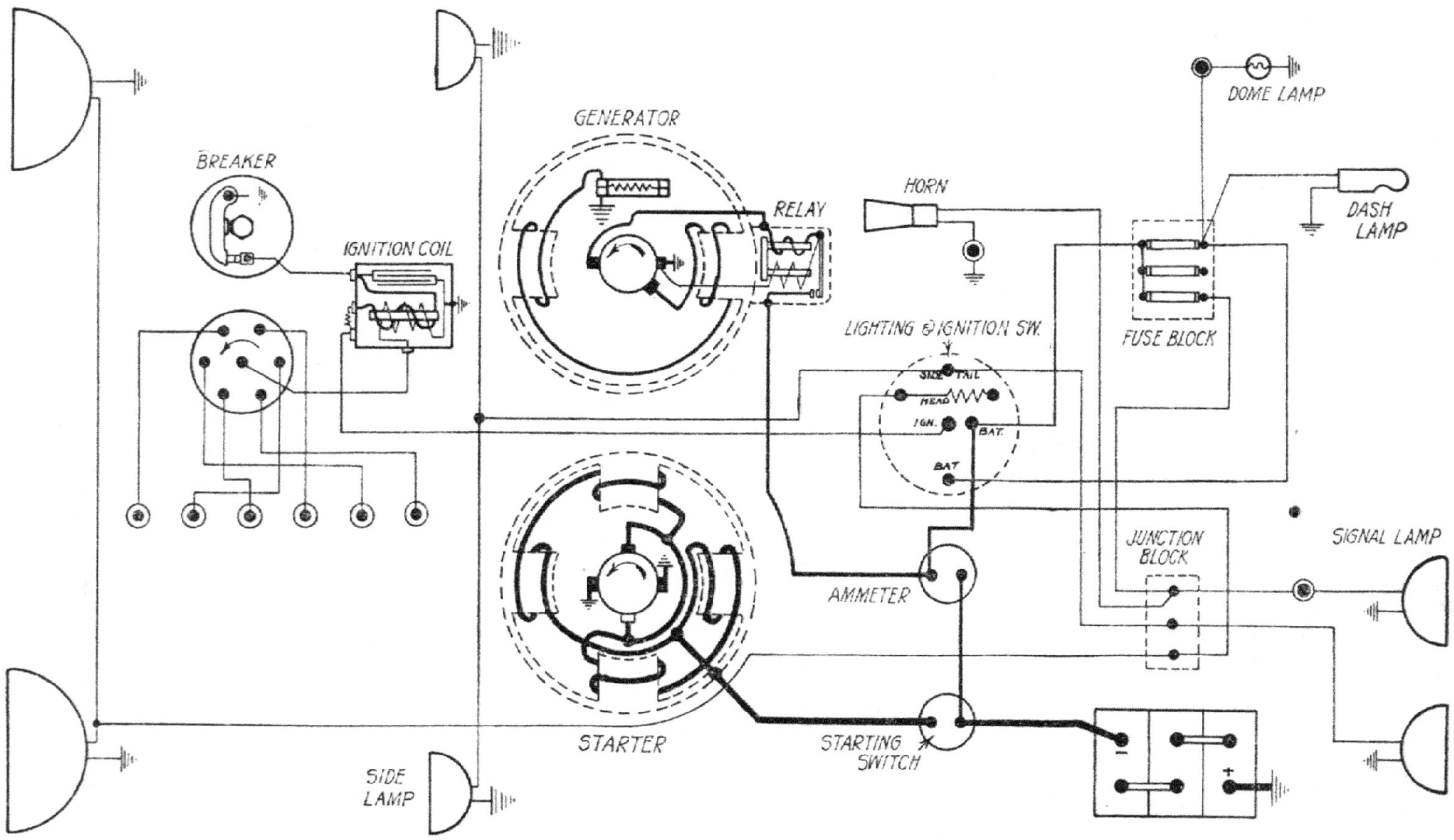

KISSEL WIRING DIAGRAM, 1923-1925, MODEL 55, REMY EQUIPMENT

K I S S E L Model 55 Year 1923-24

Remy Starter & Generator Remy Ignition

GENERATOR Model	Regulation	Max. Chg. rate
No. 912 - B	Third Brush	19 - 21 Amps.

Amps. Cut-in	Volts Warm Test	R.P.M.	Amps. Cut-in	Volts Cold Test	R.P.M.
7.0	7.2	750	0	7.0	675
11	8.2	1900	20	8.5	1800

RELAY Model	Closes	Opens	Contact Gap	Air Gap
No. 264 - C	7.2 Volts	2 Amps.	.012 - .015	.012

STARTING Motor Model	Drive	Rotation
No. 720 - D	Bendix	Anti-Clockwise

BATTERY Willard Type	Capacity	Bat. to Frame Con.
S J D - 4	6 Volts 120 Amps.	Positive

IGNITION Coil Model	DISTRIBUTOR Model	Breaker Contact Gap
No. 284 - A	No. 626 - B	.022 - .025

FIRING ORDER	Ignition Timing	SPARK PLUG
1-5-3-6-2-4	"TDC" 1¼" past	Size 7/8"L Gap .025

LIGHTING	Headlamps	Dash & Tail	Side Lamps
Single Contact	21 C.P.	6 C.P.	6 C.P.

ENGINE — Cylinder — Valve tappet clearance

Bore $3\frac{5}{16}$ Stroke $5\frac{1}{8}$ No. 6 L Head Inlet .008 Exhaust .008

CARBURETOR	COOLING	PISTON RING	OILING SYSTEM
Stromberg 1¼	Pump	3 5/16 X 3/16	Cap. 6qt. Type F&S

CLUTCH	REAR AXLE	Gear Ratio	TRANSMISSION
Plate 10"	3/4 Flt	4.42	Warner Gear

BRAKES 4-Wheel Optional Position — Brake lining width & length

Rear wheel Trans. or wheel W Service 2 X 45¼ Hand 2 X 21½

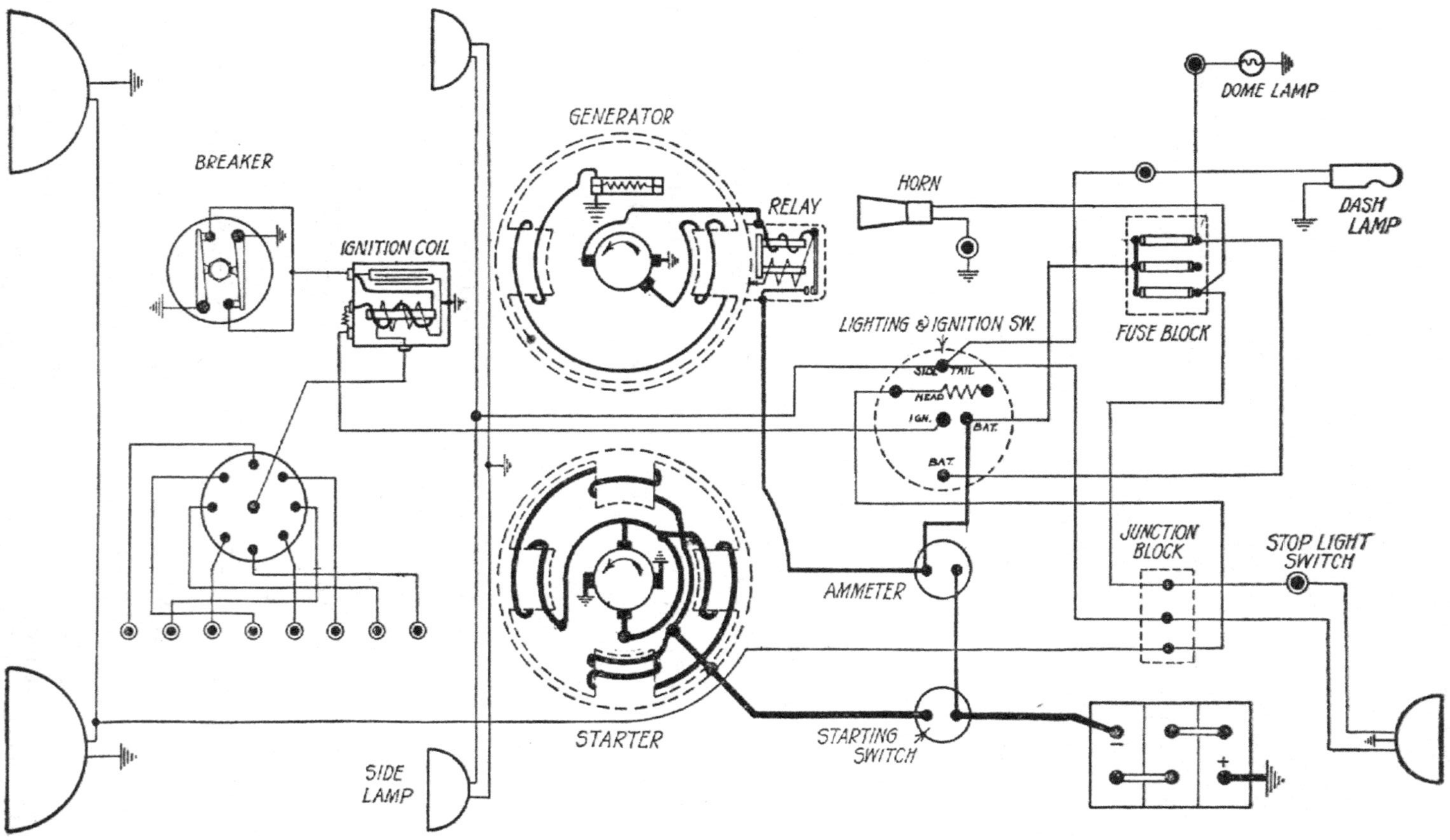

KISSEL WIRING DIAGRAM, 1925, MODEL 75, REMY SYSTEM

K I S S E L Model 75 Year 1925

Remy Starter & Generator Remy Ignition

Regulation	Max. Chg. rate and speed
3rd Brush	19 Amps. 1900

RELAY Air Gap	Contact Gap	Cut-in R.P.M.
.016"-.020"	.020"	650 or 9 M.P.H.

BATTERY Willard Type S J R - 4 Volts 6 Amps 110

Bat. to Frame Con. Pos. CONTACT BREAKER Gap .023"

Firing Order 1-6-2-5-8-3-7-4 Ignition Timing 10° A.T.C. Ret.

SPARK PLUG	ENGINE	Taxable Hp.
Size 7/8" Gap .030"	Bore 3 3/16" Stroke 4 1/2"	33

INTAKE VALVE TIMING	EXHAUST VALVE TIMING
Open 5° A.T.C. Close 35° A.B.C.	Open 55° B.B.C. Close T.D.C.

VALVE CLEARANCE Hot Intake .006" Exhaust .008"

CARBURETOR	COOLING SYSTEM	OILING SYSTEM
Schebler	Type P Cap 7½	Type Pres. Cap 8 qts.

PISTON RING: Width 1/8" Diam 3 3/16" Gap .006"-.018"

CLUTCH Plate GEAR RATIO AXLE Semi F

BRAKES

Front	Rear	Hand

Lighting	Headlights	Dash & Tail	Side Lamps
Single Contact	21 C.P.	2 C.P.	2 C.P.

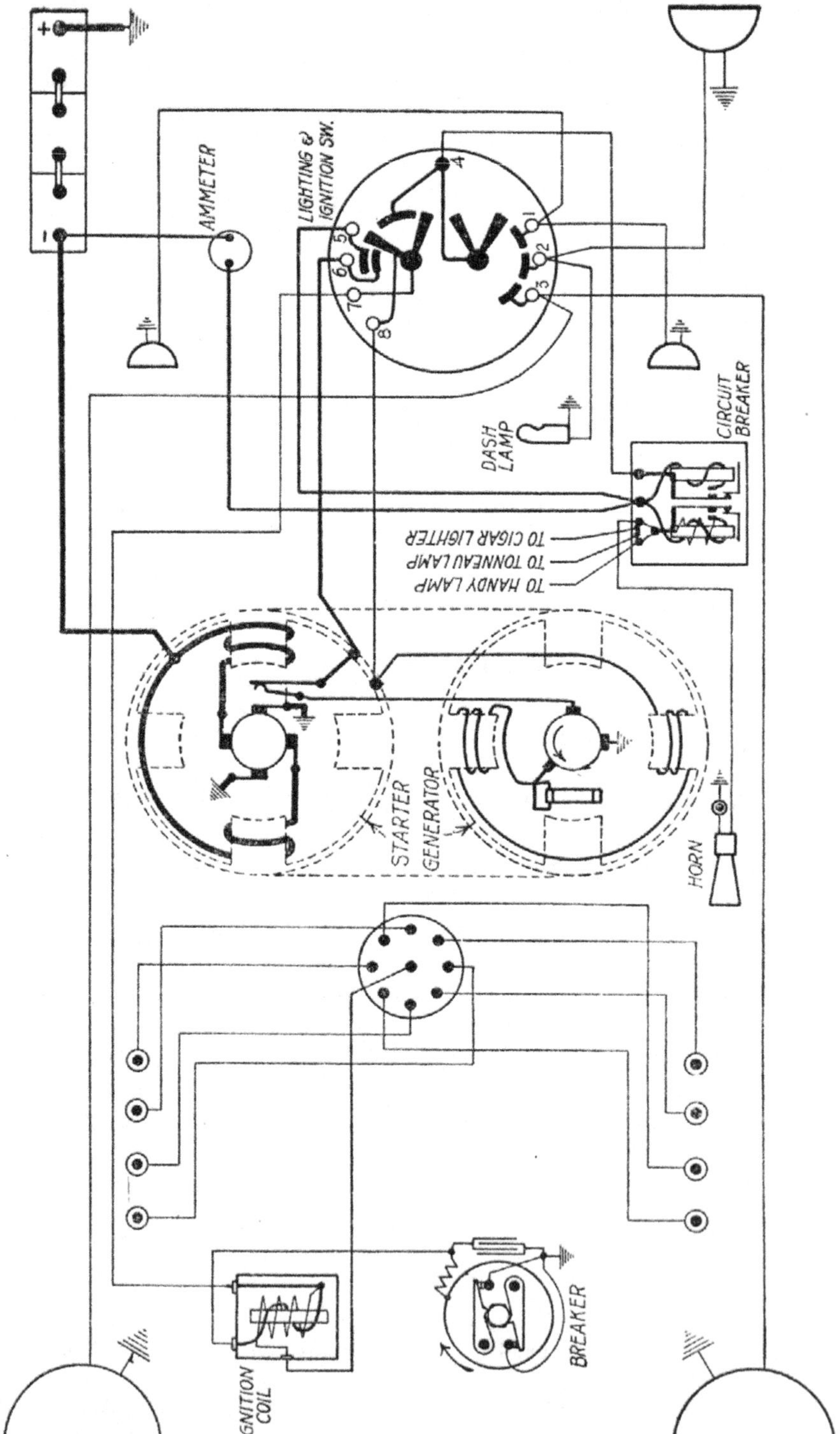

LA FAYETTE WIRING DIAGRAM, 1923-1924, MODEL 134, DELCO EQUIPMENT

L A F A Y E T T E Model 134 Year 1923-24

Delco Starter & Generator Delco Ignition

GENERATOR Model	Regulation	Max. Chg. rate
S - G - No. 187	Third Brush	15 Amps.

Amps. Cut-in	Volts Warm Test	R.P.M.	Amps. Cut-in	Volts Cold Test	R.P.M.
0	- - -	600	6	- - -	800
12	- - -	1000	15	- - -	2000

RELAY NONE USED. CONTACT MADE BY SWITCH.

STARTING Motor Model	Drive	Rotation
S - G - No. 187	Gear	Clockwise

BATTERY Exide Type — Capacity — Bat. to Frame Con.

3 - L X R V - 15 - 2 — 6 Volts 130 Amps. — Positive

IGNITION Coil Model	DISTRIBUTOR Model	Breaker Contact Gap
No. 2115	No. 5221	.018

FIRING ORDER	Ignition Timing	SPARK PLUG
1L-2R-3L-1R-4L-3R-2L-4R	TDC at 1/3 Adv.	Size 7/8" Gap .025

LIGHTING	Headlamps	Dash & Tail	Side Lamps
Single Contact	6 V 21 C.P.	3 V 2 C.P.	6 V 4 C.P.

ENGINE — Cylinder — Valve tappet clearance

Bore 3¼ Stroke 5¼ No. 8 L Head Inlet .003 Exhaust .003

CARBURETOR	COOLING	PISTON RING	OILING SYSTEM
Johnson 2"	Pump	3 1/4 X 3/16	Cap. 7 qt. Type Force

CLUTCH	REAR AXLE	Gear Ratio	TRANSMISSION
Disc 7¾	Full	4.58	Own

BRAKES	Position	Brake lining width & length	
Rear wheel	Trans. or wheel W	Service 2½ X 49	Hand 2½ X 49

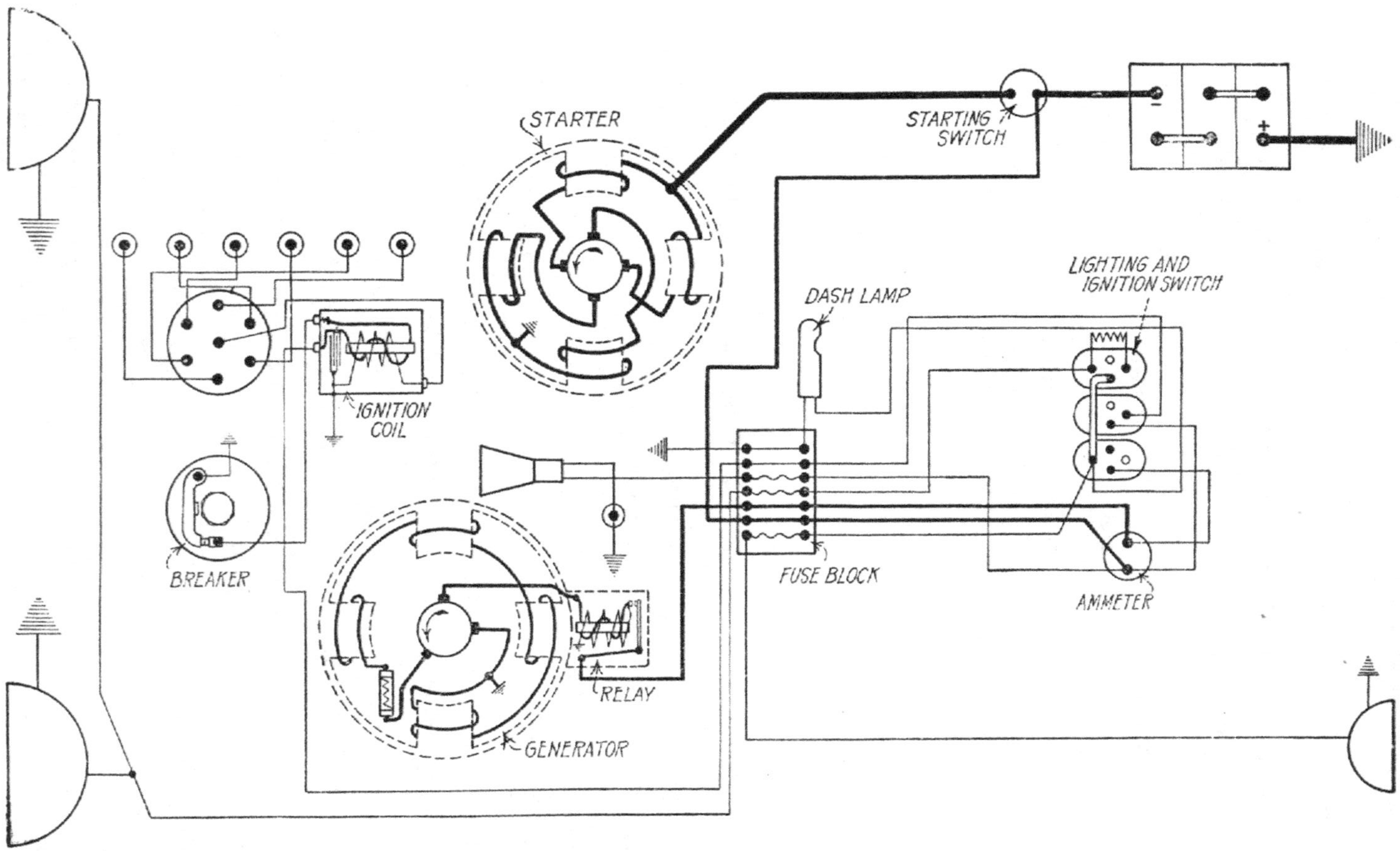

LEXINGTON WIRING DIAGRAM, 1923-1924, MODEL 23, GREY & DAVIS STARTING AND LIGHTING, CONNECTICUT IGNITION SYSTEMS

LEXINGTON Model 23 Year 1923-24

Grey & Davis Starter & Generator Connecticut Ignition

GENERATOR Model	Regulation	Max. Chg. rate
No. 647 - 650	Third Brush	15 Amps.

Amps. Cut-in	Volts Warm Test	R.P.M.	Amps. Cut-in	Volts Cold Test	R.P.M.
0	6.5	550	0	6.5	500
10	6.5	1800	14	6.5	1800

RELAY Model	Closes	Opens	Contact Gap	Air Gap
No. CR - 600	8 M P H	7 M P H	.015-.018	.012

STARTING Motor Model	Drive	Rotation
No. 532 - 534	Bendix	Anti-Clockwise

BATTERY Willard Type S J R N - 4 Capacity 6 Volts 100 Amps. Bat. to Frame Con. Positive

IGNITION Coil Model	DISTRIBUTOR Model	Breaker Contact Gap
No. 225-1-Y N	R 368 - 8 - Y	.012 - .015

FIRING ORDER	Ignition Timing	SPARK PLUG
1 - 5 - 3 - 6 - 2 - 4	TDC at retard	Size 7/8R Gap .020

LIGHTING	Headlamps	Dash & Tail	Side Lamps
Single Contact	21 C.P.	4 and 2 C.P.	4 C.P.

ENGINE Bore $3\frac{5}{16}$ Stroke $4\frac{1}{2}$ Cylinder No. 6 I Head Valve tappet clearance Inlet .010 Exhaust .010

CARBURETOR	COOLING	PISTON RING	OILING SYSTEM
Rayfield 1 1/2	Pump	3 5/16 X 3/16	Cap. 6qt. Type Force

CLUTCH	REAR AXLE	Gear Ratio	TRANSMISSION
Plate 9 1/4	Semi	5.10	Warner Gear

BRAKES Rear wheel Position Trans. or wheel T Brake lining width & length Service 45 X 2 Hand $24\frac{1}{4}$ X 2

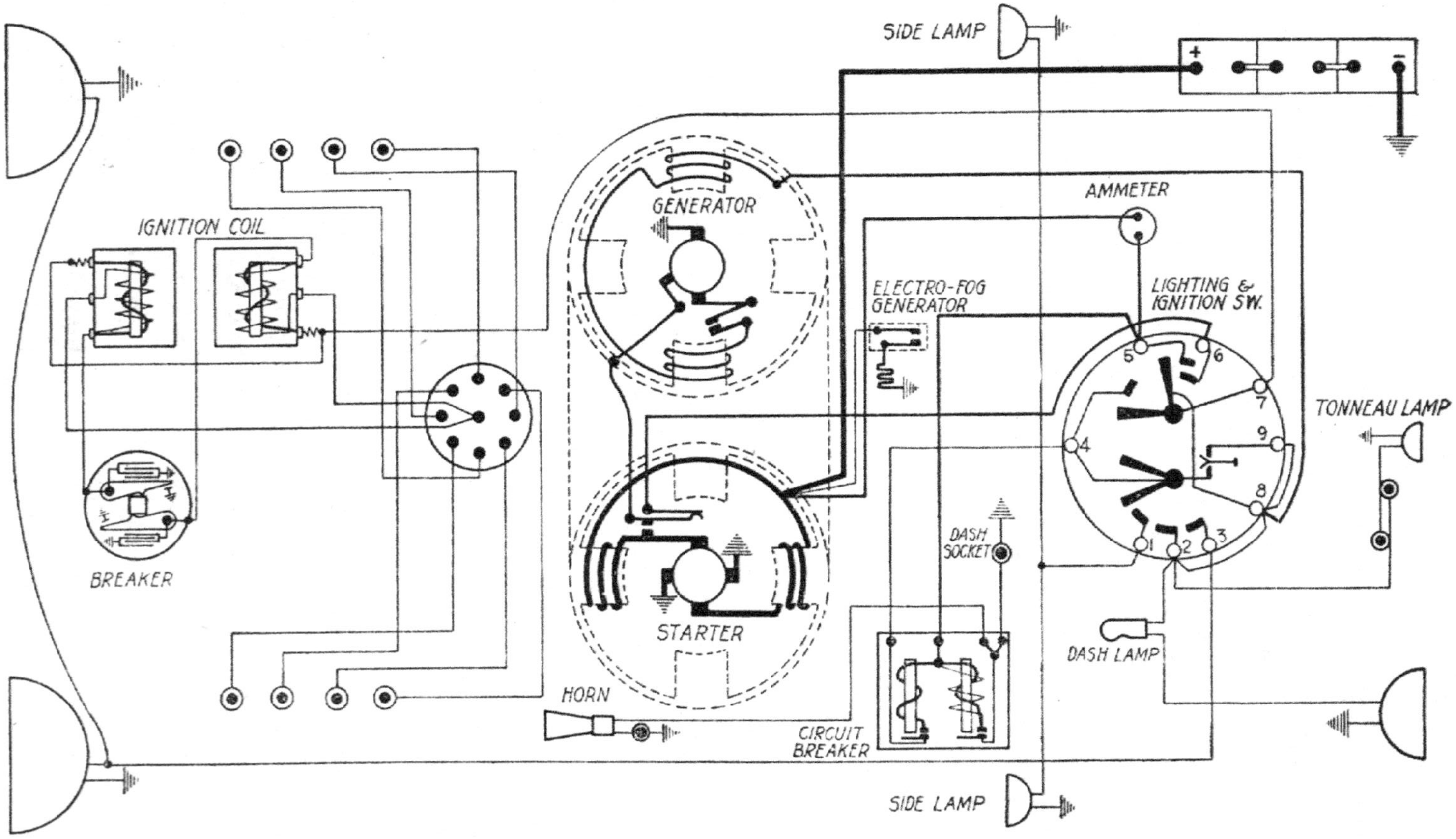

LINCOLN WIRING DIAGRAM, 1923-1925, DELCO EQUIPMENT

L.I.N.C.O.L.N. Model All. Year 1923-24

Delco Starter & Generator Delco Ignition

GENERATOR Model	Regulation	Max. Chg. rate
No. S - G - 193	Third Brush	16 amps.

Amps. Cut-in	Volts Warm Test	R.P.M.	Amps. Cut-in	Volts Cold Test	R.P.M.
3	- - -	600	6	- - -	600
16	- - -	1400	16	- - -	1200

RELAY NONE USED. CONTACT MADE BY SWITCH.

STARTING Motor Model	Drive	Rotation
No. S - G - 193	Gear	Clockwise

BATTERY Exide Type Capacity Bat. to Frame Con.

x-LRVX-15-2 6 Volts 135 Amps. Negative

IGNITION Coil Model	DISTRIBUTOR Model	Breaker Contact Gap
No. 2159	No. 5226	.020

FIRING ORDER Ignition Timing SPARK PLUG

1R-4L-2R-3L-4R-1L-3R-2L "Ret" at Retard Size 7/8" Gap .025"

LIGHTING Headlamps Dash & Tail Side Lamps

Single Contact 6V 21 C.P. 3V 2 C.P. 6V 4 C.P.

ENGINE Cylinder Valve tappet clearance

Bore $3\frac{3}{8}$ Stroke 5 No. 8 L Head Inlet .002 Exhaust .003

CARBURETOR	COOLING	PISTON RING	OILING SYSTEM
Stromberg $1\frac{1}{2}$"	Pump	3 3/8 X 1/8	Cap 10 qt. Type F&S

CLUTCH	REAR AXLE	Gear Ratio	TRANSMISSION
Disc 8 3/4"	Full	4.58	Own

BRAKES Position Brake lining width & length

Rear wheel Trans. or wheel W Service 3 X $49\frac{3}{4}$ Hand $2\frac{1}{2}$ X $46\frac{1}{2}$

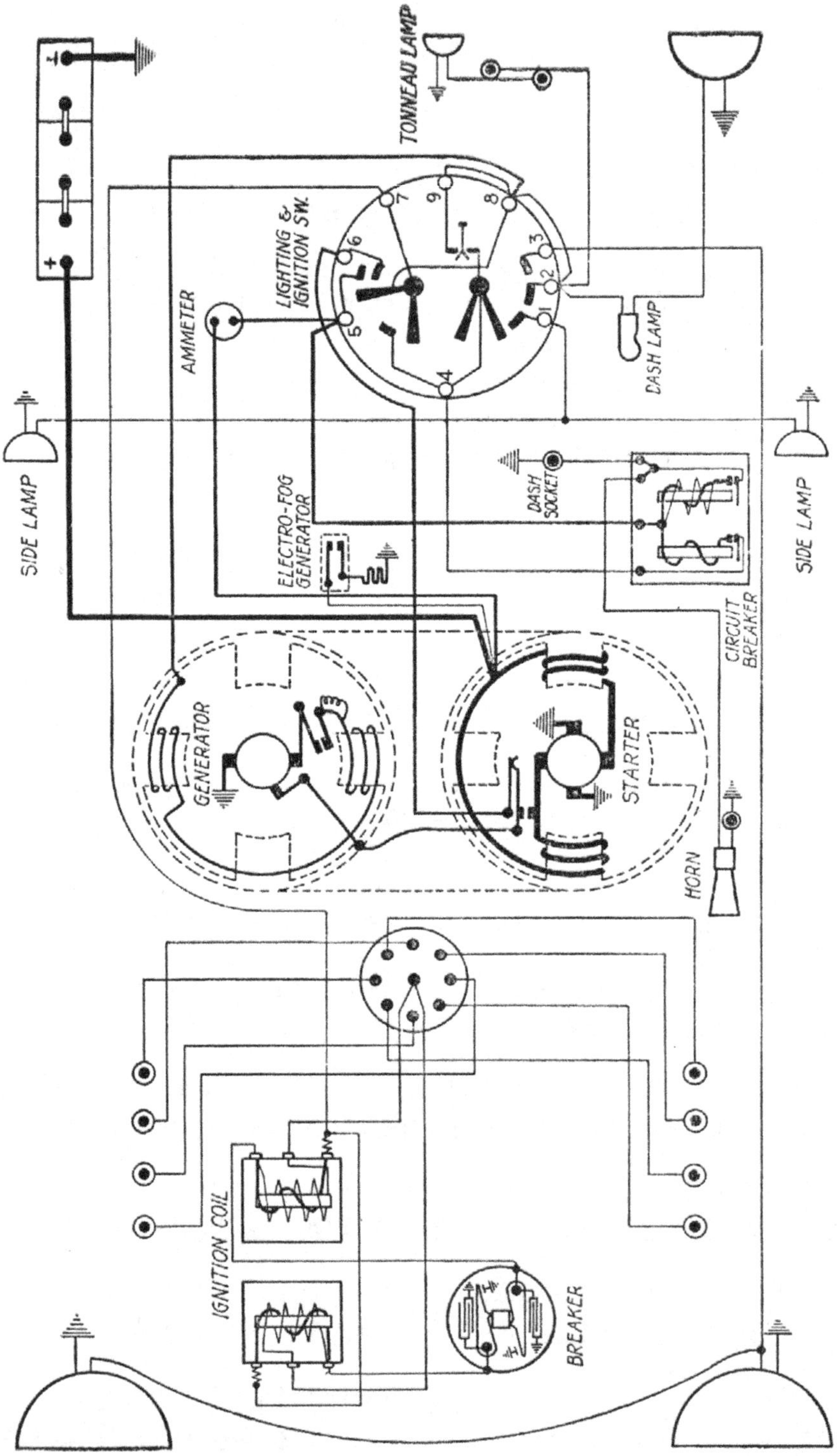

LINCOLN WIRING DIAGRAM, 1925

L I N C O L N .... Model .... Year 1925

Delco .... Starter & Generator .... Delco .... Ignition

Regulation .... Max. Chg. rate and speed

3rd Brush .... 15 Amps. 2000

RELAY Air Gap	Contact Gap	Cut-in R.P.M.
None		

BATTERY .... Exide .... Type 3 LRVX-15-2 Volts 6 Amps 135

Bat. to Frame Con. .... Neg. .... CONTACT BREAKER Gap .020"

Firing Order 1R - 4L - 2R - 3L
4R - 1L - 3R - 2L
Ignition Timing "Ret" on FL.Ret.

SPARK PLUG	ENGINE	Taxable Hp.
Size 7/8" Gap .025"	Bore 3 3/8" Stroke 5"	36

INTAKE VALVE TIMING	EXHAUST VALVE TIMING
Open $2\frac{1}{2}^{\circ}$ B.T.C. Close 46° A.B.C.	Open 48° B.B.C. Close T.D.C.

VALVE CLEARANCE .... Cold .... Intake .003" Exhaust .003"

CARBURETOR	COOLING SYSTEM	OILING SYSTEM
Stromberg	Type P Cap $7\frac{1}{2}$ gal.	Type Pres. Cap 10 qts.

PISTON RING: Width 1/8" Diam. 3.061" Gap .010"-.020"

CLUTCH Disc $8\frac{1}{4}$" X $\frac{5"}{32}$ GEAR RATIO 4-59 AXLE F F

BRAKES

Front	Rear	Hand
	$44\frac{1}{2}$" X 3" X $\frac{1}{4}$"	40 $\frac{13"}{16}$ X $2\frac{1}{2}$" X $\frac{3"}{16}$

Lighting	Headlights	Dash & Tail	Side Lamps
Single Contact	21 C.P.	2 C.P.	4 C.P.

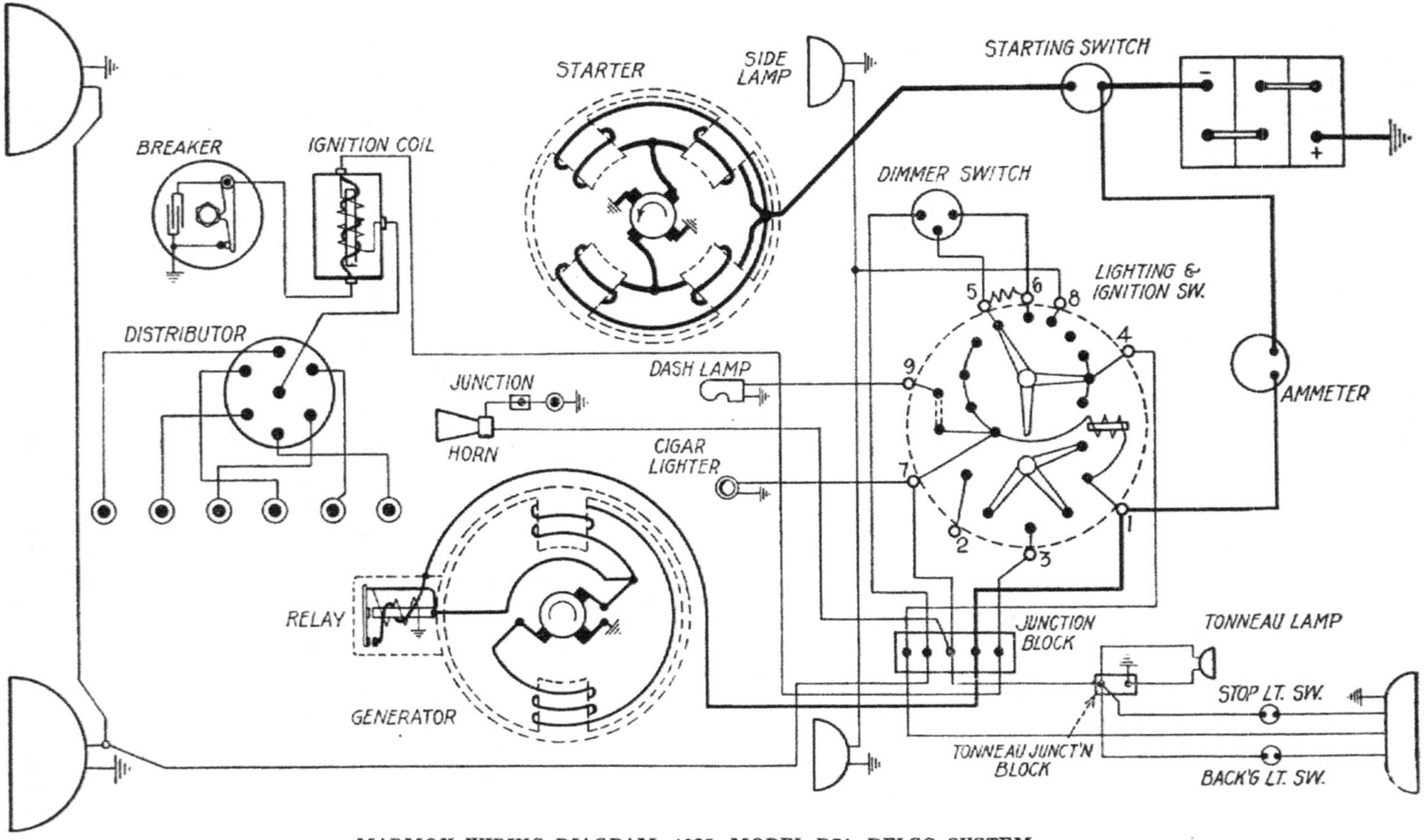

MARMON WIRING DIAGRAM, 1925, MODEL D74, DELCO SYSTEM

MARMON Model D 74 Year 1925

Delco Starter & Generator Delco Ignition

Regulation	Max. Chg. rate and speed
3rd Brush	15-18 Amps. 1600

RELAY Air Gap	Contact Gap	Cut-in R.P.M.
.025"-.035"	.025"-.035"	6-8 M.P.H.

BATTERY Prest-O-Lite Type 617 S H C Volts 6 Amps 170

Bat. to Frame Con. Pos. CONTACT BREAKER Gap .020"

Firing Order 1-5-3-6-2-4 Ignition Timing T.D.C. 5¾" Adv.

SPARK PLUG — ENGINE — Taxable Hp.

Size 7/8" Gap .025" Bore 3¾" Stroke 5 1/8" 34

INTAKE VALVE TIMING — EXHAUST VALVE TIMING

Open T.D.C. Close 35° A.B.C. Open 51° B.B.C. Close 6° A.T.C.

VALVE CLEARANCE Hot Intake .006" Exhaust .010"

CARBURETOR — COOLING SYSTEM — OILING SYSTEM

Stromberg Type P Cap 5½ gal. Type Pres. Cap 10 qts.

PISTON RING: Width .1875" Diam 3.75" Gap .002"-.007"

CLUTCH Disc. 6½" X 8½" GEAR RATIO 4.1 AXLE 3/4 F.

BRAKES

Front	Rear	Hand
27½" X 2" X 3/16"	54" X 2½" X 3/16"	47 3"/16 X 2" X 3"/16

Lighting	Headlights	Dash & Tail	Side Lamps
Single Contact	4 C.P.	2 C.P.	2 C.P.

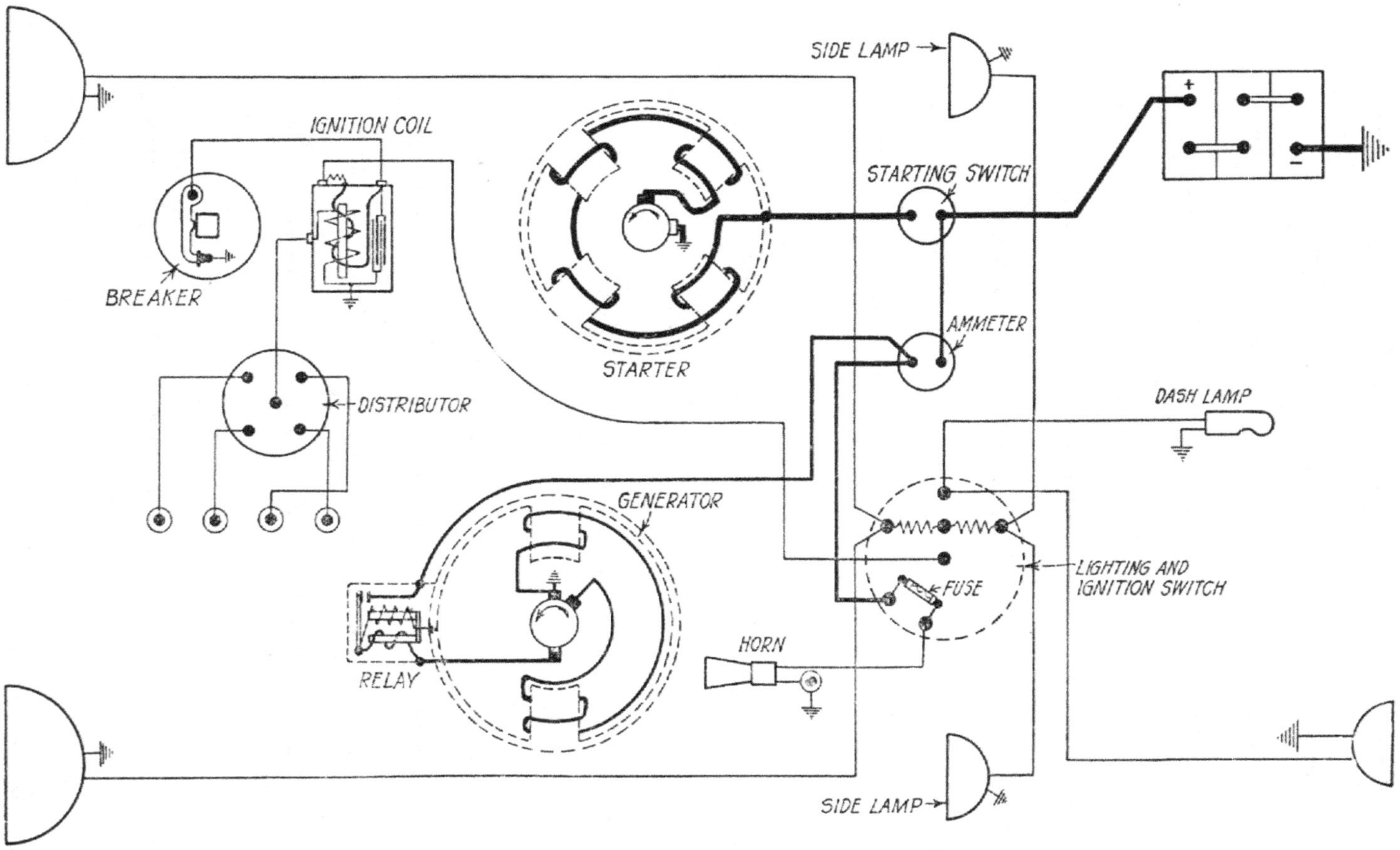

MAXWELL WIRING DIAGRAM, 1923-1924 MODELS, REMY EQUIPMENT

M A X W E L L Model All Year 1923-24

Remy Starter & Generator Remy Ignition

GENERATOR Model	Regulation	Max. Chg. rate
No. 961 - A	Third Brush	16 Amps.

Amps. Cut-in	Volts Warm Test	R.P.M.	Amps. Cut-in	Volts Cold Test	R.P.M.
0	7.2	600	11-14	8.1	1000
12	8.0	1600	16	8.1	1800

RELAY Model	Closes	Opens	Contact Gap	Air Gap
No.264B or 265A	8 MPH	7 MPH	.025	.040

STARTING Motor Model	Drive	Rotation
No. 712 - A	Bendix	Anti-Clockwise

BATTERY U. S. L. Type Capacity Bat. to Frame Con.

C D X 313 X. 6 Volts 100 Amps. Negative

IGNITION Coil Model	DISTRIBUTOR Model	Breaker Contact Gap
No. 284 - L	No. 369 A - C	.018

FIRING ORDER	Ignition Timing	SPARK PLUG
1 - 3 - 4 - 2	10°past T D C	Size 7/8R Gap .028

LIGHTING	Headlamps	Dash & Tail	Side Lamps
Single Contact	21 C.P.	2 C.P.	2 C.P.

ENGINE Cylinder Valve tappet clearance

Bore $3\frac{5}{8}$ Stroke $4\frac{1}{2}$ No. 4 L Head Inlet .004 Exhaust .006

CARBURETOR	COOLING	PISTON RING	OILING SYSTEM
Stewart 1"	Thermo	3 5/8 X 1/8	Cap. 4qt. Type F&S

CLUTCH	REAR AXLE	Gear Ratio	TRANSMISSION
Plate 9"	Semi	4.60	Own

BRAKES	Position	Brake lining width & length
Rear wheel	Trans. or wheel T	Service $1\frac{1}{2}$ X $34\frac{1}{2}$ Hand 2 X $24\frac{1}{2}$

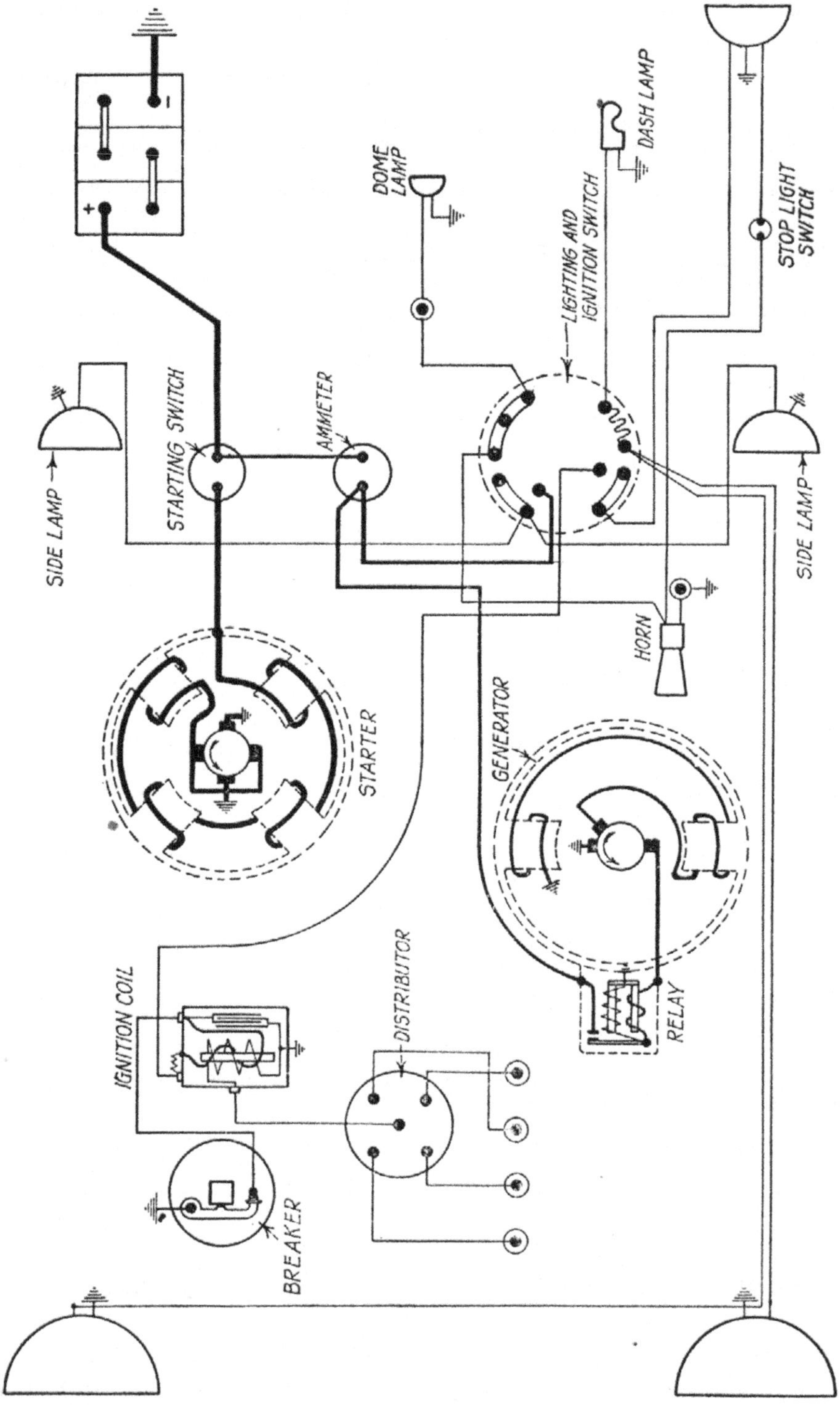

MAXWELL WIRING DIAGRAM, 1925, REMY SYSTEM

MAXWELL Model 25 Year 1925

Remy Starter & Generator Remy Ignition

Regulation	Max. Chg. rate and speed
3rd Brush	15-17 Amps. 1800

RELAY Air Gap	Contact Gap	Cut-in R.P.M.
.014"	.014"-.025"	8-10 M.P.H.

BATTERY Westinghouse Type 60.OB.13 Volts 6 Amps 105

Bat. to Frame Con. Neg. CONTACT BREAKER Gap .020"

Firing Order 1-3-4-2 Ignition Timing 10° A.T.C. Ret.

SPARK PLUG	ENGINE	Taxable Hp.
Size 7/8" Gap .027"	Bore 3 5/8" Stroke 4 1/2"	21

INTAKE VALVE TIMING	EXHAUST VALVE TIMING
Open 5° A.T.C. Close 45° A.B.C.	Open 49° B.B.C. Close 3° A.T.C.

VALVE CLEARANCE Hot Intake .004" Exhaust .006"

CARBURETOR	COOLING SYSTEM	OILING SYSTEM
Stewart	Type Ther. Cap 3½ gal.	Type Pres. Cap 4 qts.

PISTON RING: Width 1/8" Diam. 3 5/8" Gap .005"-.015"

CLUTCH Plate 8 $\frac{5"}{8}$ X $\frac{1"}{8}$ GEAR RATIO 4.6 AXLE Semi F

BRAKES

Front	Rear	Hand
	33¾" X 1¾" X $\frac{3"}{16}$	24 $\frac{5"}{8}$ X 2" X $\frac{3"}{16}$

Lighting	Headlights	Dash & Tail	Side Lamps
Single Contact	21 & 3 C.P.	3 C.P.	2 C.P.

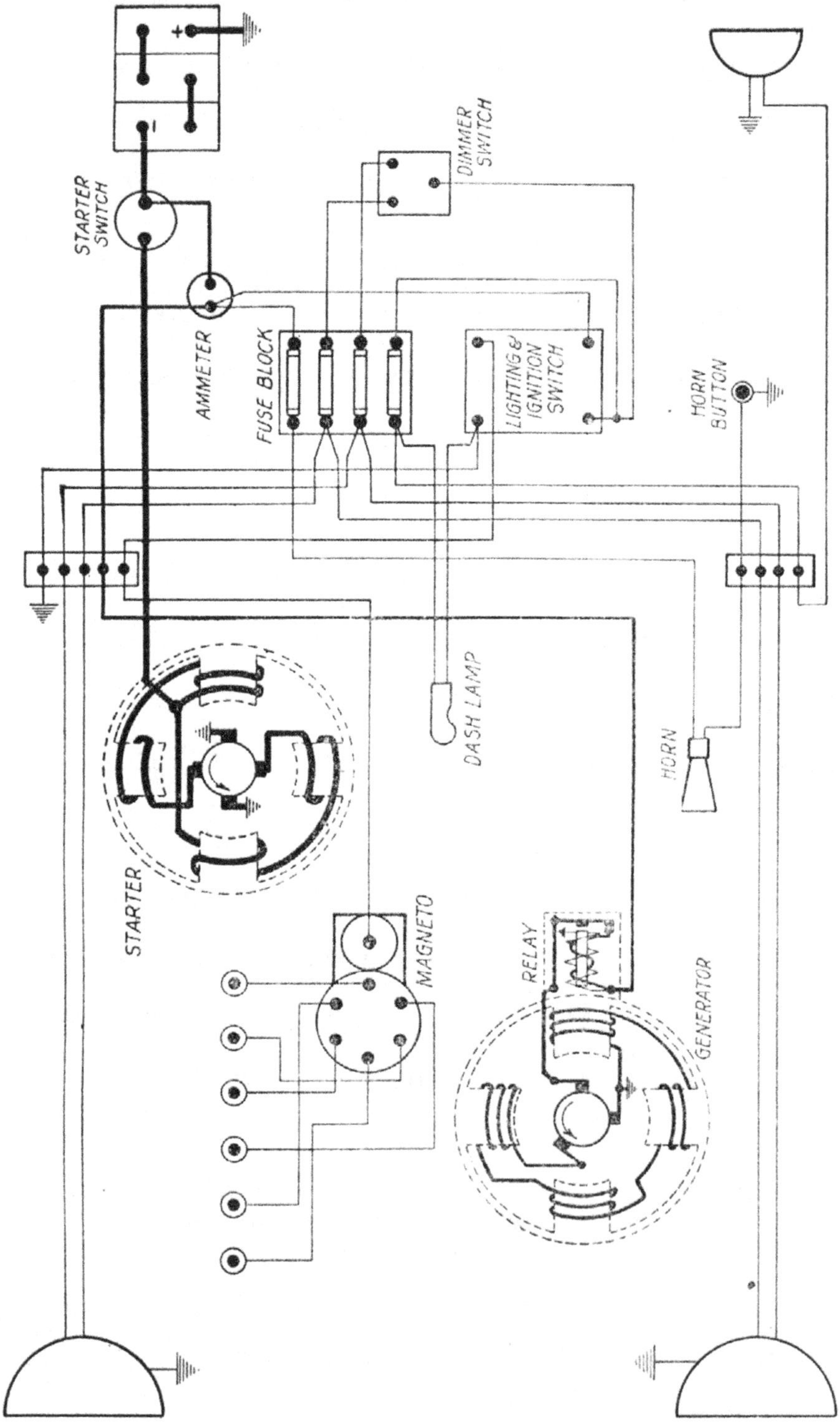

MERCER WIRING DIAGRAM, 1923, MODEL 6, WESTINGHOUSE STARTING AND LIGHTING, SPLITDORF MAGNETO IGNITION SYSTEMS

MERCER Model 6 Series 6 Year 1923

Westinghouse Starter & Generator Splitdorf Magneto Ignition

GENERATOR Model	Regulation	Max. Chg. rate
No. 56 - A T	Third Brush	17 Amps.

Amps. Cut-in	Volts Warm Test	R.P.M.	Amps. Cut-in	Volts Cold Test	R.P.M.
0	6.5	500	8	7.5	750
15	8.0	1600	17	8.0	1600

RELAY Model	Closes	Opens	Contact Gap	Air Gap
No. 301210 A	10 M P H	8 M P H	.030"-.035"	.020

STARTING Motor Model	Drive	Rotation
No. 711	Bendix	Anti-Clockwise

BATTERY Willard Type S J R - 5 Capacity 6 Volts 132 Amps. Bat. to Frame Con. Positive

IGNITION	Breaker Contact Gap
SPLITDORF, TYPE S S - 6, MAGNETO,	.020"-.024"

FIRING ORDER	Ignition Timing	SPARK PLUG
1-5-3-6-2-4	T D C at Retard	Size 7/8" Gap .020

LIGHTING	Headlamps	Dash & Tail	Side Lamps
Single Contact	21 C.P.	4 C.P.	2 C.P.

ENGINE	Cylinder	Valve tappet clearance
Bore 3¾ Stroke 6¾	No. 4 L Head	Inlet ... Exhaust ...

CARBURETOR	COOLING	PISTON RING	OILING SYSTEM
Ball & Ball 1½"	Pump	3 3/4 X 3/16	Cap. 6qt. Type Force

CLUTCH	REAR AXLE	Gear Ratio	TRANSMISSION
Disc	3/4 Flt.	4.41	Own

BRAKES	Position	Brake lining width & length
Rear wheel	Trans. or wheel T	Service 2½X17¼ Hand 2½ X 12

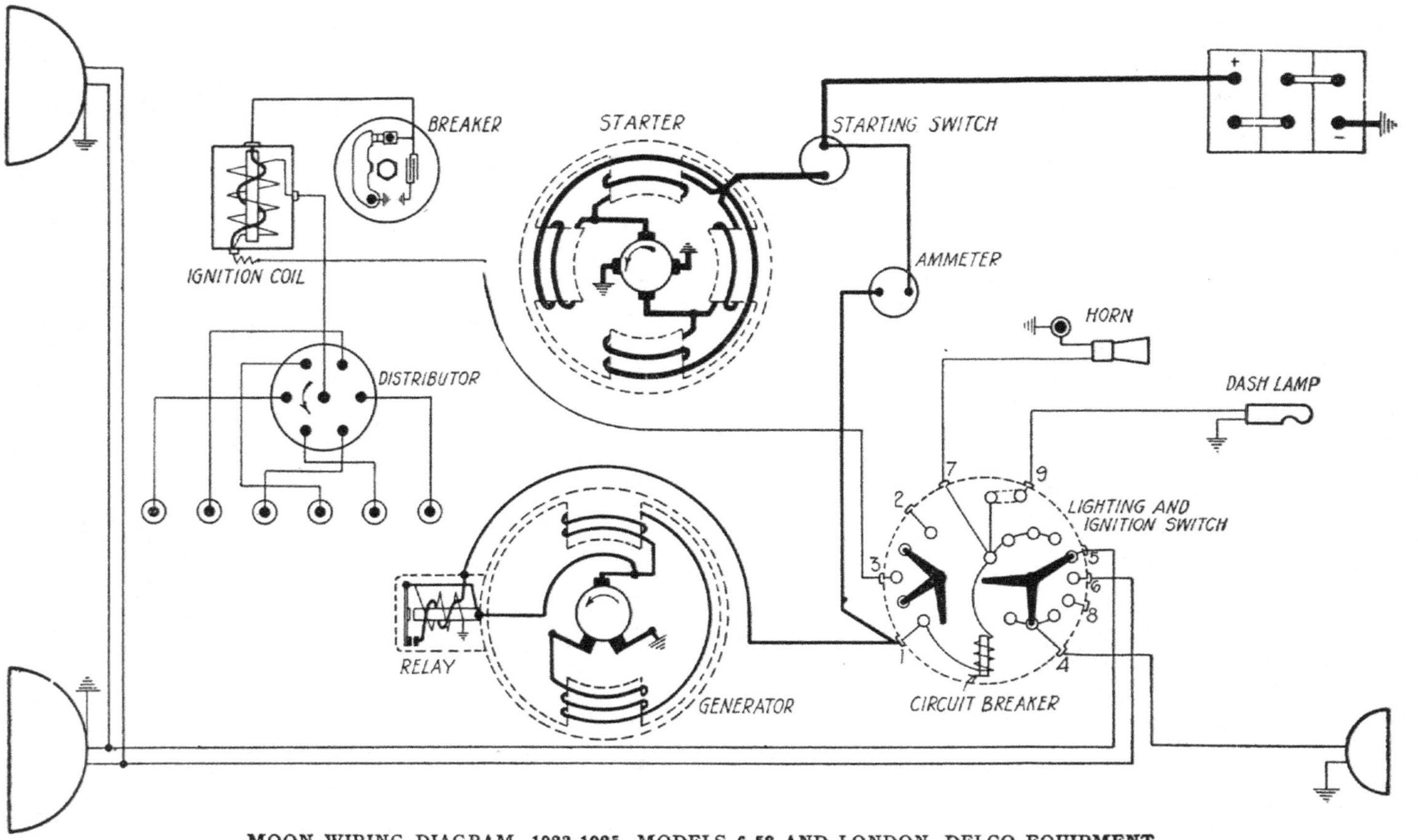

MOON WIRING DIAGRAM, 1923-1925, MODELS 6-58 AND LONDON, DELCO EQUIPMENT

M O O N   Model 6-58   Year 1923-24

Delco Starter & Generator   Delco Ignition

GENERATOR Model	Regulation	Max. Chg. rate
No. 258	Third Brush	14-16 Amps.

Amps. Cut-in	Volts Warm Test	R.P.M.	Amps. Cut-in	Volts Cold Test	R.P.M.
5	- - -	700	11	- - -	1000
14-16	- - -	1600	15	- - -	2000

RELAY Model	Closes	Opens	Contact Gap	Air Gap.
No. 258	8 M P H	6 M P H	.025"-.035"	.025

STARTING Motor Model	Drive	Rotation
No. 181	Bendix	Anti-Clockwise

BATTERY Exide Type   Capacity   Bat. to Frame Con.

3 - X C - 15 - 1   6 Volts 100 Amps.   Negative

IGNITION Coil Model	DISTRIBUTOR Model	Breaker Contact Gap
No. 2178	No. 5256	.020"-.027"

FIRING ORDER	Ignition Timing	SPARK PLUG
1-5-3-6-2-4	"U D" at Retard	Size 7/8 Gap .030

LIGHTING	Headlamps	Dash & Tail	Side Lamps
Single Contact	21 C.P.	2 C.P.	4 C.P.

ENGINE   Cylinder   Valve tappet clearance

Bore $3\frac{3}{8}$ Stroke $4\frac{1}{2}$ No. 6   Head   Inlet .004 Exhaust .006

CARBURETOR	COOLING	PISTON RING	OILING SYSTEM
Stromberg $1\frac{1}{4}$"	Pump	3 3/8 X 3/16	Cap. 6 qt. Type F&S

CLUTCH	REAR AXLE	Gear Ratio	TRANSMISSION
Plate 10"	Semi	5.09	Warner-Gear

BRAKES   Position   Brake lining width & length

Rear wheel   Trans. or wheel W   Service 2 X 22   Hand $1\frac{3}{4}$ X 44

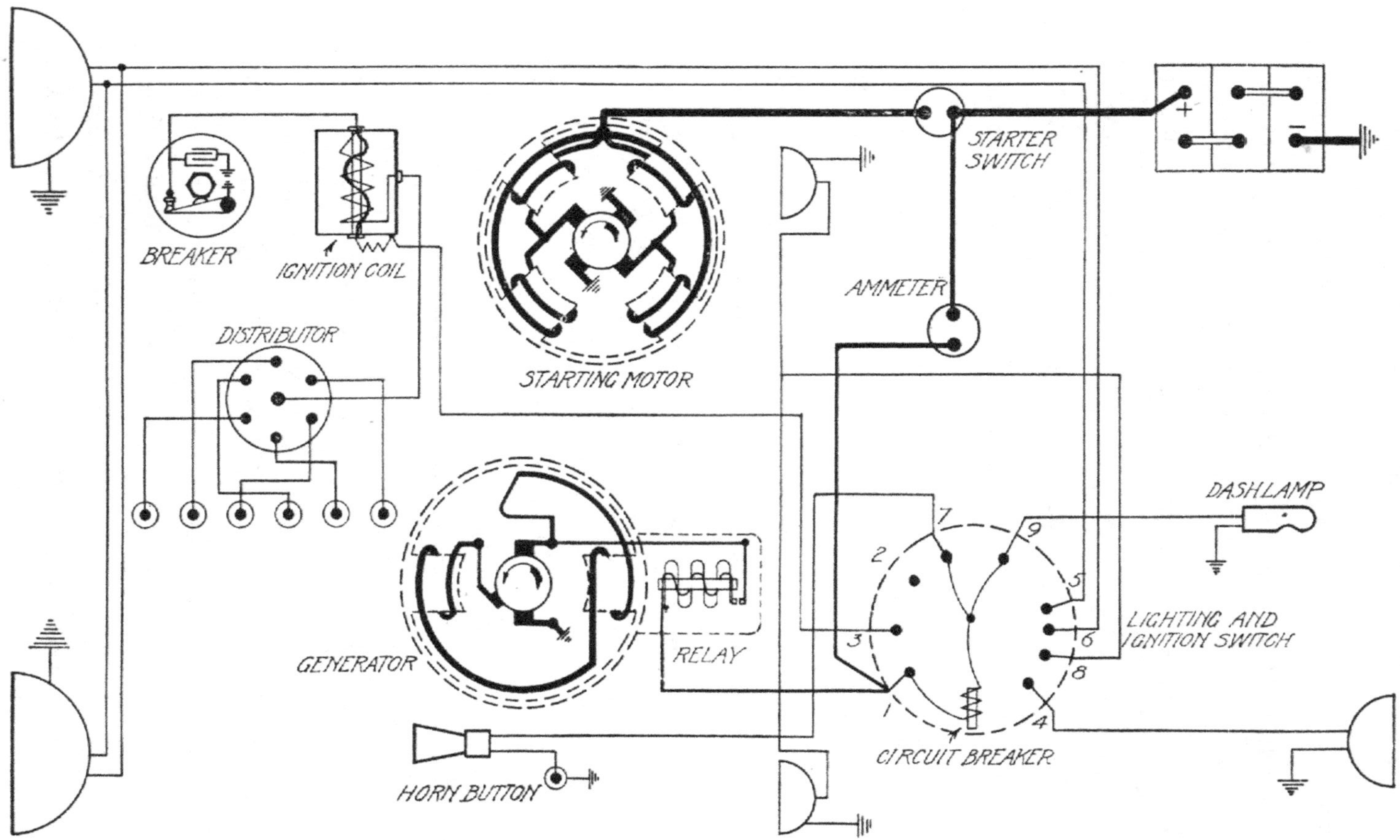

MOON WIRING DIAGRAM, 1925, NEWPORT MODEL, DELCO SYSTEM

M O O N Model Newport Year 1925

Delco Starter & Generator Delco Ignition

Regulation	Max. Chg. rate and speed
3rd Brush	17 Amps. 1400

RELAY Air Gap	Contact Gap	Cut-in R.P.M.
.025"	.046"	600 or 8 M.P.H.

BATTERY Exide Type 3-X3-13-1 Volts 6 Amps 85

Bat. to Frame Con. Neg. CONTACT BREAKER Gap .020"

Firing Order 1-5-3-6-2-4 Ignition Timing T.D.C. Ret.

SPARK PLUG	ENGINE	Taxable Hp.
Size 7/8" Gap .025"	Bore 3 1/8" Stroke 4 1/4"	23

INTAKE VALVE TIMING	EXHAUST VALVE TIMING
Open 4° A.T.C. Close 46° A.B.C.	Open 41° B.B.C. Close 1° A.T.C.

VALVE CLEARANCE Hot Intake .004" Exhaust .006"

CARBURETOR	COOLING SYSTEM	OILING SYSTEM
Stromberg	Type P Cap 4½ gal.	Type Pres. Cap

PISTON RING: Width 3/16" Diam. 3 1/8" Gap .002"

CLUTCH Plate GEAR RATIO 5.1 AXLE Semi F

BRAKES

Front	Rear	Hand
27 1/4" x 1 3/4" x 3/16"	27 1/4" x 1 3/4" x 3/16"	24" x 2" x 5/32"

Lighting	Headlights	Dash & Tail	Side Lamps
Single Contact	21 & 4 C.P.	2 C.P.	2 C.P.

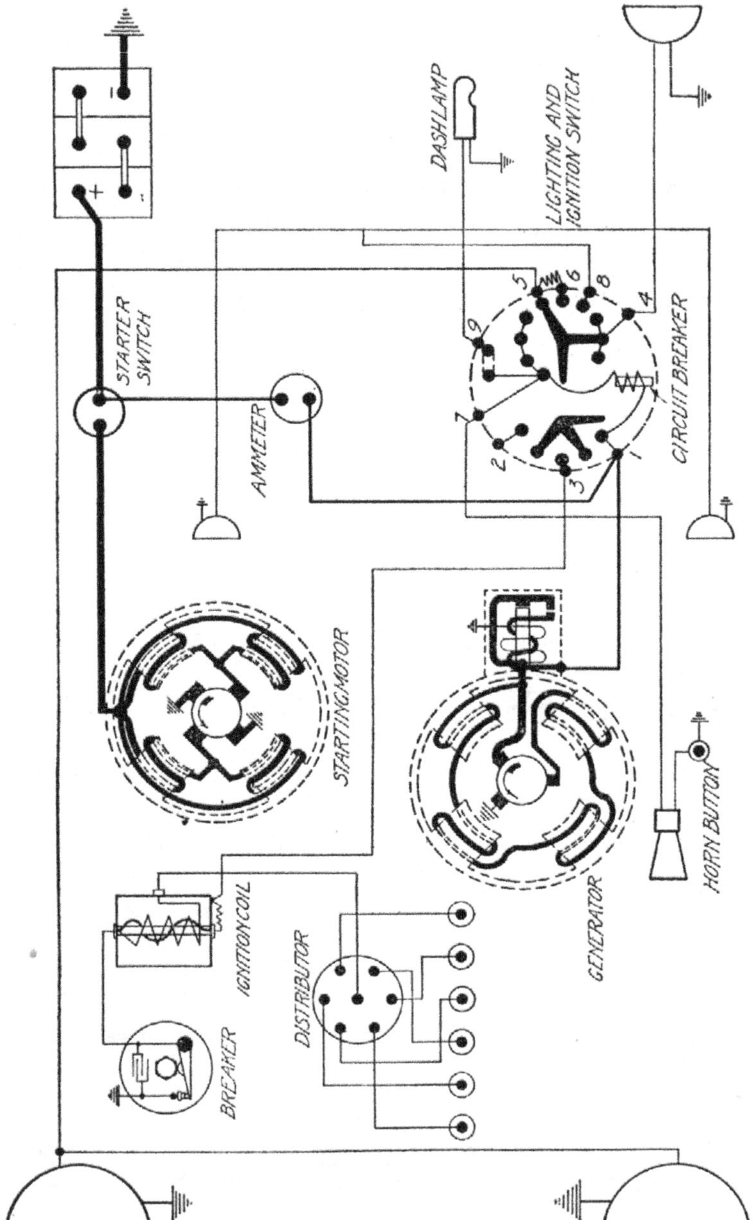

MOON, 1925, SERIES A

M O O N Model A Year 1925

Delco Starter & Generator Delco Ignition

Regulation	Max. Chg. rate and speed
3rd Brush	12 Amps. 1600

RELAY Air Gap	Contact Gap	Cut-in R.P.M.
.015"-.020"	.030"	8 M.P.H.

BATTERY Exide Type 3-XC-31-1 Volts 6 Amps 90

Bat. to Frame Con. Neg. CONTACT BREAKER Gap .014"

Firing Order 1-5-3-6-2-4 Ignition Timing T.D.C. Ret.

SPARK PLUG	ENGINE	Taxable Hp.
Size 7/8" Gap .025"	Bore 3 1/8" Stroke 4 1/4"	23

INTAKE VALVE TIMING	EXHAUST VALVE TIMING
Open 4°A.T.C. Close 46°A.B.C.	Open Close 1°A.T.C.

VALVE CLEARANCE Hot Intake .004" Exhaust .004"

CARBURETOR	COOLING SYSTEM	OILING SYSTEM
Stromberg	Type P Cap $4\frac{1}{2}$ gal.	Type Pres. Cap To Level.

PISTON RING: Width Diam Gap

CLUTCH Plate GEAR RATIO 4.7 AXLE Semi F.

BRAKES

Front	Rear	Hand
$27\frac{1}{2}$" X $1\frac{1}{2}$" X $\frac{3}{16}$"	$27\frac{1}{2}$" X $1\frac{1}{2}$" X $\frac{3}{16}$"	18" X 2" X $\frac{5}{32}$"

Lighting	Headlights	Dash & Tail	Side Lamps
Single Contact	21 C.P.	2 C.P.	2 C.P.

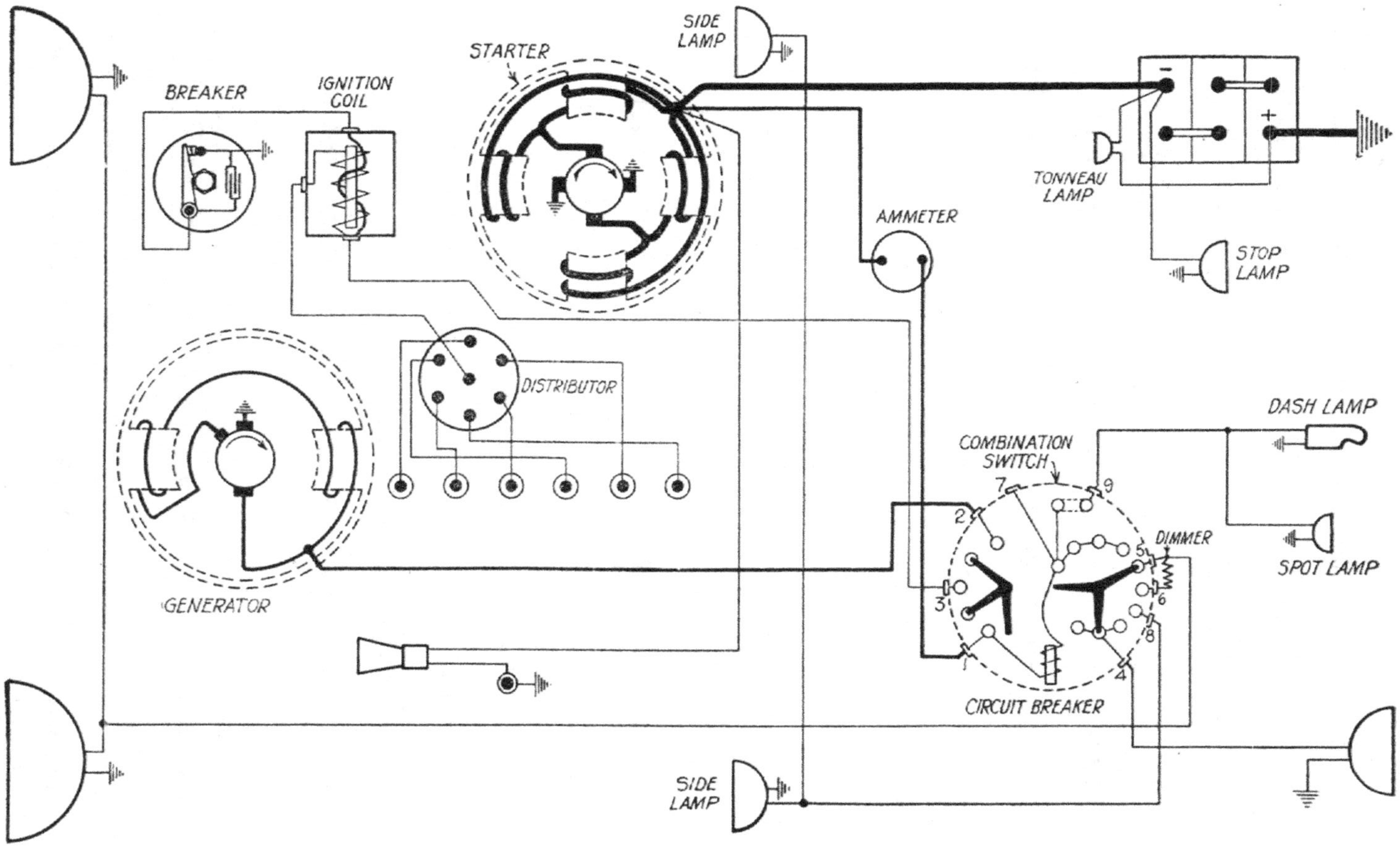

NASH WIRING DIAGRAM, 1923-1925, MODELS 691-698 AND 161-169, DELCO EQUIPMENT

N A S H   Model 691 - 698   Year 1923-24

Delco Starter & Generator   Delco Ignition

GENERATOR Model	Regulation	Max. Chg. rate
No. 229	Third Brush	19 Amps.

Amps. Cut-in	Volts Warm Test	R.P.M.	Amps. Cut-in	Volts Cold Test	R.P.M.
5	- - -	800	8	- - -	800
17	- - -	1500	19	- - -	1500

RELAY   NONE USED. CONTACT MADE BY SWITCH.

STARTING Motor Model	Drive	Rotation
No. 230	Gears	Clockwise

BATTERY Willard Type	Capacity	Bat. to Frame Con.
S J R N - 4	6 Volts 110 Amps.	Positive

IGNITION Coil Model	DISTRIBUTOR Model	Breaker Contact Gap
No. 2178	No. 5244	.020"-.025"

FIRING ORDER	Ignition Timing	SPARK PLUG
1-5-3-6-2-4	T.D.C. at Retard	Size 7/8" Gap .025"

LIGHTING	Headlamps	Dash & Tail	Side Lamps
Single Contact	21 C.P.	2 C.P.	2 C.P.

ENGINE	Cylinder	Valve tappet clearance
Bore 3¼ Stroke 5	No. 6 I Head	Inlet .015 Exhaust .015

CARBURETOR	COOLING	PISTON RING	OILING SYSTEM
Marvel	Pump	3 1/4 X 3/16	Cap. 6qt. Type C S

CLUTCH	REAR AXLE	Gear Ratio	TRANSMISSION
Disc	Semi	4.50	Own

BRAKES	Position	Brake lining width & length
Rear wheel	Trans. or wheel T	Service 2½ X 48¾ Hand 2½X18¾

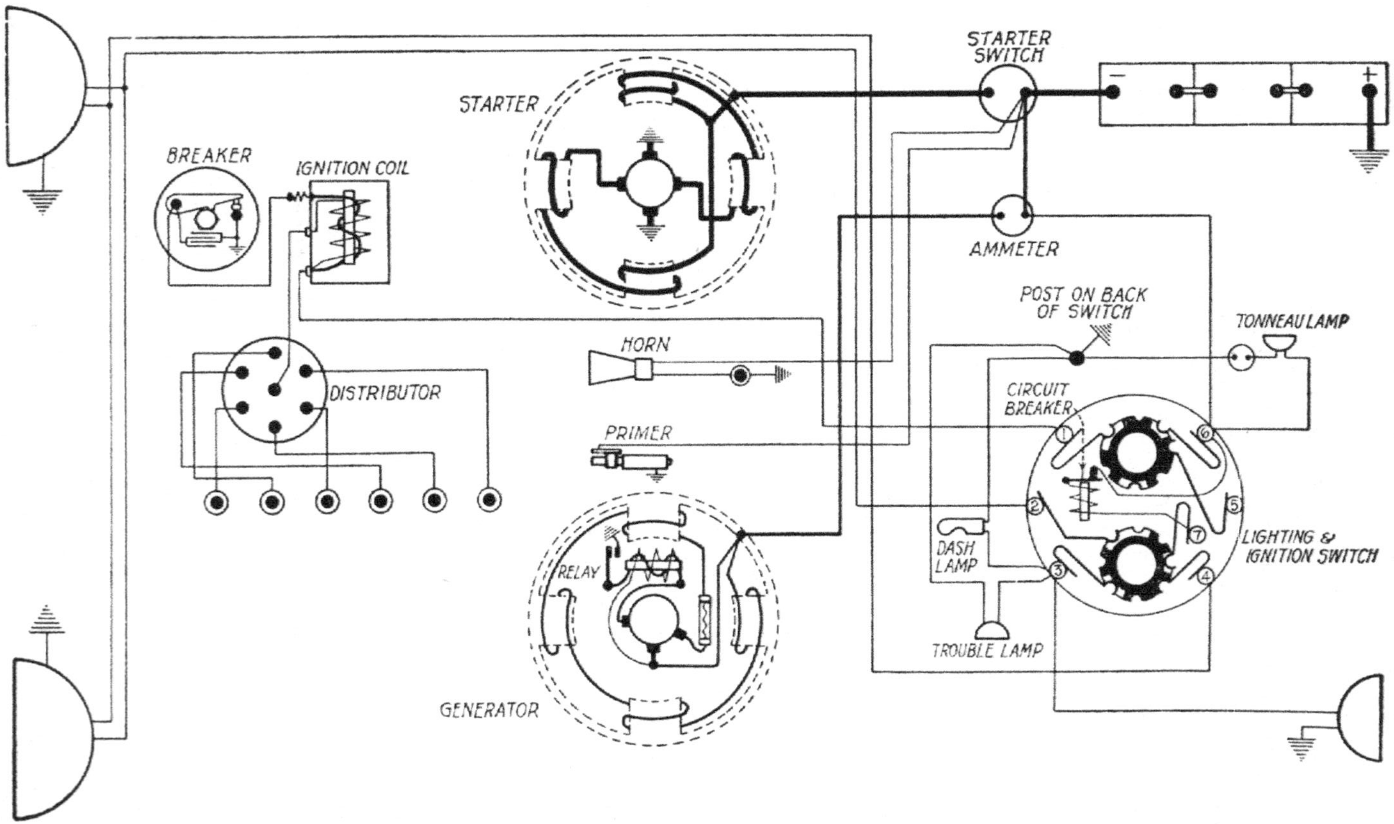

NATIONAL WIRING DIAGRAM, 1923-1924, MODEL BB, WESTINGHOUSE STARTING AND LIGHTING, DELCO IGNITION SYSTEMS

N A T I O N A L Model B B Year 1923-24

Westinghouse Starter & Generator Delco Ignition

GENERATOR Model	Regulation	Max. Chg. rate
No. 760	Third Brush	17 Amps.

Amps. Cut-in	Volts Warm Test	R.P.M.	Amps. Cut-in	Volts Cold Test	R.P.M.
0	6.5	500	8	6.5	675
13	7.5	1250	17	7.5	1250

RELAY Model	Closes	Opens	Contact Gap	Air Gap
No. 262982	8 M P H	6 M P H	.030"-.035"	.010

STARTING Motor Model	Drive	Rotation
No. 571	Bendix	Anti-Clockwise

BATTERY Prest-O-L Type Capacity Bat. to Frame Con.

615 - R H N 6 Volts 125 Amps. Positive

IGNITION Coil Model	DISTRIBUTOR Model	Breaker Contact Gap
No. 2159	No. 5211	.018

FIRING ORDER	Ignition Timing	SPARK PLUG
1-5-3-6-2-4	3/4" past T D C	Size 7/8" Gap .025

LIGHTING	Headlamps	Dash & Tail	Side Lamps
Single Contact	6 V 21 C.P.	3 V 2 C.P.	6 V 4 C.P.

ENGINE Cylinder Valve tappet clearance

Bore $3\frac{1}{2}$ Stroke $5\frac{1}{4}$ No. 6 I Head Inlet .003 Exhaust .005

CARBURETOR	COOLING	PISTON RING	OILING SYSTEM
Rayfield 1 1/2"	Pump	3 1/2 X 3/16	Cap. 10qt Type F & S

CLUTCH	REAR AXLE	Gear Ratio	TRANSMISSION
Plate 12"	Full	4.08	Brown Lipe

BRAKES Position Brake lining width & length

Rear wheel Trans. or wheel W Service $2\frac{1}{2}$ X 48 Hand $2\frac{1}{4}$ X 46

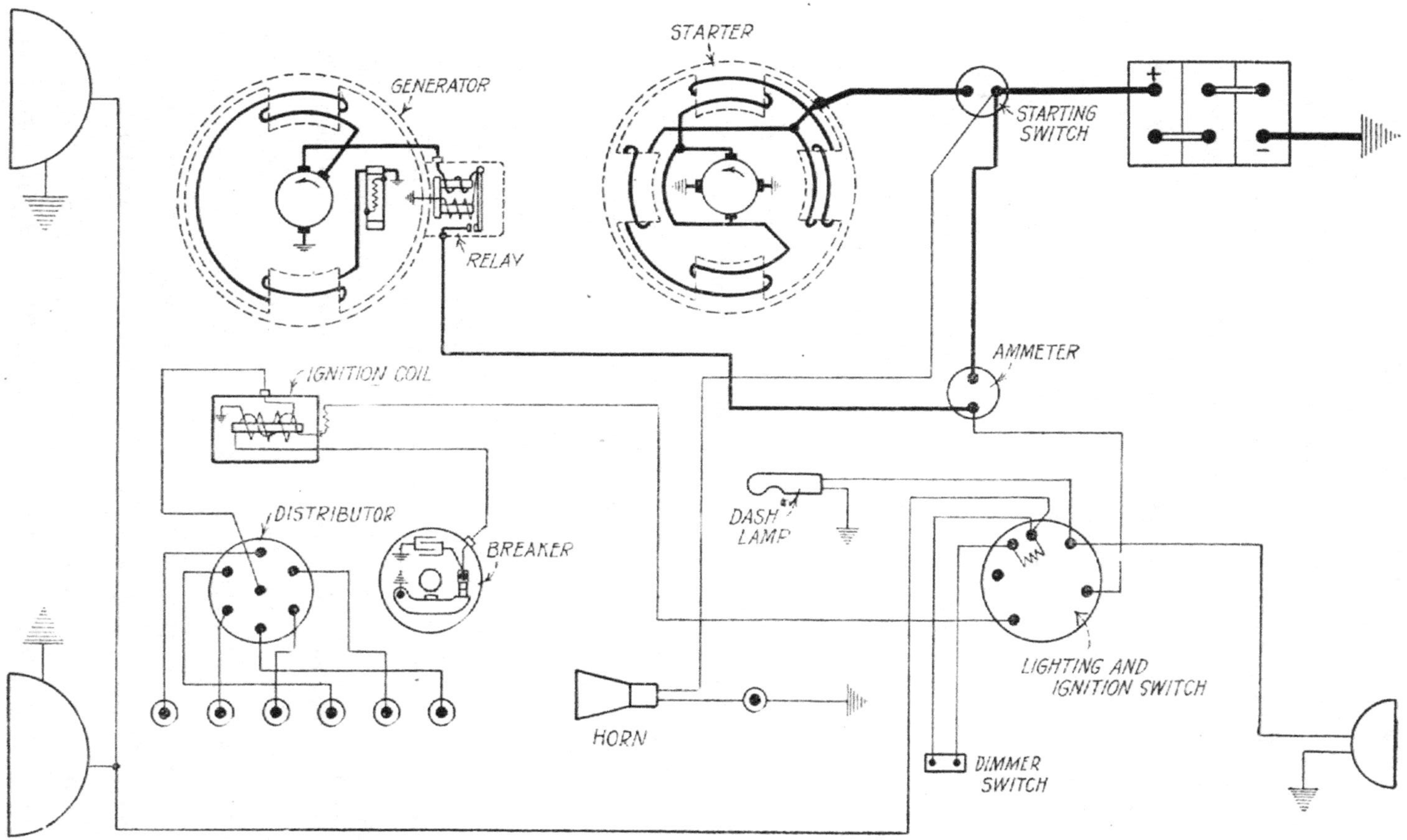

OAKLAND WIRING DIAGRAM, 1922-1923, MODEL 6-44, REMY EQUIPMENT

O A K L A N D Model 6 - 44 Year 1922-23

Remy Starter & Generator Remy Ignition

GENERATOR Model No. 917 - F | Regulation Third Brush | Max. Chg. rate 16 - 18 Amps.

Amps. Cut-in	Volts Warm Test	R.P.M.	Amps. Cut-in	Volts Cold Test	R.P.M.
0	7.2	820	0	7.3	600
7.0	7.2	1400	18	8.35	1560

RELAY Model	Closes	Opens	Contact Gap	Air Gap
No. 264 - B	10 M P H	8 M P H	.012 - .015	.012

STARTING Motor Model	Drive	Rotation
No. 701 - A	Bendix	Anti-Clockwise

BATTERY Prest-O-L. Type 613 - R H N | Capacity 6 Volts 100 Amps. | Bat. to Frame Con. Negative

IGNITION Coil Model	DISTRIBUTOR Model	Breaker Contact Gap
No. 284 - M	No. 606-D or 616-C	.023 - .025

FIRING ORDER	Ignition Timing	SPARK PLUG
1-5-3-6-2-4	"UDC 1-6" Retard	Size 7/8L Gap .025

LIGHTING	Headlamps	Dash & Tail	Side Lamps
Single Contact	21 C.P.	2 C.P.	4 C.P.

ENGINE Bore $2\frac{13}{16}$ Stroke $4\frac{3}{4}$ | Cylinder No. 6 I Head | Valve tappet clearance Inlet .010 Exhaust .010

CARBURETOR	COOLING	PISTON RING	OILING SYSTEM
Marvel 1"	Pump	2 13/16 X 1/8	Cap. 4qt. Type F. & S

CLUTCH	REAR AXLE	Gear Ratio	TRANSMISSION
Plate 9	Semi	4.70	Muncie

BRAKES Rear wheel | Position Trans. or wheel W | Brake lining width & length Service $1\frac{7}{8}$ X 35 Hand $1\frac{5}{8}$ X 35

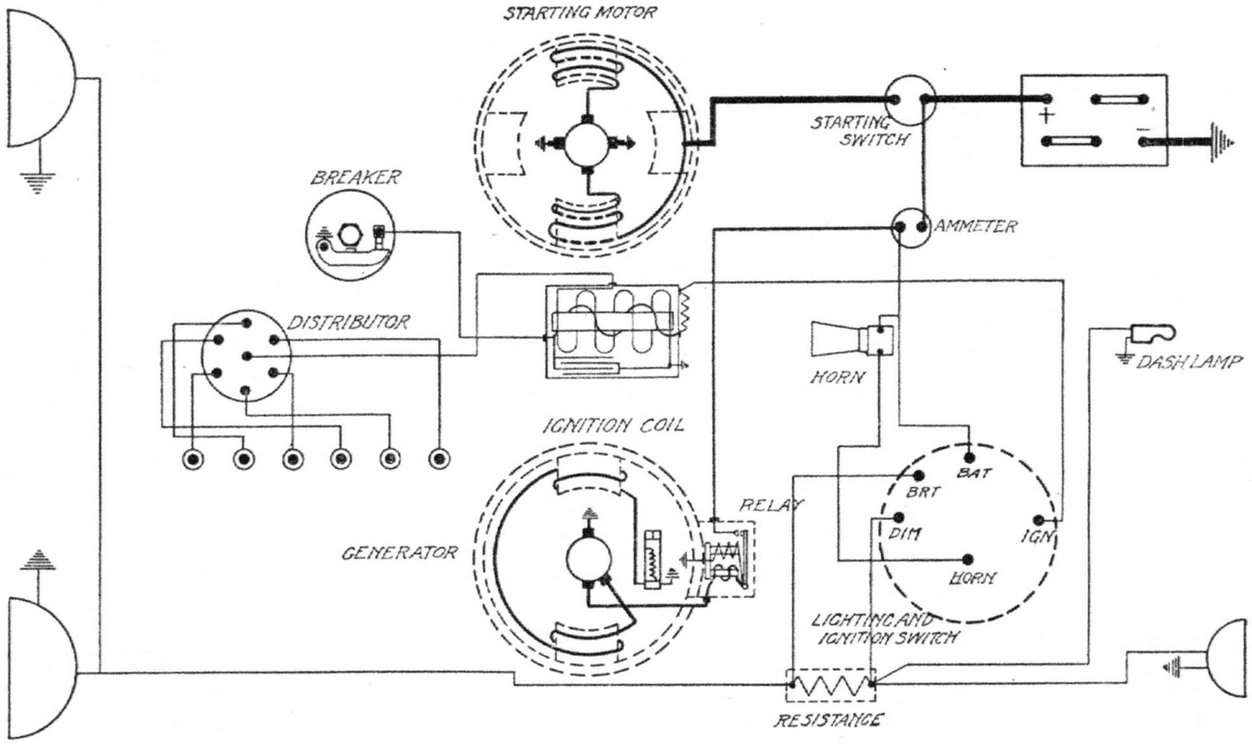

OAKLAND, 1925, MODEL 6-54

OAKLAND Model 6-54 Year 1925

Remy Starter & Generator Remy Ignition

Regulation	Max. Chg. rate and speed
3rd Brush	18-20 Amps. 1900

RELAY Air Gap	Contact Gap	Cut-in R.P.M.
.014"-.018"	.015"-.025"	675

BATTERY Prest-O-Lite Type 613 R.H.N. Volts 6 Amps 100

Bat. to Frame Con. Neg. CONTACT BREAKER Gap .020"

Firing Order 1-5-3-6-2-4 Ignition Timing 5° B.T.C. "0" in center of clamp scale on Housing

SPARK PLUG		ENGINE		Taxable Hp.
Size 7/8"	Gap .025"	Bore 2 7/8"	Stroke 4 3/4"	20

INTAKE VALVE TIMING		EXHAUST VALVE TIMING	
Open 5°A.T.C.	Close 40°A.B.C.	Open 45°B.B.C.	Close 5°A.T.C.

VALVE CLEARANCE Hot Intake .004" Exhaust .004"

CARBURETOR	COOLING SYSTEM		OILING SYSTEM	
Stromberg	Type P	Cap 3 gal.	Type Pres.	Cap 4 qts.

PISTON RING: Width 1/8" Diam. 2 7/8" Gap .006"

CLUTCH Disc 8 $\frac{7"}{8}$ X $\frac{5"}{32}$ GEAR RATIO 5.1 AXLE Semi F

BRAKES

Front	Rear	Hand
34" X 1 $\frac{7"}{8}$ X $\frac{5"}{32}$	34 $\frac{11"}{16}$ X 1 $\frac{7"}{8}$ X $\frac{5"}{32}$	16½" X 2½" X $\frac{3"}{16}$

Lighting	Headlights	Dash & Tail	Side Lamps
Single Contact	21 C.P.	2 C.P.	2 C.P.

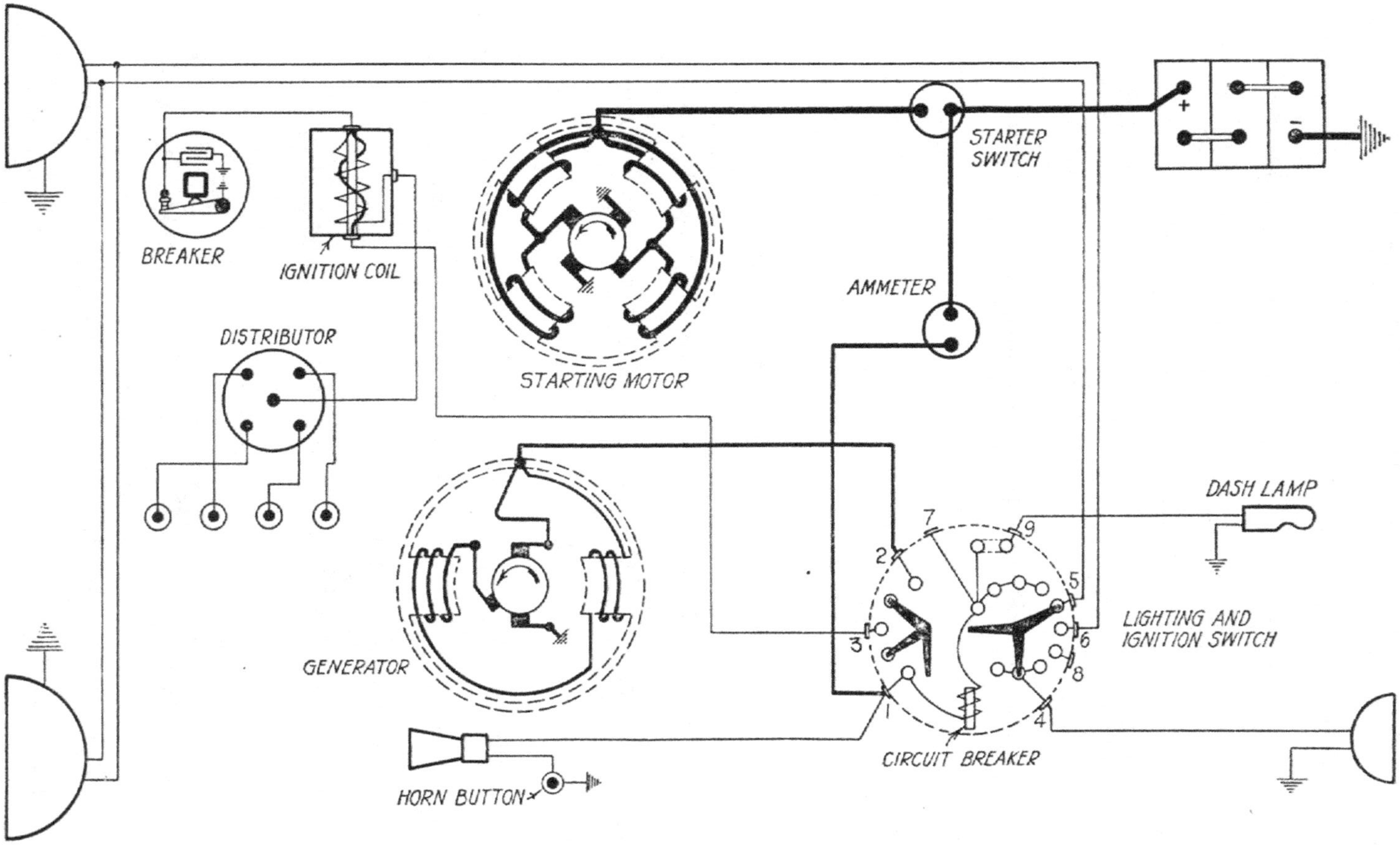

OLDSMOBILE WIRING DIAGRAM, 1923, MODEL 43A AND 1924 MODEL 6-30, DELCO EQUIPMENT

O L D S M O B I L E Model 30 Year 1923-24

Delco Starter & Generator Delco Ignition

GENERATOR Model Regulation Max. Chg. rate

No. 273 Third Brush 12 Amps.

Amps. Cut-in	Volts Warm Test	R.P.M.	Amps. Cut-in	Volts Cold Test	R.P.M.
8	- - -	1000	11	- - -	1000
12	- - -	1600	15	- - -	1600

RELAY NONE USED. CONTACT MADE BY SWITCH.

STARTING Motor Model Drive Rotation

No. 274 Bendix Clockwise

BATTERY Willard Type Capacity Bat. to Frame Con.

X W - 13 6 Volts 80 Amps. Negative

IGNITION Coil Model DISTRIBUTOR Model Breaker Contact Gap

No. 2178 No. 273 .020" - .025"

FIRING ORDER Ignition Timing SPARK PLUG

1-5-3-6-2-4 T.D.C. at Retard Size 7/8" Gap .025

LIGHTING Headlamps Dash & Tail Side Lamps

Single Contact 21 C.P 2 C.P - - - C.P

ENGINE Cylinder Valve tappet clearance

Bore $2\frac{3}{4}$ Stroke $4\frac{3}{4}$ No. 6 L Head Inlet .004 Exhaust .006

CARBURETOR COOLING PISTON RING OILING SYSTEM

Zenith 1 3/16" Pump 2 3/4 X 3/16 Cap. 6qt. Type Force

CLUTCH REAR AXLE Gear Ratio TRANSMISSION

Plate 9" Semi 5.10 Muncie

BRAKES Position Brake lining width & length

Rear wheel Trans. or wheel T Service $1\frac{1}{2}$ X 34 Hand 2 X $18\frac{1}{4}$

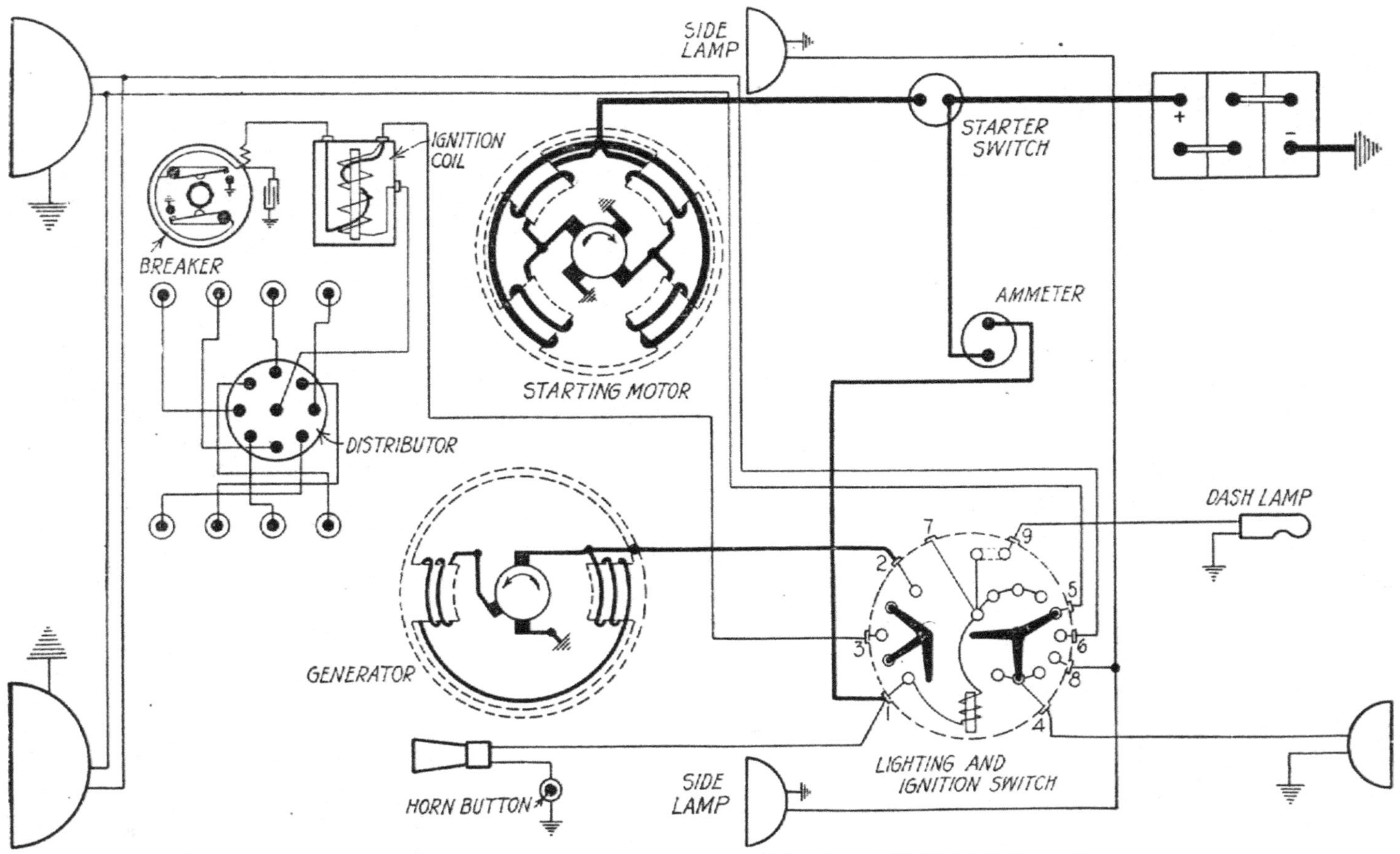

OLDSMOBILE WIRING DIAGRAM, 1923, MODEL 8-47, DELCO EQUIPMENT

O.L.D.S.M.O.B.I.L.E. Model 47 Light 8 Year 1923

Delco Starter & Generator Delco Ignition

GENERATOR Model — No. 148 | Regulation — Third Brush | Max. Chg. rate — 19 Amps.

Amps. Cut-in	Volts Warm Test	R.P.M.	Amps. Cut-in	Volts Cold Test	R.P.M.
0	- - -	550	10	- - -	900
17-19	- - -	1400	13-17	- - -	1800

RELAY NONE USED. CONTACT MADE BY SWITCH.

STARTING Motor Model — No. 219 | Drive — Bendix | Rotation — Clockwise

BATTERY Willard Type S R - 4 | Capacity 6 Volts 80 Amps. | Bat. to Frame Con. Negative

IGNITION Coil Model — No. 2127 | DISTRIBUTOR Model — No. 5210 | Breaker Contact Gap — .020"

FIRING ORDER — 1R-4L-3R-2L 4R-1L-2R-3L | Ignition Timing — "1&4" TDC at Retard | SPARK PLUG Size 7/8L Gap .030

LIGHTING — Single Contact | Headlamps 15 C.P. | Dash & Tail 2 C.P. | Side Lamps 2 C.P

ENGINE Bore $2\frac{7}{8}$ Stroke $4\frac{1}{2}$ | Cylinder No. 8 L Head | Valve tappet clearance Inlet .004 Exhaust .004

CARBURETOR Johnson 1 1/4 | COOLING Pump | PISTON RING 2 7/8 X 3/16 | OILING SYSTEM Cap. 6 Type Force

CLUTCH Disc 10" | REAR AXLE Full | Gear Ratio 4.67 | TRANSMISSION Muncie

BRAKES Rear wheel | Position Trans. or wheel W | Brake lining width & length Service $1\frac{7}{8}$ X $38\frac{1}{4}$ Hand $1\frac{5}{8}$ X $35\frac{1}{4}$

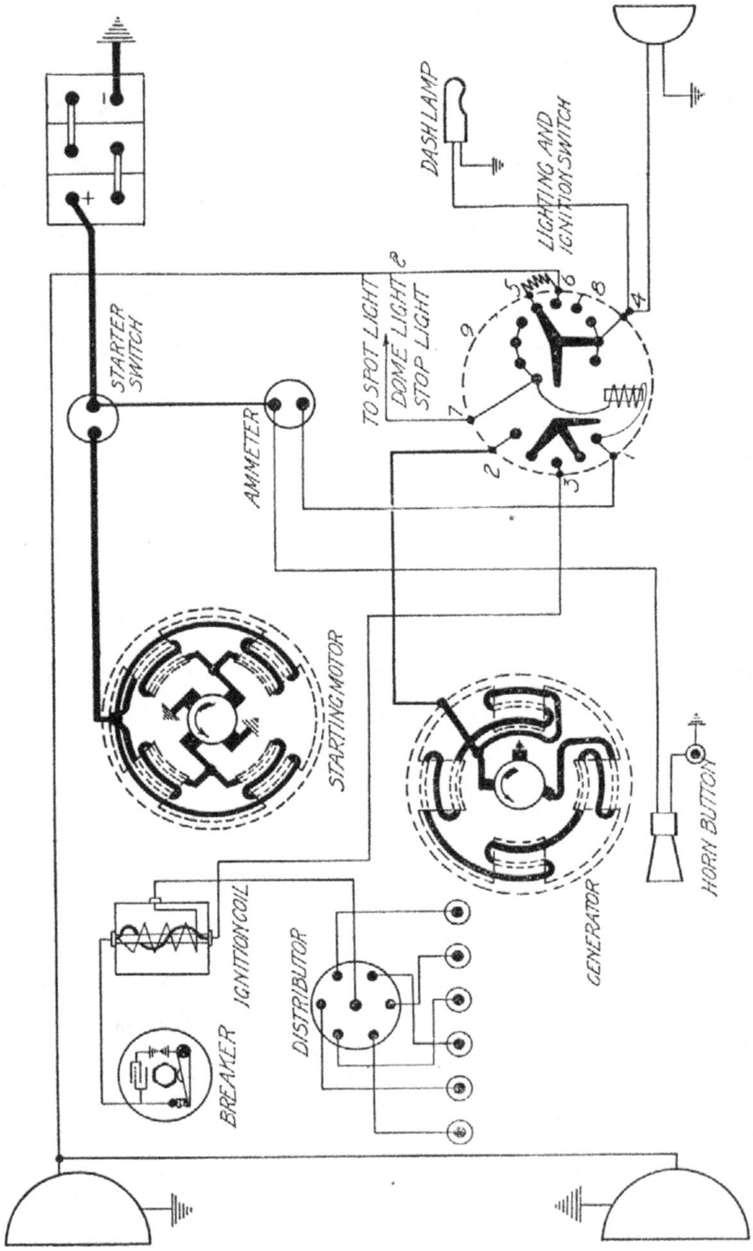

OLDSMOBILE, 1925, MODEL 30

O L D S M O B I L E Model 30 Year 1925

Delco Starter & Generator Delco Ignition

Regulation	Max. Chg. rate and speed
3rd Brush	12 Amps. 1600

RELAY Air Gap	Contact Gap	Cut-in R.P.M.
None		

BATTERY Willard Type X W - 13 Volts 6 Amps. 80

Bat. to Frame Con. Neg. CONTACT BREAKER Gap .020"

Firing Order 1-5-3-6-2-4 Ignition Timing T.D.C. Ret.

SPARK PLUG	ENGINE	Taxable Hp.
Size 7/8" Gap .030"	Bore 2 3/4" Stroke 4 3/4"	18

INTAKE VALVE TIMING	EXHAUST VALVE TIMING
Open 5°A.T.C. Close 50°A.B.C.	Open 42°B.B.C. Close 5°A.T.C.

VALVE CLEARANCE Hot Intake .004" Exhaust .004"

CARBURETOR	COOLING SYSTEM	OILING SYSTEM
Zenith	Type P Cap 3¼ gal.	Type Pres. Cap 6 qts.

PISTON RING: Width 3/16" Diam. 2.75" Gap .003"-.013"

CLUTCH Plate 9" X 1/8" GEAR RATIO 5.1 AXLE Semi F

BRAKES

Front	Rear	Hand
	37 5"/8 X 1¼" X 5"/32	Trans 18 1"/8 X 2" X 3"/16

Lighting	Headlights	Dash & Tail	Side Lamps
Single Contact	21 C.P.	2 C.P.	2 C.P.

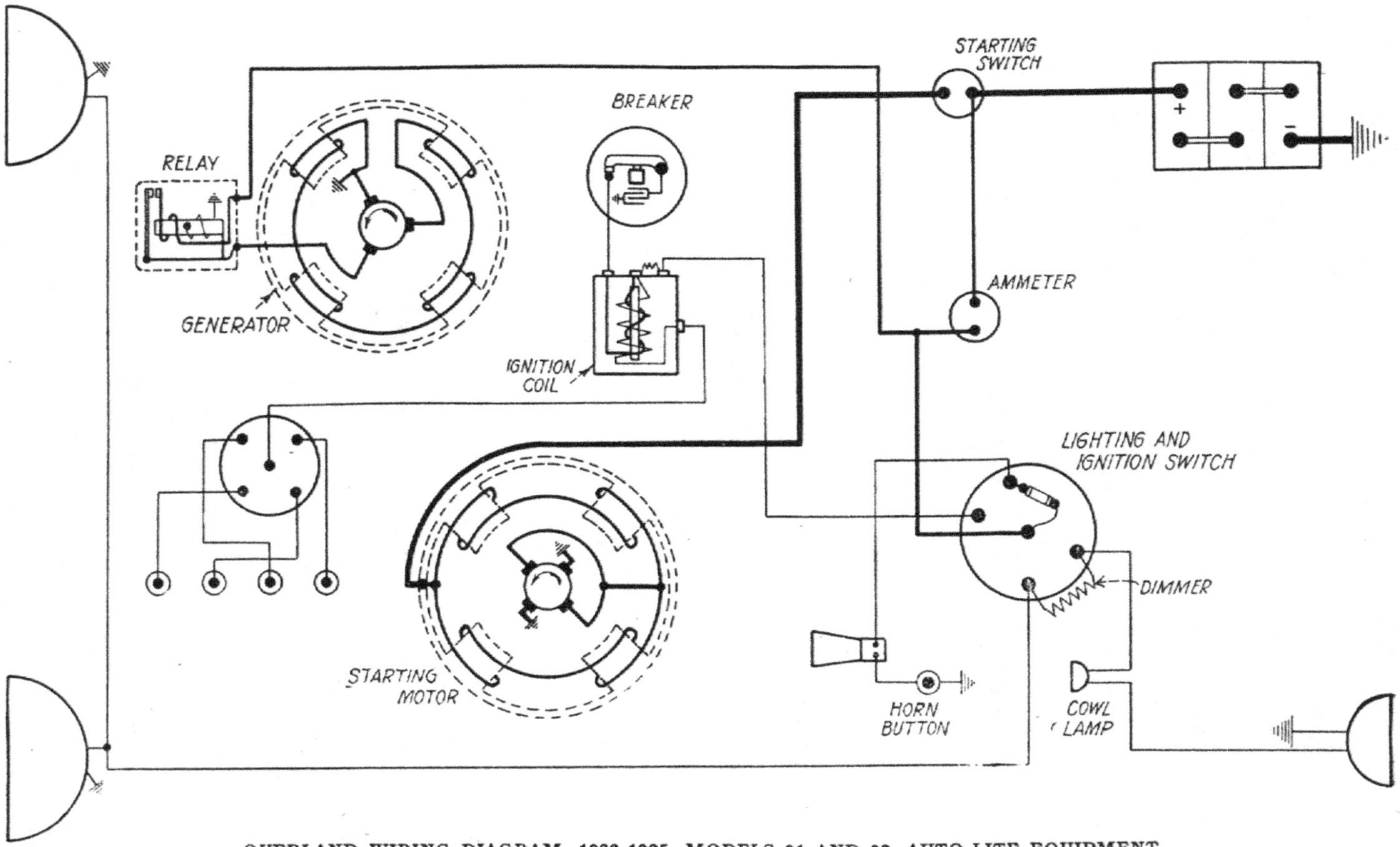

OVERLAND WIRING DIAGRAM, 1923-1925, MODELS 91 AND 92, AUTO-LITE EQUIPMENT

OVERLAND	Model 91 & 92	Year 1923-24

Auto-Lite	Starter & Generator	Auto-Lite	Ignition

GENERATOR Model	Regulation	Max. Chg. rate
G P	Third Brush	15 Amps.

Amps. Cut-in	Volts Warm Test	R.P.M.	Amps. Cut-in	Volts Cold Test	R.P.M.
0	7.05	475	0	6.5	390
10	8.0	1200	15	7.5	1200

RELAY Model	Closes	Opens	Contact Gap	Air Gap
G P	9 MPH	7 MPH	.025-.035	.012

STARTING Motor Model	Drive	Rotation
M G	Bendix	Anti-Clockwise

BATTERY U S L Type	Capacity	Bat. to Frame Con.
CDM-311-X	6 Volts 85 Amps.	Negative

IGNITION Coil Model	DISTRIBUTOR Model	Breaker Contact Gap
I G	I G	.018

FIRING ORDER	Ignition Timing	SPARK PLUG
1-3-4-2	2½" before TDC	Size 1/2" Gap .025

LIGHTING	Headlamps	Dash & Tail	Side Lamps
Single Contact	6 V 21 C.P.	3 V 2 C.P.	6 V 4 C.P.

ENGINE	Cylinder	Valve tappet clearance
Bore 3½ Stroke 4	No. 4 L Head	Inlet .003 Exhaust .004

CARBURETOR	COOLING	PISTON RING	OILING SYSTEM
Tillotson	Thermo	3½ X $\frac{3}{16}$	Cap 6 qt Type C S

CLUTCH	REAR AXLE	Gear Ratio	TRANSMISSION
Plate 8"	Semi	4.50	Own

BRAKES	Position	Brake lining width & length
Rear wheel	Trans. or wheel W	Service 1½ X 31½ Hand 1¼ X 10

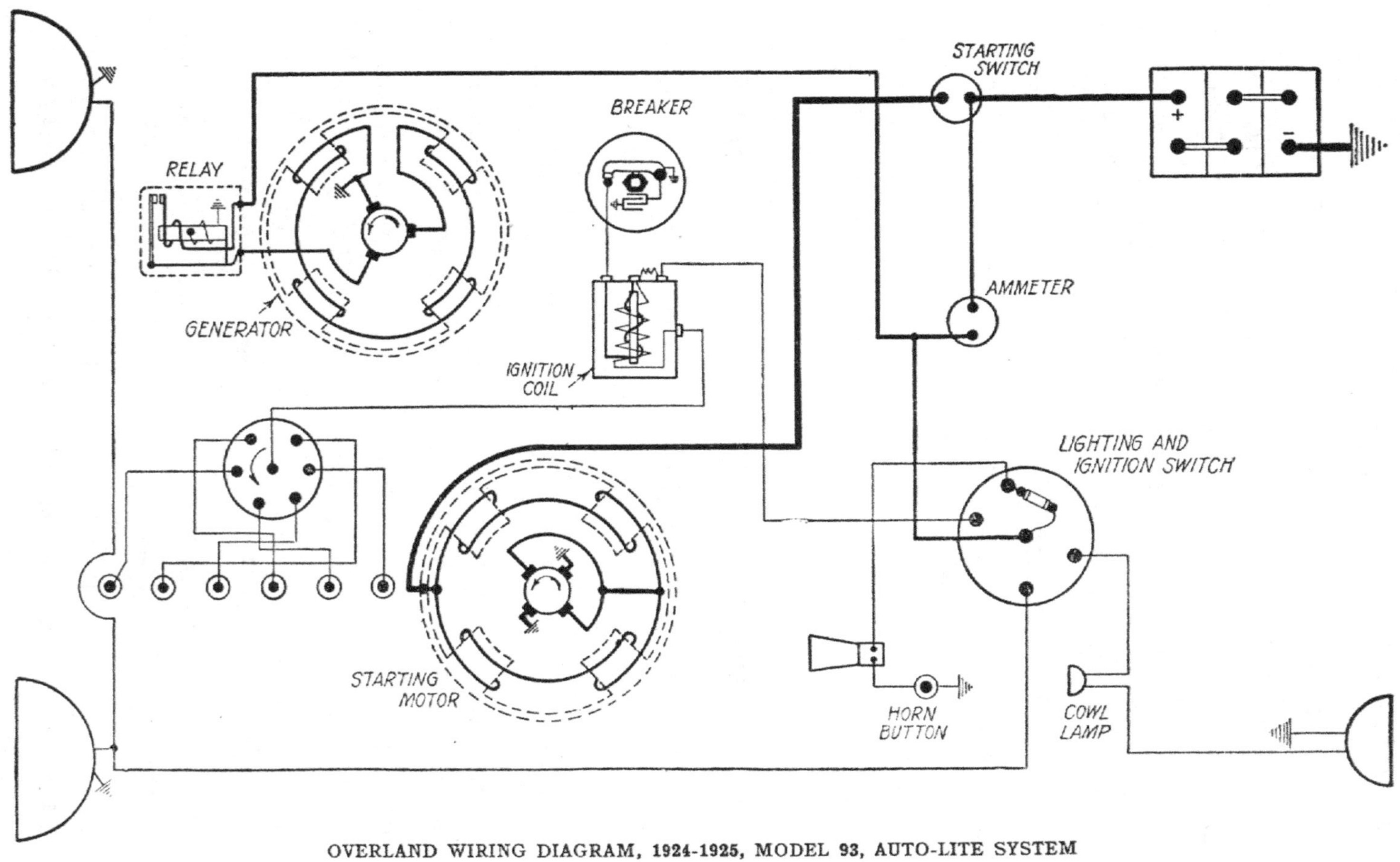

OVERLAND WIRING DIAGRAM, 1924-1925, MODEL 93, AUTO-LITE SYSTEM

O V E R L A N D Model 93 Year 1925

Auto-Lite Starter & Generator Auto-Lite Ignition

Regulation	Max. Chg. rate and speed
3rd Brush	16 Amps. 1600

RELAY Air Gap	Contact Gap	Cut-in R.P.M.
.010"-.015"	.025"-.035"	350

BATTERY U.S.L. Type 3-CVX-5X Volts 6 Amps 96

Bat. to Frame Con. Neg. CONTACT BREAKER Gap .018"

Firing Order 1-5-3-6-2-4 Ignition Timing 20° B.T.C. Adv.

SPARK PLUG	ENGINE	Taxable Hp.
Size 7/8" Gap .025"	Bore 3" Stroke 4"	22

INTAKE VALVE TIMING	EXHAUST VALVE TIMING
Open 5° A.T.C. Close 45° A.B.C.	Open 39° B.B.C. Close 5° A.T.C.

VALVE CLEARANCE Hot Intake .004" Exhaust .004"

CARBURETOR	COOLING SYSTEM	OILING SYSTEM
Tillotson	Type P Cap	Type Cap

PISTON RING: Width Diam. Gap

CLUTCH Disc GEAR RATIO 5.1 AXLE Semi F

BRAKES

Front	Rear	Hand

Lighting	Headlights	Dash & Tail	Side Lamps
Contact	C.P.	C.P.	C.P.

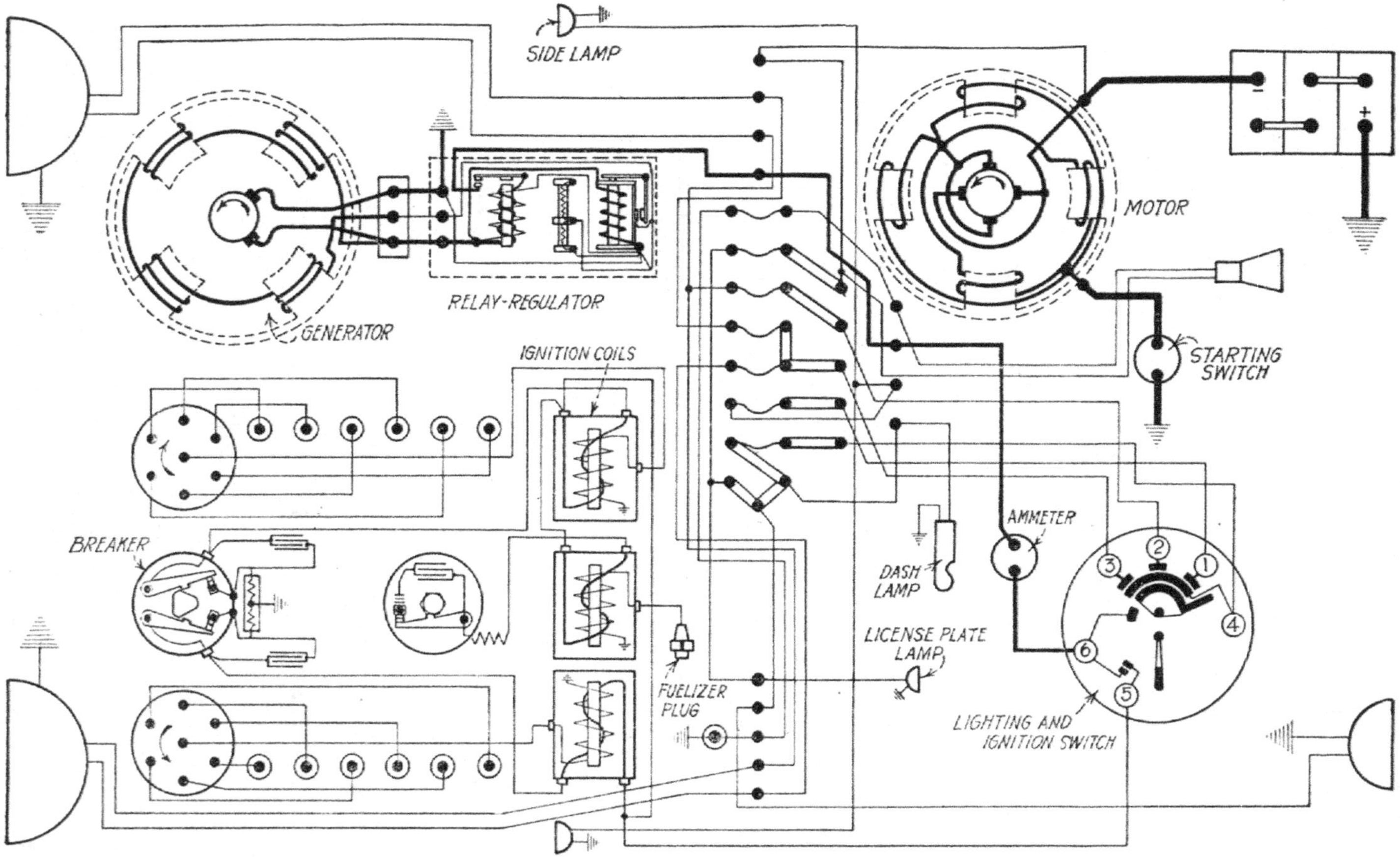

PACKARD WIRING DIAGRAM, 1922-1923, MODELS 3-25 AND 3-35, BIJUR STARTING AND LIGHTING, DELCO IGNITION SYSTEMS

PACKARD Model 3-25 & 3-35 Year 1922-23

Bijur Starter & Generator Delco Ignition

GENERATOR Model	Regulation	Max. Chg. rate
No. M-718, M-718A, M-1426.	Vibrator	15-18 Amps.

Amps. Cut-in	Volts Warm Test	R.P.M.	Amps. Cut-in	Volts Cold Test	R.P.M.
0	6.5	350	0	6.5	350
15-18	8.0	750	15-18	8.0	750

RELAY Model	Closes	Opens	Contact Gap	Air Gap
No. K - 646	8 M.P.H.	7 M.P.H	Adj. to 6.5V	Fixed

STARTING Motor Model	Drive	Rotation
No. M-800 M-1041	Pinion	Anti-Clockwise

BATTERY Willard Type Capacity Bat. to Frame Con.

S M W - 7 6 Volts 120 Amps. Positive

IGNITION Coil Model	DISTRIBUTOR Model	Breaker Contact Gap
No. 2150	No. 5161	.015 - .020

FIRING ORDER	Ignition Timing	SPARK PLUG
1R-6L-4R-3L-2R-5L-6R-1L-3R-4L-5R-2L.	"SR" at advance	Size 7/8R Gap .030

LIGHTING	Headlamps	Dash & Tail	Side Lamps
Single Contact	24 C.P.	4 C.P.	4 C.P.

ENGINE Cylinder Valve tappet clearance

Bore 3 Stroke 5 No. 12 L Head Inlet .005 Exhaust .005

CARBURETOR	COOLING	PISTON RING	OILING SYSTEM
Own	Pump	3 X 3/16	Cap. 10½ q. Type F&S

CLUTCH	REAR AXLE	Gear Ratio	TRANSMISSION
Disc 8 7/8	Semi	4.36	Own

BRAKES Position Brake lining width & length

Rear wheel Trans. or wheel W Service 1¼ X 42½ Hand 2½ X 53¼

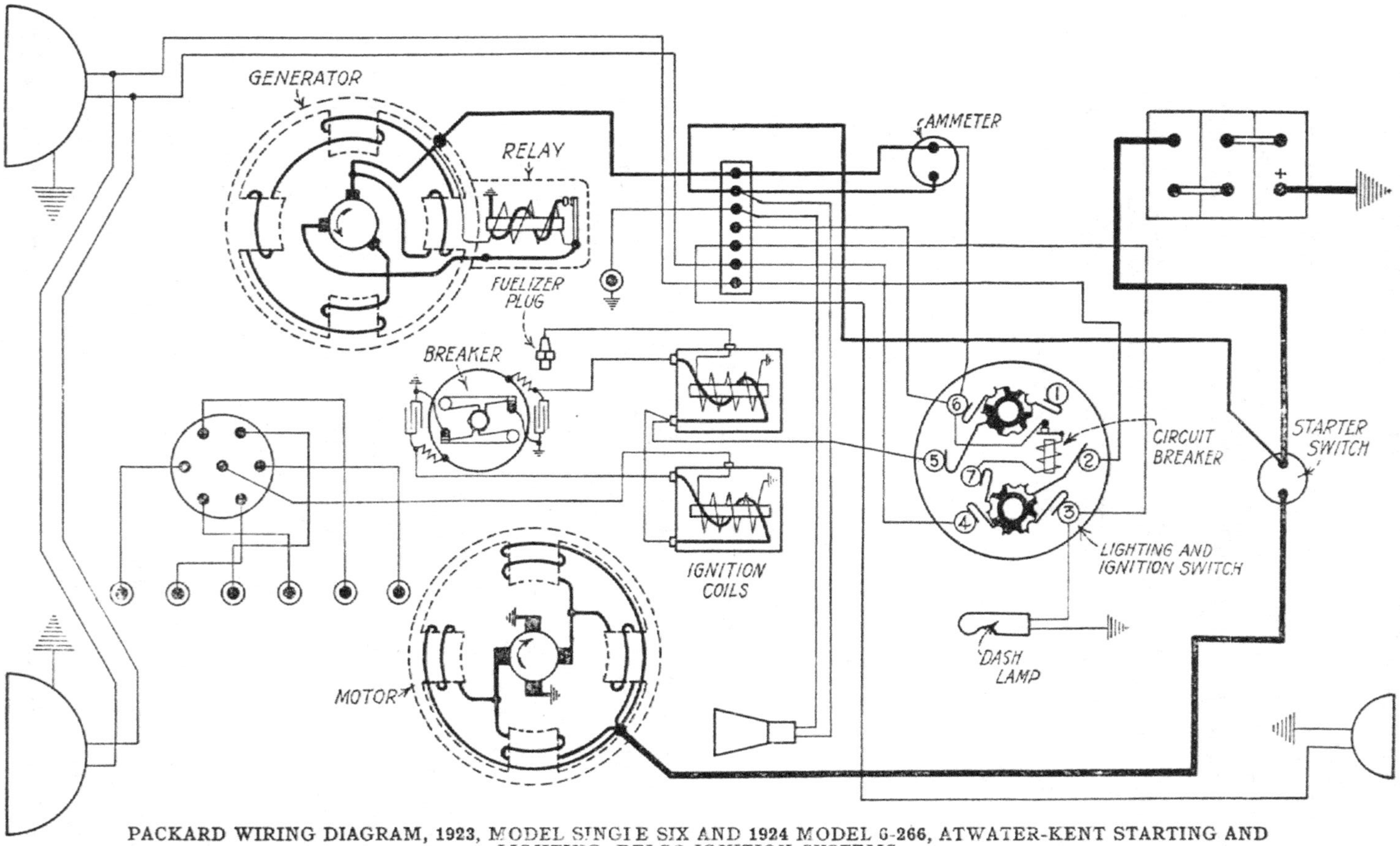

PACKARD WIRING DIAGRAM, 1923, MODEL SINGLE SIX AND 1924 MODEL 6-266, ATWATER-KENT STARTING AND LIGHTING, DELCO IGNITION SYSTEMS

PACKARD Model 6-266-233 Year 1924

Atwater-Kent Starter & Generator Delco Ignition

GENERATOR Model	Regulation	Max. Chg. rate
No. 5870	Third Brush	11 Amps.

Amps. Cut-in	Volts Warm Test	R.P.M.	Amps. Cut-in	Volts Cold Test	R.P.M.
0	6.5	475	5	7.0	615
11	8.0	1000	9	8.0	1750

RELAY Model	Closes	Opens	Contact Gap	Air Gap
No. 5870	10 MPH	8 MPH	.018-.020	.035

STARTING Motor Model	Drive	Rotation
No. 6484	Bendix	Anti-Clockwise

BATTERY Willard Type or Prest-O-Lite Capacity 6 Volts 100 Amps. Bat. to Frame Con. Positive

IGNITION Coil Model	DISTRIBUTOR Model	Breaker Contact Gap
No. 2175	No. 5249	.020-.025

FIRING ORDER	Ignition Timing	SPARK PLUG
1-5-3-6-2-4	2" to TDC at Adv.	Size 7/8 Gap .030

LIGHTING Single Contact Headlamps 21 C.P. Dash & Tail 2 C.P. Side Lamps 4 C.P.

ENGINE Bore $3\frac{3}{8}$ Stroke 5 Cylinder No. 6 L Head Valve tappet clearance Inlet .004 Exhaust .004

CARBURETOR	COOLING	PISTON RING	OILING SYSTEM
Own $1\frac{1}{2}$	Pump	$3\frac{3}{8}$ X $\frac{1}{8}$	Cap 6 qt. Type Force

CLUTCH	REAR AXLE	Gear Ratio	TRANSMISSION
Disc $8\frac{1}{8}$	Semi	4.66	Own

BRAKES Four wheel Position Trans. or wheel W. Brake lining width & length Service 2 X $41\frac{1}{4}$ Hand $1\frac{1}{2}$ X $39\frac{1}{2}$

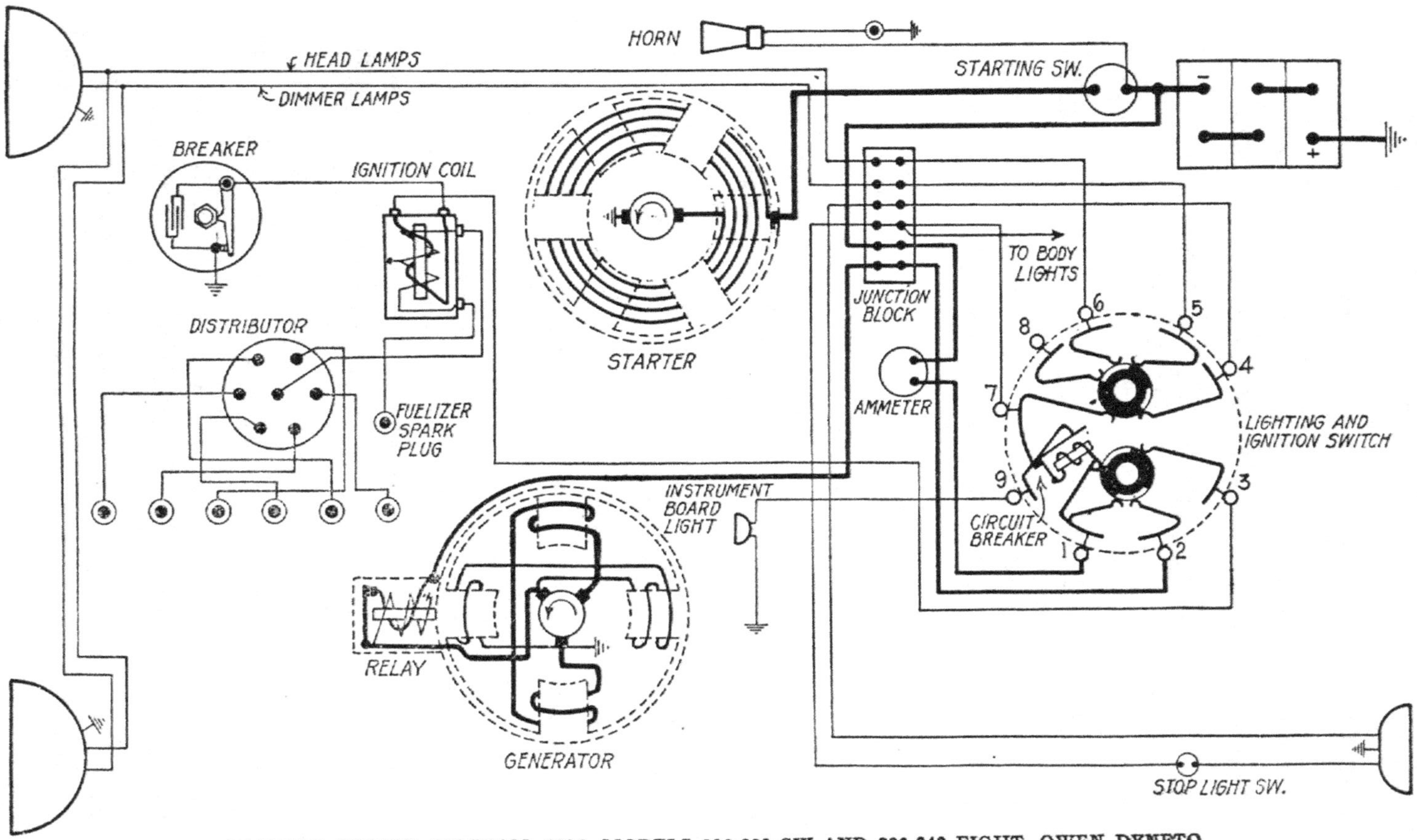

PACKARD WIRING DIAGRAM, 1925, MODELS 326-333 SIX AND 336-343 EIGHT, OWEN DYNETO GENERATING AND STARTING AND LIGHTING, DELCO IGNITION

P A C K A R D (Six) Model 326-333 Year 1925

Owen-Dyneto Starter & Generator Delco Ignition

Regulation	Max. Chg. rate and speed
3rd Brush	14-6 Amps. 900

RELAY Air Gap	Contact Gap	Cut-in R.P.M.
		8 M.P.H.

BATTERY Prest-O-Lite Type 613 S H C Volts 6 Amps. 112

Bat. to Frame Con. Pos. CONTACT BREAKER Gap .020"

Firing Order 1-5-3-6-2-4 Ignition Timing T.D.C. 1-6 2" B.T.C. Adv.

SPARK PLUG	ENGINE	Taxable Hp.
Size 7/8" Gap 1/32"	Bore 3 3/8" Stroke 5"	27

INTAKE VALVE TIMING	EXHAUST VALVE TIMING
Open T.D.C. Close 45° A.B.C.	Open 45° B.B.C. Close T.D.C.

VALVE CLEARANCE Hot Intake .004" Exhaust .004"

CARBURETOR	COOLING SYSTEM	OILING SYSTEM
Own	Type P Cap 4 7/8 gal.	Type Pres. Cap 9 qts.

PISTON RING: Width 1/8" Diam. 3 3/8" Gap

CLUTCH Disc GEAR RATIO 4.66 AXLE Semi F

BRAKES

Front	Rear	Hand
	41½" X 2"	39 3/32" X 1 3/4"

Lighting	Headlights	Dash & Tail	Side Lamps
Single Contact	21 & 6 C.P.	2 C.P.	2 C.P.

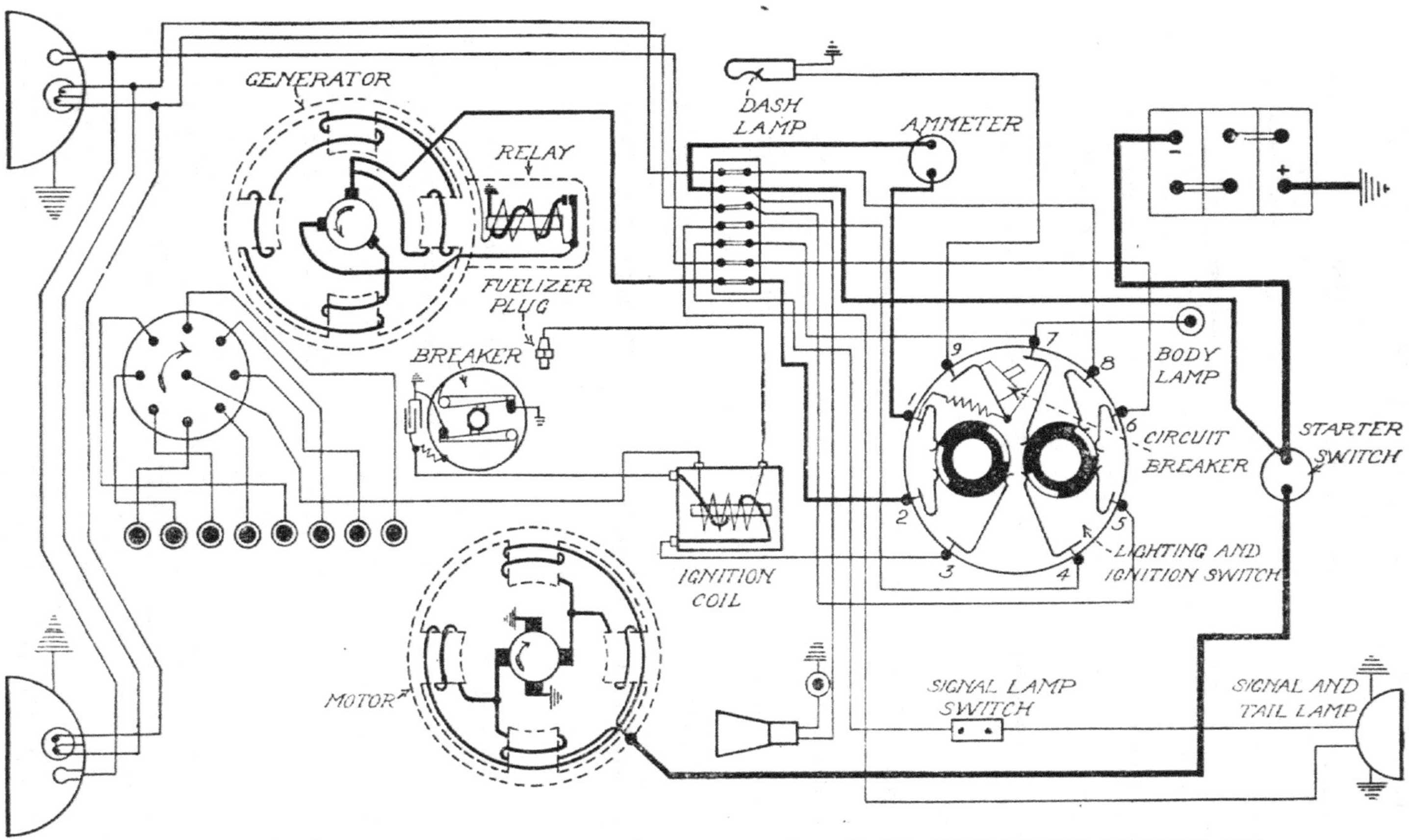

PACKARD WIRING DIAGRAM, 1924-1925, MODEL 336-343, SINGLE EIGHT, OWEN-DYNETO STARTING AND LIGHTING, DELCO IGNITION SYSTEMS

P A C K A R D (Eight) Model 336-343 Year 1925

Owen-Dyneto Starter & Generator Delco Ignition

Regulation: 3rd Brush

Max. Chg. rate and speed: 14-16 Amps.-900

RELAY Air Gap ______ Contact Gap ______ Cut-in R.P.M. 8 M.P.H.

BATTERY Prest-O-Lite Type 617 S.H.C. Volts 6 Amps 160

Bat. to Frame Con. Pos. CONTACT BREAKER Gap .015"

Firing Order 1-3-2-5-8-6-7-4 Ignition Timing A.D.C. 1-8 1 5/8" T.D.C. Adv.

SPARK PLUG		ENGINE		Taxable Hp.
Size 7/8"	Gap 3 1/2"	Bore 3 3/8"	Stroke 5"	37

INTAKE VALVE TIMING		EXHAUST VALVE TIMING	
Open T.D.C.	Close 45° A.B.C.	Open 45° B.B.C.	Close T.D.C.

VALVE CLEARANCE Hot Intake .004" Exhaust .004"

CARBURETOR	COOLING SYSTEM		OILING SYSTEM	
Own	Type P	Cap	Type Pres.	Cap 6 qts.

PISTON RING: Width 1/8" Diam. 3 3/8" Cap

CLUTCH Disc GEAR RATIO 4.7 AXLE Semi F

BRAKES

Front	Rear	Hand
	41 3/8" X 2"	31¾" X 2"

Lighting		Headlights	Dash & Tail	Side Lamps
Single	Contact	21 C.P.	4 C.P.	4 C.P.

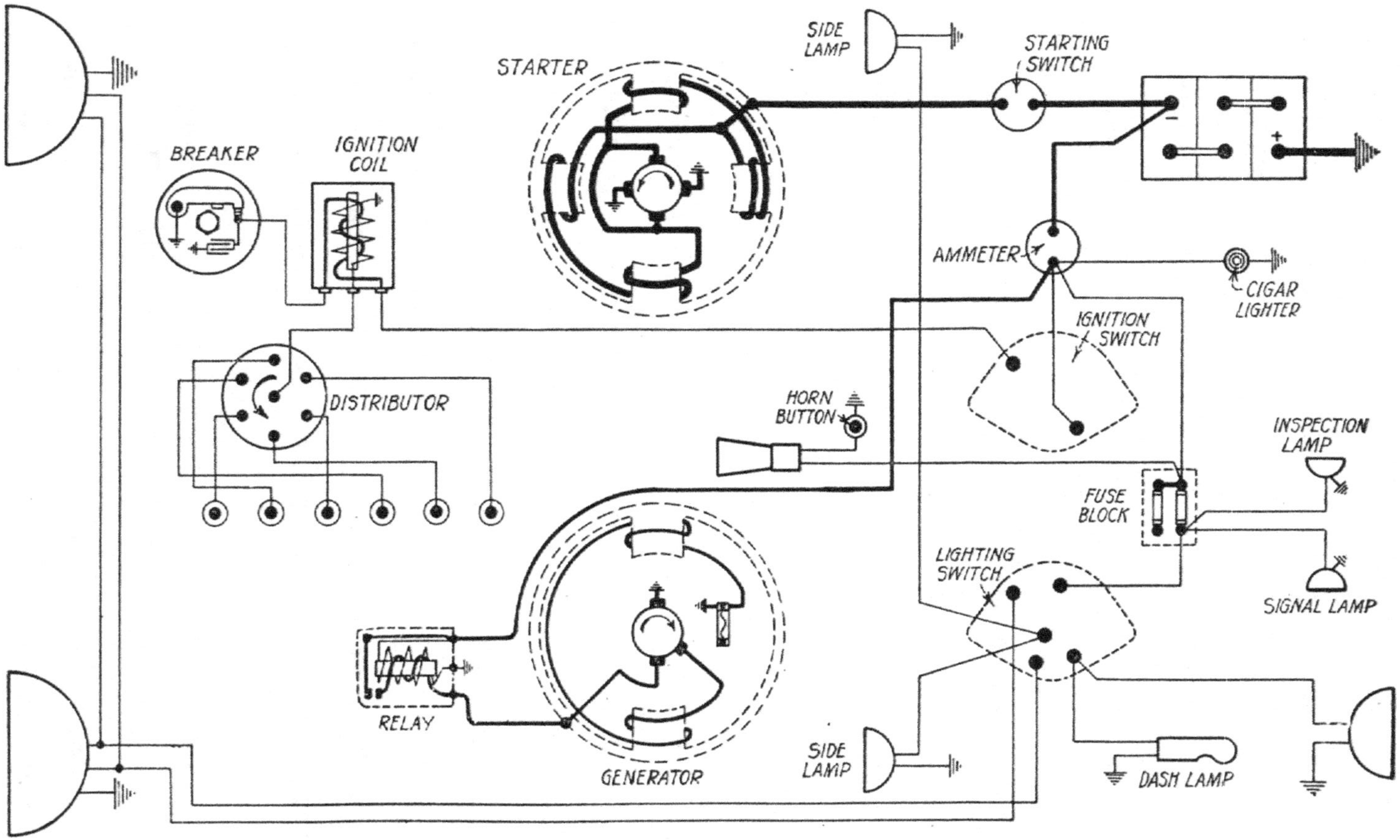

PAIGE WIRING DIAGRAM, 1923-1925, MODELS 6-70 AND 21-24, REMY STARTING AND LIGHTING, ATWATER KENT IGNITION SYSTEMS

P A I G E Model 6 - 70 Year 1924

Remy Starter & Generator Atwater-Kent Ignition

GENERATOR Model	Regulation	Max. Chg. rate
917 - U	Third Brush	19- 21 Amps.

Amps. Cut-in	Volts Warm Test	R.P.M.	Amps. Cut-in	Volts Cold Test	R.P.M.
0	7.2	550	7.	7.2	675
1	8.2	1800	20	8.5	1650

RELAY Model	Closes	Opens	Contact Gap	Air Gap
No. 266 - F	9 M P H	7 M P H	.014	.014

STARTING Motor Model	Drive	Rotation
No. 722 - A	Bendix	Clockwise

BATTERY Westing-house Type 6-WO - 17 Capacity 6 Volts 150 Amps. Bat. to Frame Con. Positive

IGNITION Coil Model	DISTRIBUTOR Model	Breaker Contact Gap
No. N - 3.	No. R A	.015"

FIRING ORDER	Ignition Timing	SPARK PLUG
1-5-3-6-2-4	"IN-OP-6" at Retard	Size 7/8" Gap .025"

LIGHTING	Headlamps	Dash & Tail	Side Lamps
Single Contact	2 C.P.	4 C.P.	4 C.P.

ENGINE — Cylinder — Valve tappet clearance

Bore $3\frac{3}{4}$ Stroke 5 No. 6 L Head Inlet .004 Exhaust .005

CARBURETOR	COOLING	PISTON RING	OILING SYSTEM
Rayfield 1 3/8"	Pump	3 3/4 X 3/16	Cap. 7qt. Type F&S

CLUTCH	REAR AXLE	Gear Ratio	TRANSMISSION
Plate 9 3/4	3/4 Flt	4.60	Warner-Gear

BRAKES	Position	Brake lining width & length
Rear wheel	Trans. or wheel T	Service $2\frac{1}{2}$ X $24\frac{1}{4}$ Hand $2\frac{1}{2}$ X 20

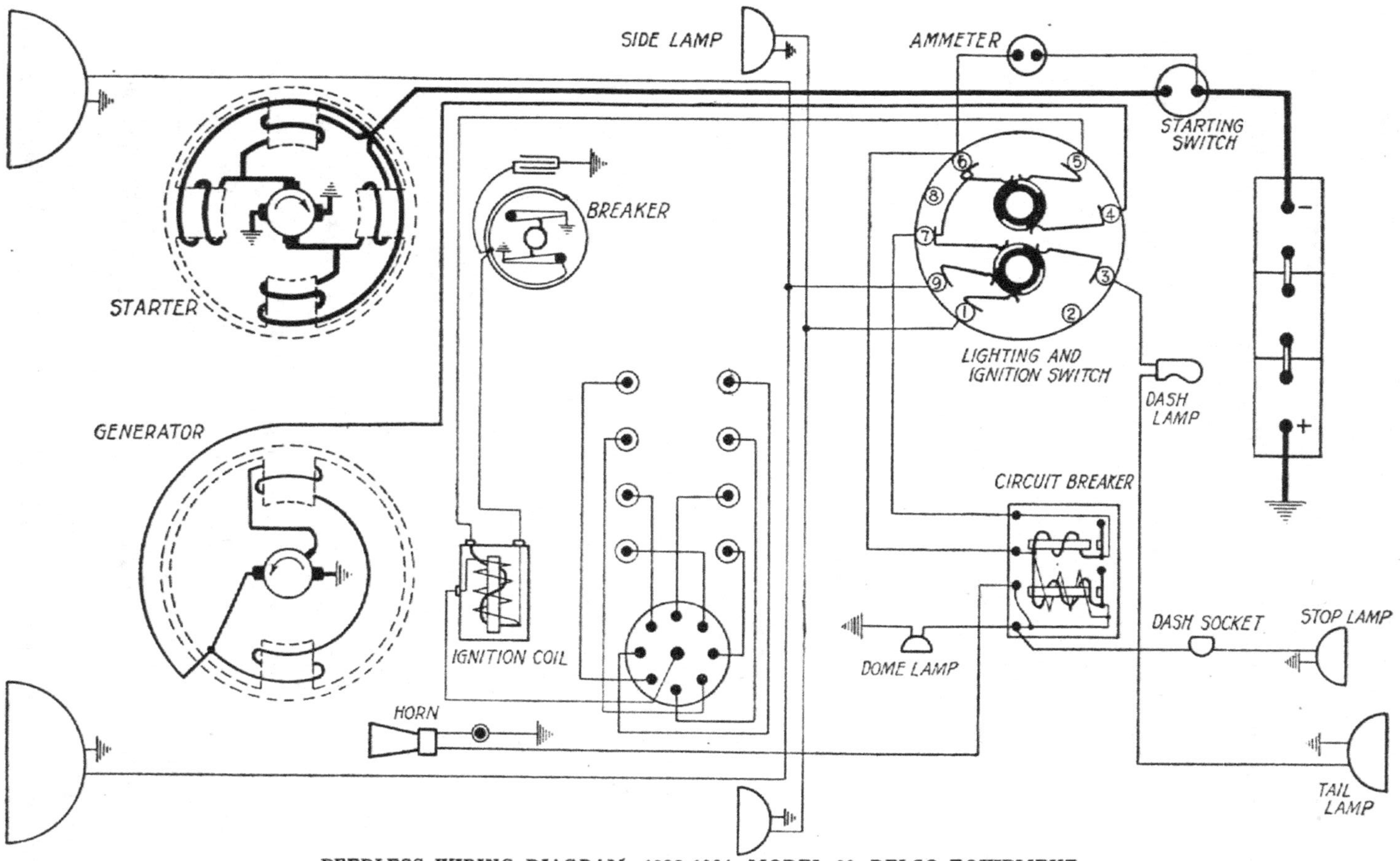

PEERLESS WIRING DIAGRAM, 1923-1924, MODEL 66, DELCO EQUIPMENT

PEERLESS Model 66 Year 1923-24

Delco Starter & Generator Delco Ignition

GENERATOR Model	Regulation	Max. Chg. rate
No. 247	Third Brush	17 Amps.

Amps. Cut-in	Volts Warm Test	R.P.M.	Amps. Cut-in	Volts Cold Test	R.P.M.
5	- - -	800	9	- - -	1000
15	- - -	1500	17	- - -	2000

RELAY NONE USED. CONTACT MADE BY SWITCH.

STARTING Motor Model	Drive	Rotation
No. 248	Bendix	Clockwise

BATTERY Exide Type Capacity Bat. to Frame Con.

3-XXV-19-2 6 Volts 120 Amps. Positive

IGNITION Coil Model	DISTRIBUTOR Model	Breaker Contact Gap
No. 2115	No. 5250	.015"-.022"

FIRING ORDER	Ignition Timing	SPARK PLUG
1R-4L-3R-2L-4R-1L-2R-3L	TDC at Retard	Size 7/8" Gap .025"

LIGHTING	Headlamps	Dash & Tail	Side Lamps
Single Contact	6 V 21 C.P.	3 V 2 C.P.	6 V 2 C.P.

ENGINE Cylinder Valve tappet clearance

Bore 3¼ Stroke 5 No. 8 L Head Inlet .008 Exhaust .008

CARBURETOR	COOLING	PISTON RING	OILING SYSTEM
Ball 1 1/8	Pump	3 1/4 X 1/8	Cap 10 qt Type Force

CLUTCH	REAR AXLE	Gear Ratio	TRANSMISSION
Disc 8 7/16	Semi	4.90	Own

BRAKES 4-Wheel Optional Position Brake lining width & length

Rear wheel Trans. or wheel W Service 2½ X 42¾ Hand 2½ X 44

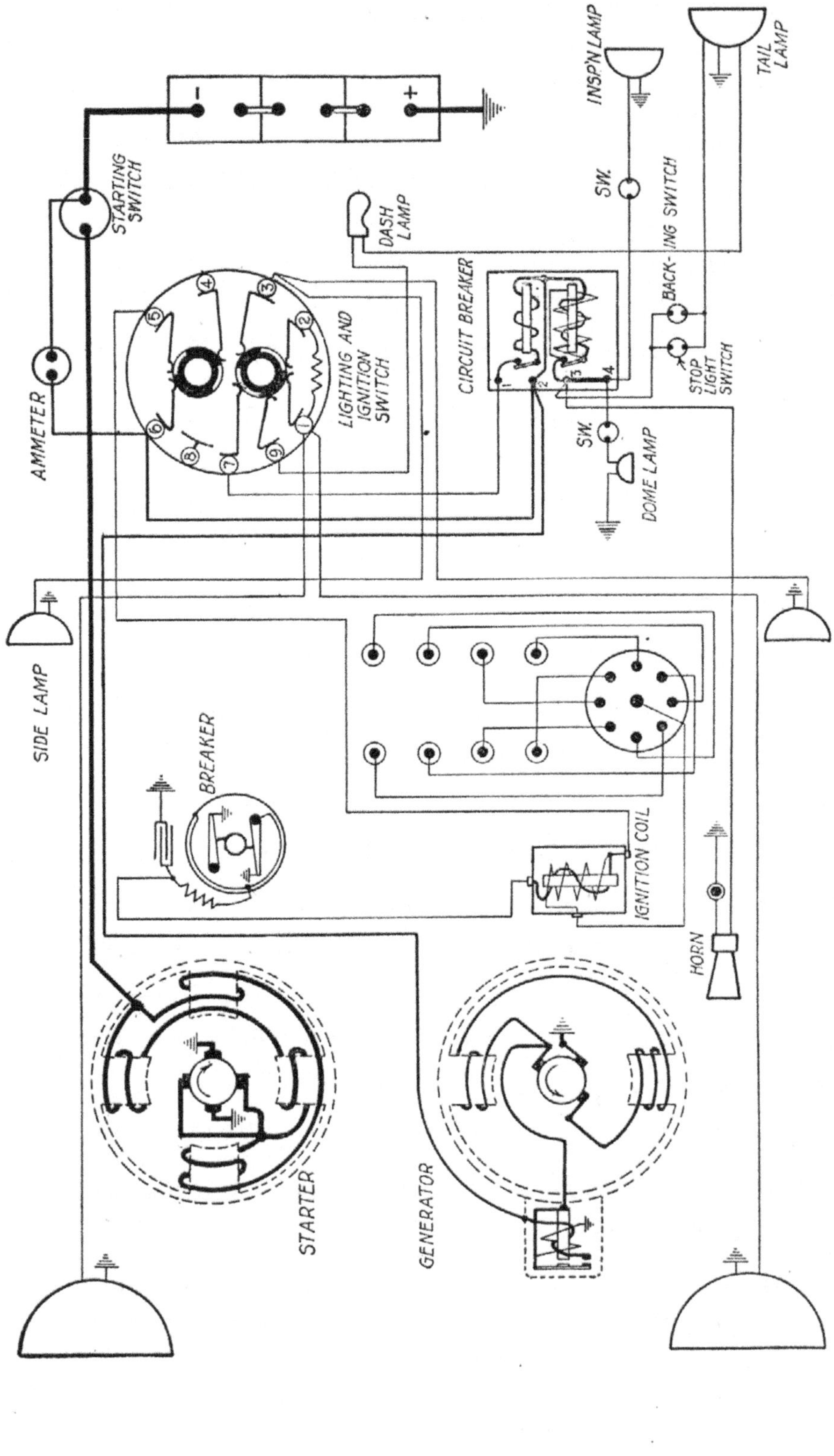

PEERLESS WIRING DIAGRAM, 1925, MODEL 67, SERIAL No. 329501 AND UP, DELCO SYSTEM

PEERLESS Model 67 Year 1925

Delco Starter & Generator Delco Ignition

Regulation	Max. Chg. rate and speed
3rd Brush	20 Amps. 1600

RELAY Air Gap	Contact Gap	Cut-in R.P.M.
.025"-.030"	.015"-.020"	

BATTERY Exide Type 3-XX-V-19-2 Volts 6 Amps 135

Bat. to Frame Con. Pos. CONTACT BREAKER Gap .015"

Firing Order 1R-1L-4L-4L 2L-3R-3L-2R Ignition Timing T.D.C. Ret.

SPARK PLUG		ENGINE		Taxable Hp.
Size 7/8"	Gap .020"	Bore $3\frac{1}{4}$"	Stroke 5"	34

INTAKE VALVE TIMING		EXHAUST VALVE TIMING	
Open T.D.C.	Close 47° A.B.C.	Open 47° B.B.C.	Close T.D.C.

VALVE CLEARANCE Hot Intake .008" Exhaust .008"

CARBURETOR	COOLING SYSTEM		OILING SYSTEM	
Stromberg	Type P	Cap 7 gals.	Type Pres.	Cap 10 qts.

PISTON RING: Width 1/8" Diam $3\frac{1}{4}$" Gap .001"

CLUTCH Disc 8 $\frac{7"}{16}$ X $\frac{5"}{32}$ GEAR RATIO 4.9 AXLE Semi F

BRAKES

Front	Rear	Hand
20 $\frac{3"}{16}$ X 2" X $\frac{3"}{16}$	20 $\frac{3"}{16}$ X 2" X $\frac{3"}{16}$	44" X $2\frac{1}{4}$" X $\frac{3"}{16}$

Lighting	Headlights	Dash & Tail	Side Lamps
Single Contact	21 & 4 C.P.	2 C.P.	4 C.P.

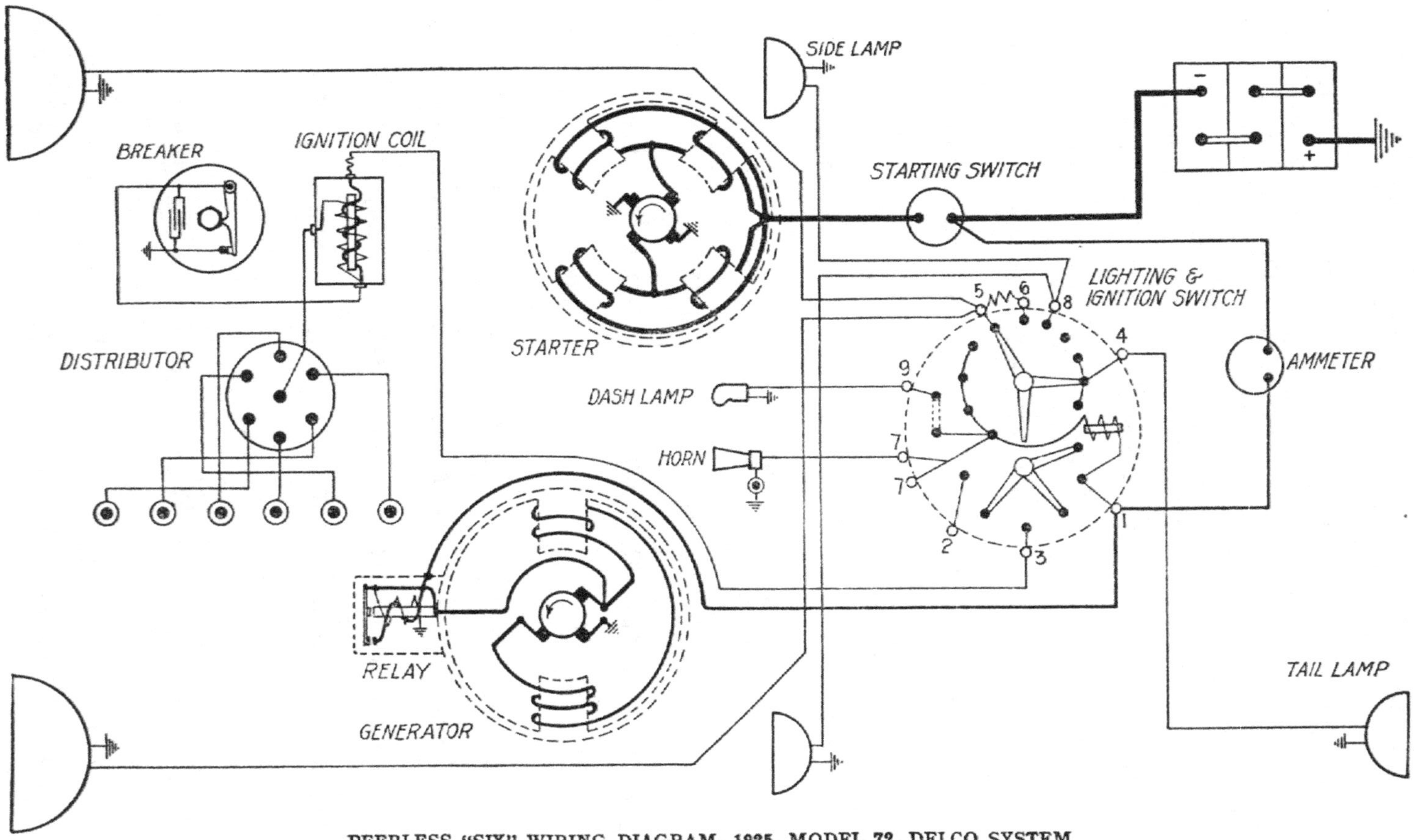

PEERLESS "SIX" WIRING DIAGRAM, 1925, MODEL 72, DELCO SYSTEM

P E E R L E S S Model 72 Year 1925

Delco Starter & Generator Delco Ignition

Regulation	Max. Chg. rate and speed
3rd Brush	18 Amps. 1600

RELAY Air Gap	Contact Gap	Cut-in R.P.M.
.015"-.020"	.030"	8 M.P.H.

BATTERY Exide Type 3-XX-15-1 Volts 6 Amps 105

Bat. to Frame Con. Pos. CONTACT BREAKER Gap .020"

Firing Order 1-5-3-6-2-4 Ignition Timing 15° on Fl. B.T.C. Adv.

SPARK PLUG		ENGINE		Taxable Hp.
Size 7/8"	Gap .025"	Bore $3\frac{1}{2}$"	Stroke 5"	29

INTAKE VALVE TIMING		EXHAUST VALVE TIMING	
Open T.D.C.	Close 50° A.B.C.	Open 50° B.B.C.	Close T.D.C.

VALVE CLEARANCE Hot Intake .007" Exhaust .007"

CARBURETOR	COOLING SYSTEM		OILING SYSTEM	
Johnson	Type P	Cap $4\frac{1}{2}$ gal.	Type Pres.	Cap 10 qts.

PISTON RING: Width 1/8" Diam. $3\frac{1}{2}$" Gap .015"

CLUTCH Disc. $6\frac{1}{4}$"X8 $\frac{7"}{16}$ GEAR RATIO 4.45 AXLE Semi F.

BRAKES

Front	Rear	Hand
14 $\frac{11"}{32}$ X 2" X $\frac{3"}{16}$	18" X 2" X $\frac{3"}{16}$	40 $\frac{13"}{16}$ X $1\frac{3}{4}$" X $\frac{3}{16}$

Lighting	Headlights	Dash & Tail	Side Lamps
Single Contact	21 C.P.	2 C.P.	2 C.P.

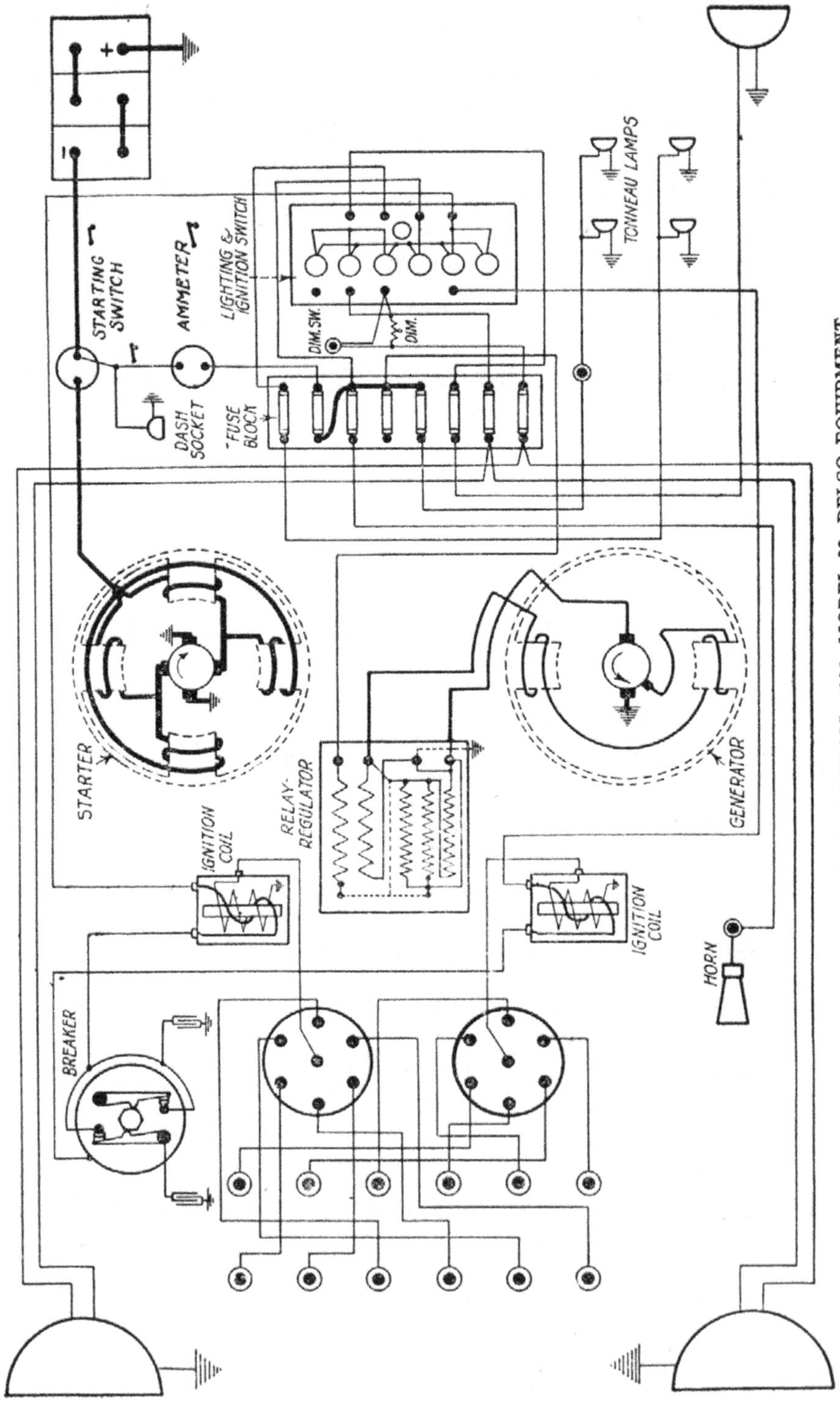

PIERCE-ARROW WIRING DIAGRAM, 1923, MODEL 33, DELCO EQUIPMENT

P I E R C E - A R R O W Model Series 33 Year 1923

Delco Starter & Generator Delco Ignition

GENERATOR Model	Regulation	Max. Chg. rate
No. 238	Third Brush	20 Amps.

Amps. Cut-in	Volts Warm Test	R.P.M.	Amps. Cut-in	Volts Cold Test	R.P.M.
3	- - -	600	10	- - -	800
16	- - -	1200	20	- - -	1600

RELAY Model	Closes	Opens	Contact Gap	Air Gap
No. 5513	8 M P H	6 M P H	Fixed	.030"

STARTING Motor Model	Drive	Rotation
No. 191 - 252	Gear	Clockwise

BATTERY Willard Type Capacity Bat. to Frame Con.

S J R N - 6 6 Volts 160 Amps. Positive

IGNITION Coil Model	DISTRIBUTOR Model	Breaker Contact Gap
No. 2171	No. 5216	.020" - .027"

FIRING ORDER	Ignition Timing	SPARK PLUG
1-5-3-6-2-4	1¼" past T.D.C.	Size 7/8" Gap .030

LIGHTING	Headlamps	Dash & Tail	Side Lamps
Single Contact	21 C.P.	4 C.P.	4 C.P.

ENGINE Cylinder Valve tappet clearance

Bore 4 Stroke 5½ No. 6 T Head Inlet .003 Exhaust .003

CARBURETOR	COOLING	PISTON RING	OILING SYSTEM
Own.	Pump	4 X 3/16	Cap. 10 qt. Type F&S.

CLUTCH	REAR AXLE	Gear Ratio	TRANSMISSION
Disc	Semi	4.29	Own

BRAKES 4-Wheel Optional Position Brake lining width & length

Rear wheel Trans. or wheel W Service 3¼ X 51 Hand 2¼ X 33½

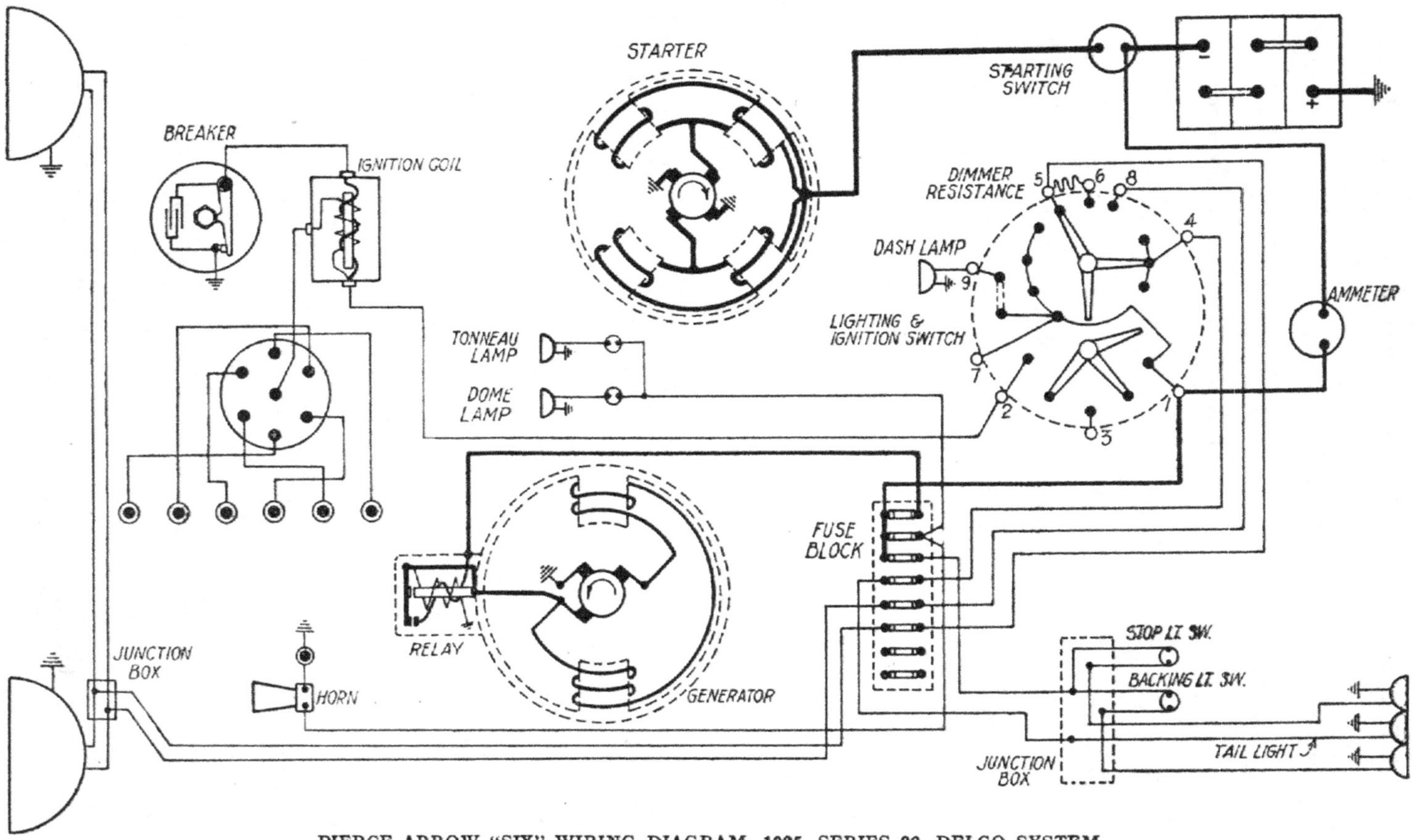

PIERCE ARROW "SIX" WIRING DIAGRAM, 1925, SERIES 80, DELCO SYSTEM

P I E R C E A R R O W 6 Model 80 Year 1925

Delco Starter & Generator Delco Ignition

Regulation	Max. Chg. rate and speed
3rd Brush	18 Amps. 1500

RELAY Air Gap	Contact Gap	Cut-in R.P.M.
.015"-.020"	.030"	8 M.P.H.

BATTERY Willard Type SJR-4 Volts 6 Amps. 110

Bat. to Frame Con. Pos. CONTACT BREAKER Gap .020"

Firing Order 1-5-3-6-2-4 Ignition Timing Ignition on Fl. 2¼" A.T.C. Ret.

SPARK PLUG	ENGINE	Taxable Hp.
Size 7/8" Gap .030"	Bore 3½" Stroke 5"	29

INTAKE VALVE TIMING	EXHAUST VALVE TIMING
Open T.D.C. Close 50°A.B.C.	Open 45°B.B.C. Close 5°A.T.C.

VALVE CLEARANCE Hot Intake .003" Exhaust .003"

CARBURETOR	COOLING SYSTEM	OILING SYSTEM
Own	Type P Cap 5¼ gal.	Type Pres. Cap 9 qts.

PISTON RING: Width 3/16" Diam 3 1/2" Gap

CLUTCH Disc. 7¼"X11 7"/8 GEAR RATIO 4.45 AXLE Semi Fl.

BRAKES

Front	Rear	Hand
29" X 2" X 3"/16	29" X 2" X 3"/16	29" X 2" X 3"/16

Lighting	Headlights	Dash & Tail	Side Lamps
Single	Contact 21 - 4 C.P.	4 C.P.	4 C.P.

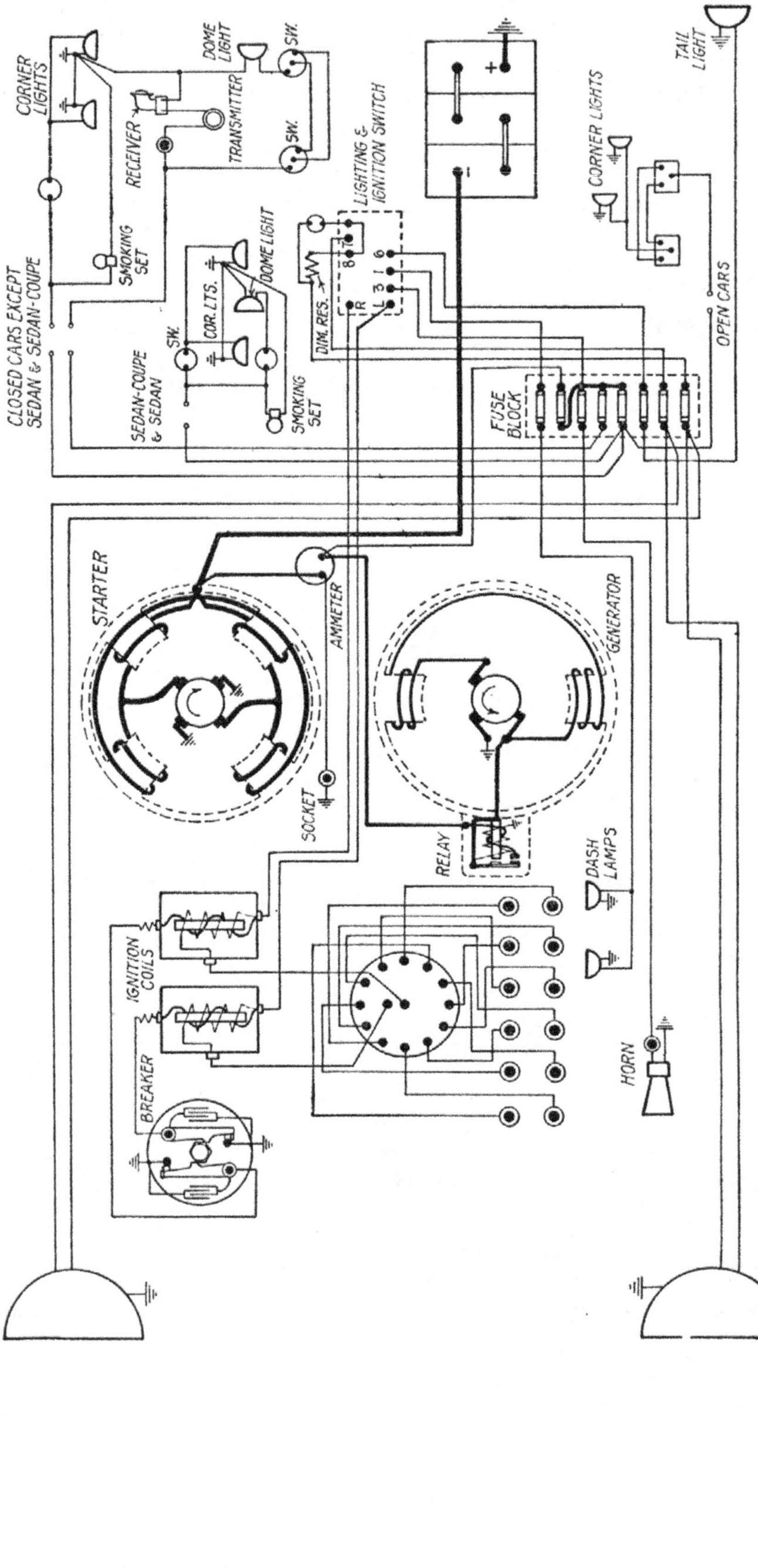

PIERCE ARROW "EIGHT" WIRING DIAGRAM, 1925, SERIES 33, DELCO SYSTEM

PIERCE ARROW 8 Model 33 Year 1925

Delco Starter & Generator Delco Ignition

Regulation — Max. Chg. rate and speed

3rd Brush — 22 Amps. 1600

RELAY Air Gap	Contact Gap	Cut-in R.P.M.
.030"	.015"-.020"	8 M.P.H.

BATTERY Willard Type SJRN-6 Volts 6 Amps. 162.5

Bat. to Frame Con. Pos CONTACT BREAKER Gap .020"

Firing Order 1-5-3-6-2-4 Ignition Timing Ign. on Fl. 1¼" A.T.C. Ret.

SPARK PLUG	ENGINE	Taxable Hp.
Size 7/8" Gap .030"	Bore 4" Stroke 5½"	38

INTAKE VALVE TIMING	EXHAUST VALVE TIMING
Open 5°B.T.C. Close 45°A.T.C.	Open 45°B.B.C. Close 5°A.T.C.

VALVE CLEARANCE Hot Intake .003" Exhaust .003"

CARBURETOR	COOLING SYSTEM	OILING SYSTEM
Own	Type P Cap 6½ gal.	Type Pres. Cap 10½ qts.

PISTON RING: Width 3/16" Diam. 4" Gap

CLUTCH Disc 8¼"X10¼" GEAR RATIO 4.29 AXLE Semi F

BRAKES

Front	Rear	Hand
	51 1/8" X 3¼" X 3/16"	— X 2¾" X 3/16"

Lighting	Headlights	Dash & Tail	Side Lamps
Single	Contact 21 & 4 C.P.	4 C.P.	4 C.P.

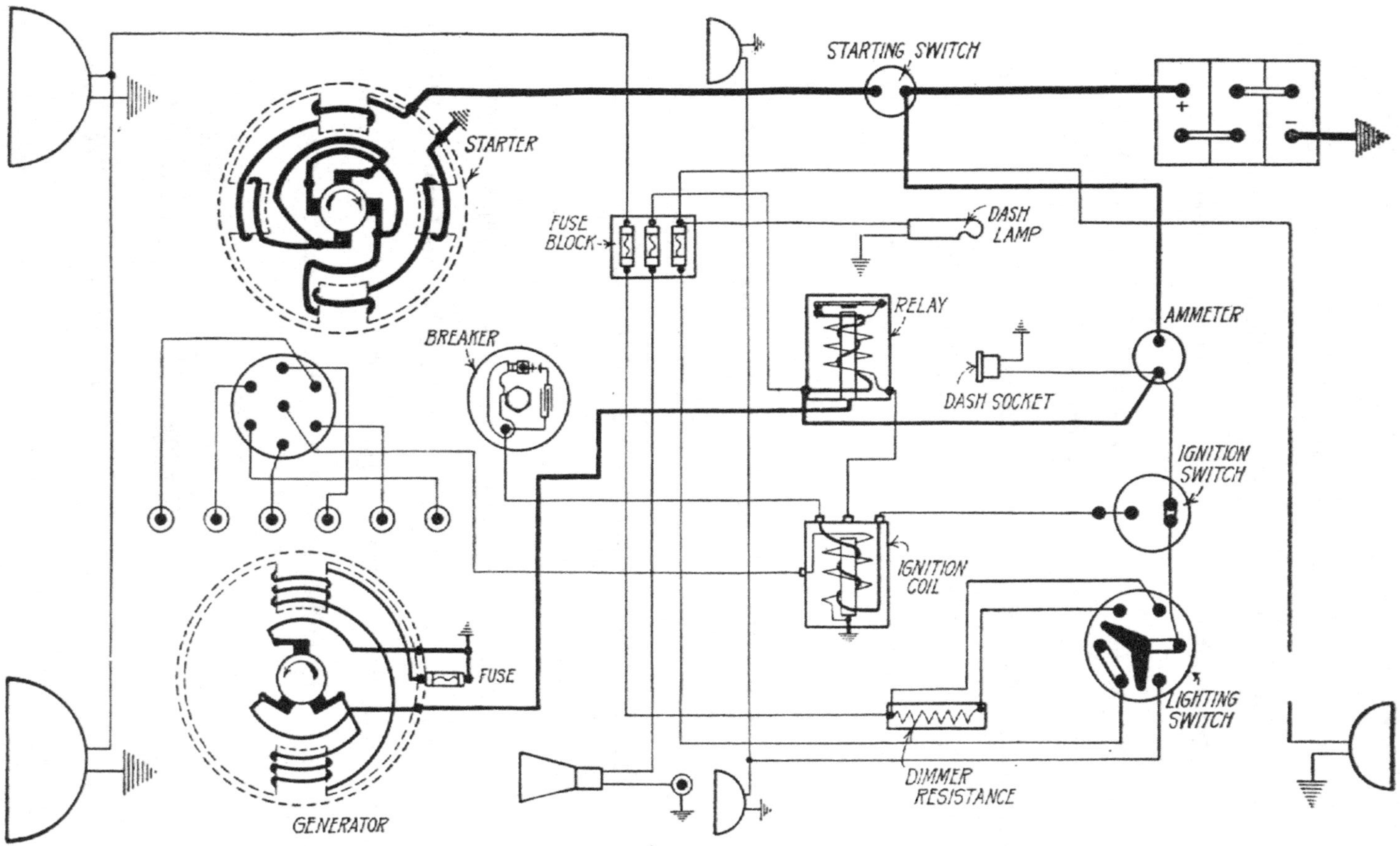

REO WIRING DIAGRAM, 1923-1925, MODELS T AND T6, SERIES B AND C, NORTH EAST EQUIPMENT

R E O Model T Series B Year 1923-24

North East Starter & Generator North East Ignition

GENERATOR Model	Regulation	Max. Chg. rate
No. 3929 - L - 6	Third Brush	18 - 19 Amps.

Amps. Cut-in	Volts Warm Test	R.P.M.	Amps. Cut-in	Volts Cold Test	R.P.M.
15.6	6.5	1250	18	7.0	1320
17.6	8.0	1600	19	7.5	1440

RELAY Model	Closes	Opens	Contact Gap	Air Gap
8212 or 11690	10 M P R	8 M P R	.025 - .040	.010

STARTING Motor Model	Drive	Rotation
No. 3581 - A	Chain	Clockwise

BATTERY Willard Type Capacity Bat. to Frame Con.

S J R N - 4 6 Volts 111 Amps. Negative

IGNITION Coil Model	DISTRIBUTOR Model	Breaker Contact Gap
No. 10481	10560	.020

FIRING ORDER	Ignition Timing	SPARK PLUG
1-4-2-6-3-5	"Spark" at Rt. Sup.	Size 1/2 L Gap .030

LIGHTING	Headlamps	Dash & Tail	Side Lamps
Single Contact	21 C.P.	2 C.P.	2 C.P.

ENGINE Cylinder Valve tappet clearance

Bore $3\frac{3}{16}$ Stroke 5 No. 6 F Head Inlet .003 Exhaust .003

CARBURETOR	COOLING	PISTON RING	OILING SYSTEM
Rayfield 1 1/4	Pump	3 3/16X3/16	Cap. 7 qt. Type F&S

CLUTCH	REAR AXLE	Gear Ratio	TRANSMISSION
Disc. 6 5/8	Semi	4.70	Own

BRAKES Position Brake lining width & length

Rear wheel Trans. or wheel W Service 2 X $45\frac{1}{2}$ Hand 2 X $40\frac{1}{2}$

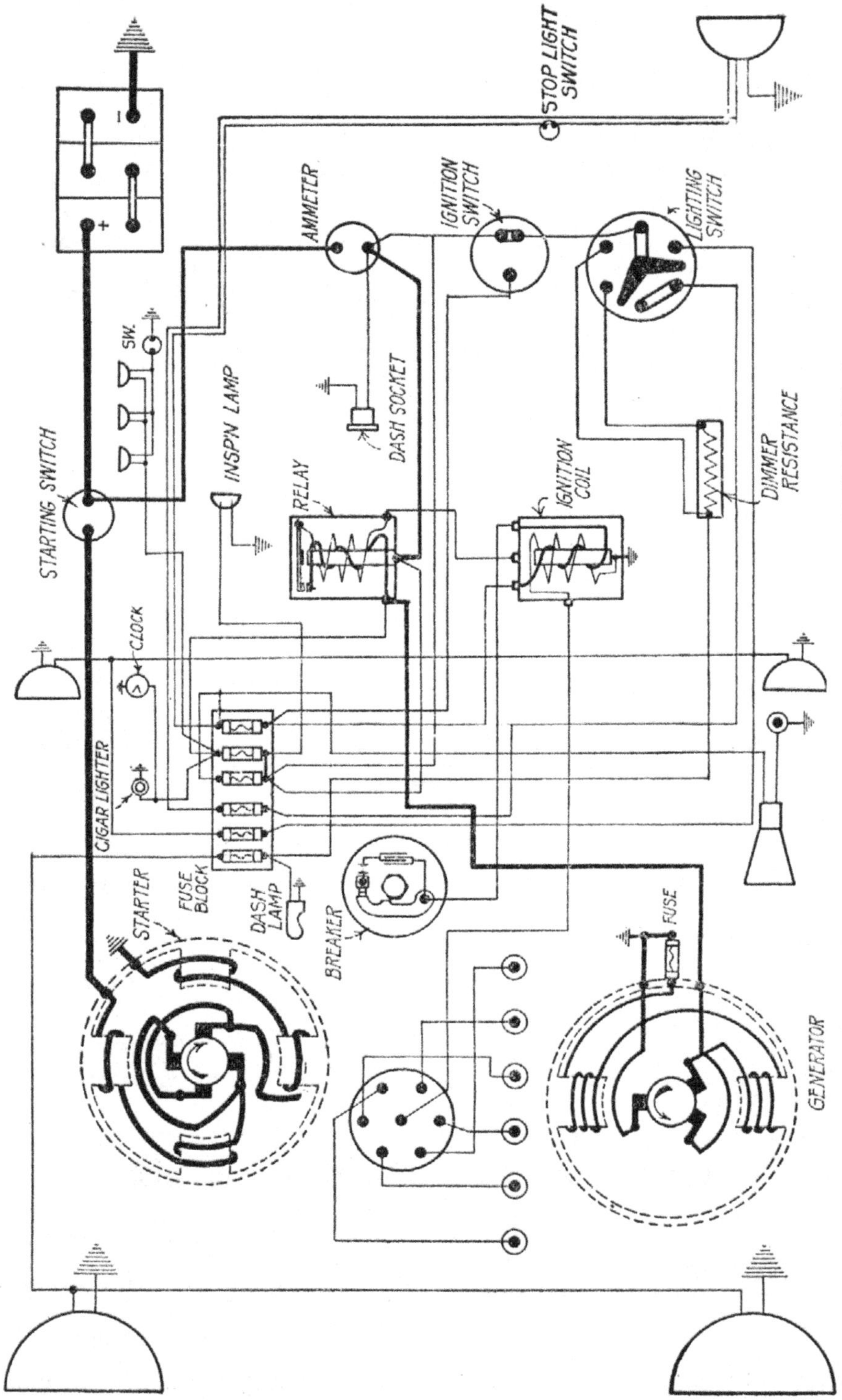

REO WIRING DIAGRAM, LATE 1925, MODEL T6, SERIES E AND G

R E O Model T 6 Ser. E & G Year Late 1925

North East Starter & Generator North East Ignition

Regulation	Max. Chg. rate and speed
3rd Brush	20 Amps. 1600

RELAY Air Gap	Contact Gap	Cut-in R.P.M.
.010"	.025"-.040"	525 or 10 M.P.H.

BATTERY Willard Type S J R - 4 Volts 6 Amps 125

Bat. to Frame Con. Neg. CONTACT BREAKER Gap .020"

Firing Order 1-4-2-6-3-5 Ignition Timing "Spark" on FL.Ret.

SPARK PLUG	ENGINE	Taxable Hp.
Size Gap .025"	Bore 3 3/16" Stroke 5"	24

INTAKE VALVE TIMING	EXHAUST VALVE TIMING
Open T.D.C. Close 69°-30 A.B.C.	Open 50°-30B.B.C. Close

VALVE CLEARANCE Hot Intake .002" Exhaust .004"

CARBURETOR	COOLING SYSTEM	OILING SYSTEM
Schebler	Type P Cap 3 gal.	Type Pres. Cap 7 qts.

PISTON RING: Width Diam Gap

CLUTCH Disc 6 5/8" GEAR RATIO 4.7 AXLE Semi F

BRAKES

Front	Rear	Hand

Lighting	Headlights	Dash & Tail	Side Lamps
Single Contact	21 C.P.	2 C.P.	2 C.P.

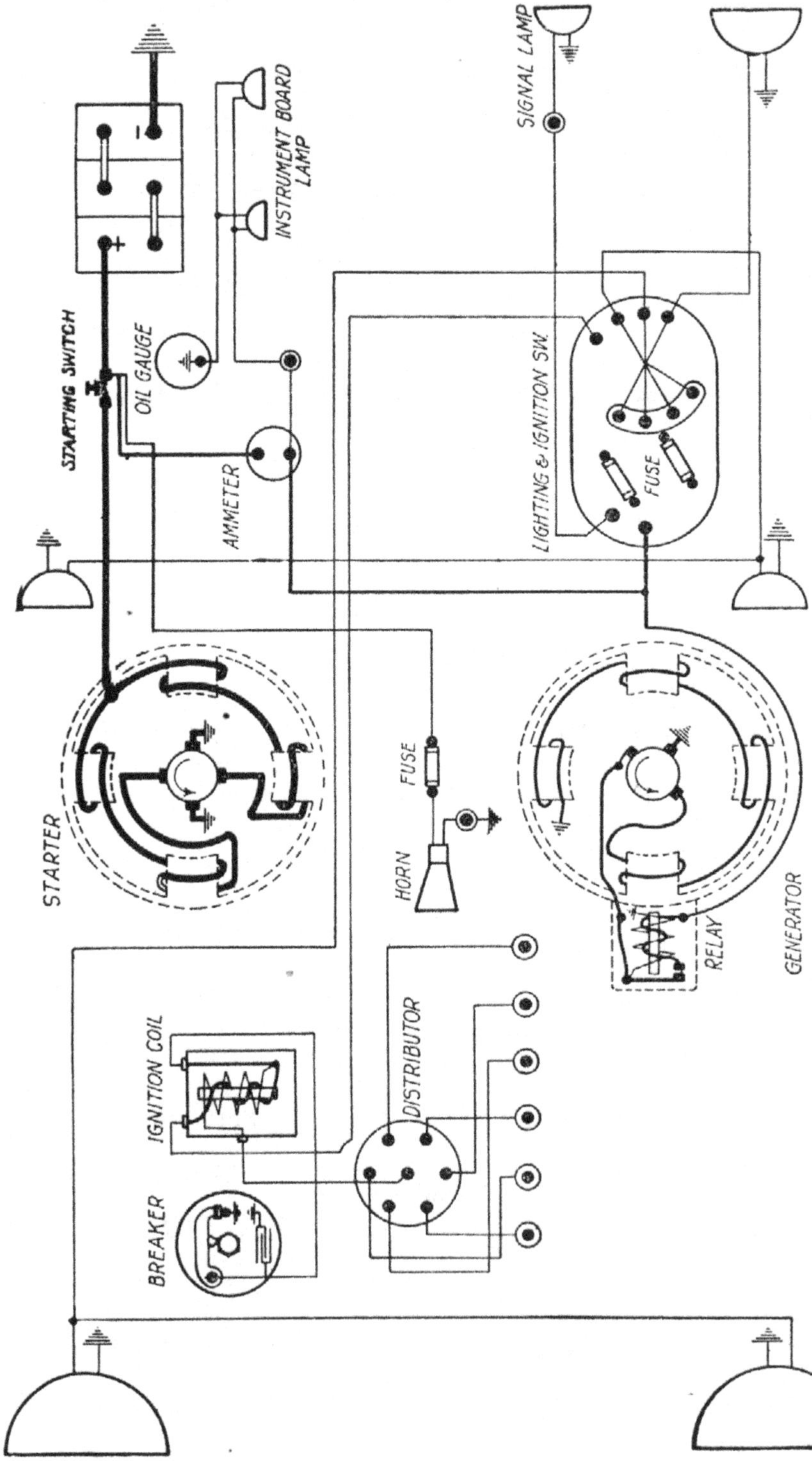

RICKENBACKER WIRING DIAGRAM, 1923-1925, MODELS B AND D, BOSCH EQUIPMENT

RICKENBACKER Model B Year 1923-24

Bosch Starter & Generator Bosch Ignition

GENERATOR Model	Regulation	Max. Chg. rate
No. 1032	Third Brush	15 Amps.

Amps. Cut-in	Volts Warm Test	R.P.M.	Amps. Cut-in	Volts Cold Test	R.P.M.
0	6.8	600	4	7.5	640
10	8.5	1600	15	8.5	1600

RELAY Model	Closes	Opens	Contact Gap	Air Gap
No. 1032	7 M.P.H.	6 M.P.H	.015"-.020"	.010"

STARTING Motor Model	Drive	Rotation
No. 933	Bendix	Anti-Clockwise

BATTERY Prest-O-L. Type Capacity Bat. to Frame Con.

613 - R H N 6 Volts 108 Amps. Negative

IGNITION Coil Model	DISTRIBUTOR Model	Breaker Contact Gap
No. T C - 30	No. T - 6148	.020

FIRING ORDER	Ignition Timing	SPARK PLUG
1-4-2-6-3-5	T.D.C. at Retard	Size 7/8" Gap .025"

LIGHTING	Headlamps	Dash & Tail	Side Lamps
Single Contact	21 C.P.	2 C.P.	5 C.P.

ENGINE Cylinder Valve tappet clearance

Bore $3\frac{1}{8}$ Stroke $4\frac{3}{4}$ No. 6 L Head Inlet .005" Exhaust .007"

CARBURETOR	COOLING	PISTON RING	OILING SYSTEM
Stromberg 1 1/4	Pump	3 1/8 X 3/16	Cap. 7 qt. Type F. & S.

CLUTCH	REAR AXLE	Gear Ratio	TRANSMISSION
Plate 9 7/8	Semi	5.00	Warner-Gear

BRAKES	Position	Brake lining width & length
Four wheel	Trans. or wheel T	Service $1\frac{5}{8}$ X 33 Hand 2 X $24\frac{1}{2}$

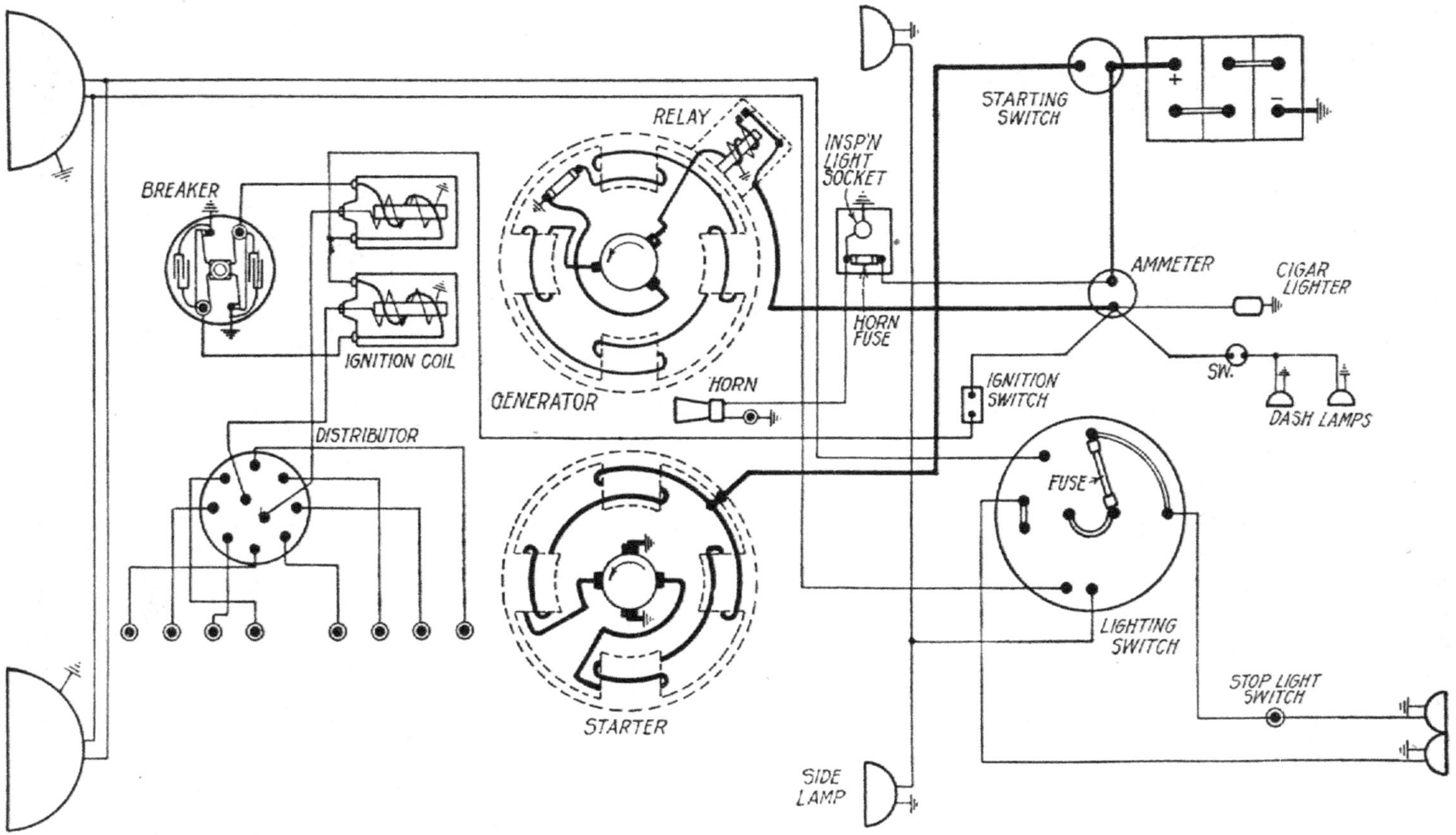

RICKENBACKER WIRING DIAGRAM, 1925, MODEL A-8, 1924-25, SERIAL NO. 25000 UP, AMERICAN BOSCH GENERATING AND STARTING AND LIGHTING, DELCO IGNITION

RICKENBACKER Model A 8 Year 1925

American Bosch Starter & Generator Delco Ignition

Regulation	Max. Chg. rate and speed
3rd Brush	12-14 Amps. 1600

RELAY Air Gap	Contact Gap	Cut-in R.P.M.
.010"	.015"-.020"	500

BATTERY U.S.L. Type 3-HVX-8X Volts 6 Amps 166

Bat. to Frame Con. Neg. CONTACT BREAKER Gap .020"

Firing Order 1-6-2-5-8-3-7-4 Ignition Timing T.D.C. Ret.

SPARK PLUG	ENGINE	Taxable Hp.
Size 7/8" Gap .025"	Bore 3 3/16" Stroke 4 3/4"	33

INTAKE VALVE TIMING	EXHAUST VALVE TIMING
Open ... Close ...	Open ... Close ...

VALVE CLEARANCE Hot Intake .008" Exhaust .008"

CARBURETOR	COOLING SYSTEM	OILING SYSTEM
Zenith	Type P Cap 6½ gal.	Type Pres. Cap

PISTON RING: Width 1/8" Diam 3 3/16" Gap

CLUTCH Disc. GEAR RATIO 4.6 AXLE

BRAKES

Front	Rear	Hand

Lighting	Headlights	Dash & Tail	Side Lamps
Single Contact	21 & 2 C.P.	2 C.P.	2 C.P.

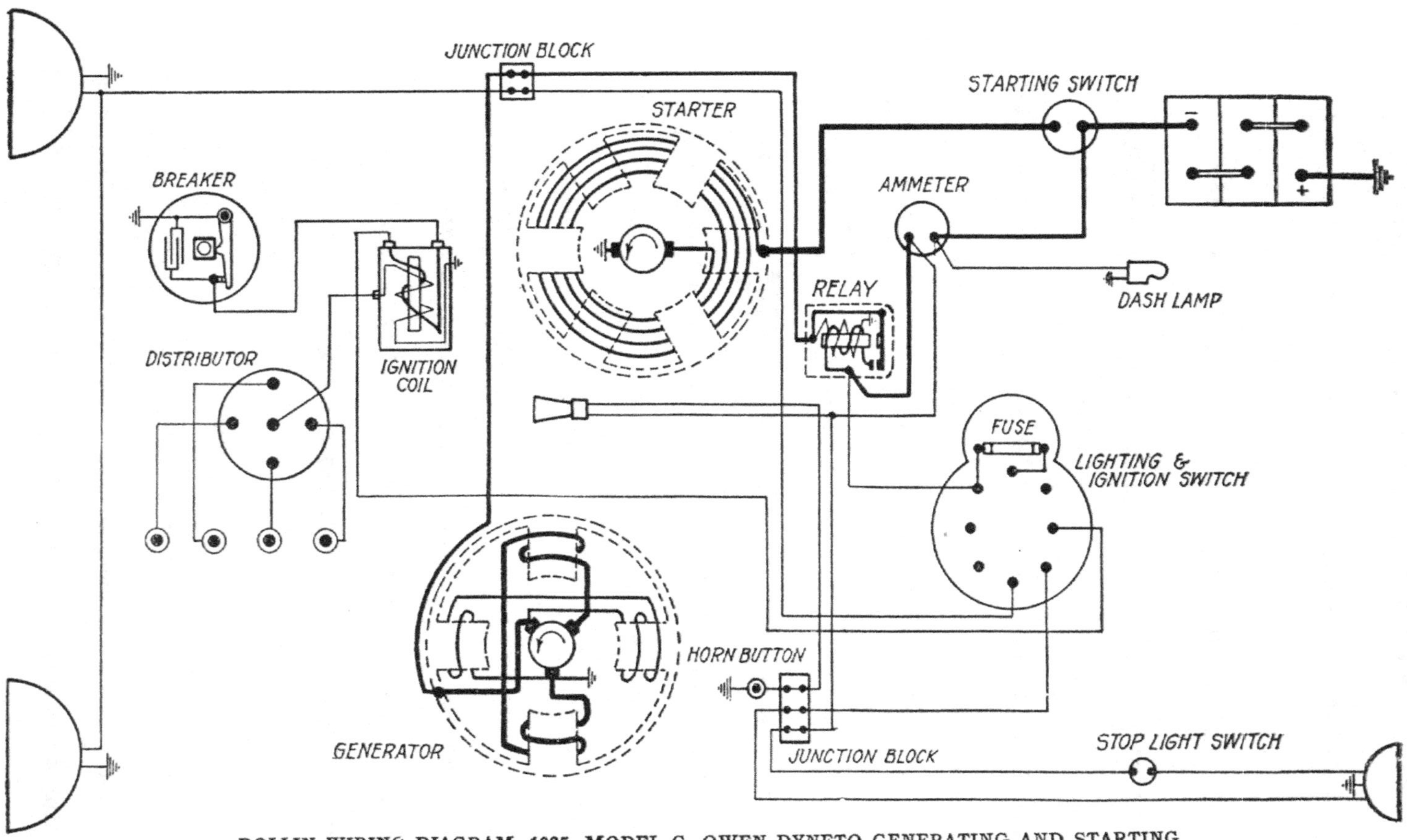

ROLLIN WIRING DIAGRAM, 1925, MODEL G, OWEN DYNETO GENERATING AND STARTING AND LIGHTING, CONNECTICUT IGNITION

ROLLIN Model G Year 1925

Owen-Dyneto Starter & Generator Connecticut Ignition

Regulation	Max. Chg. rate and speed
3rd Brush	13-15 Amps. 1000

RELAY Air Gap	Contact Gap	Cut-in R.P.M.

BATTERY U.S.L. Type 3-CVX-6X Volts 6 Amps 15

Bat. to Frame Con. Pos. CONTACT BREAKER Gap .020"

Firing Order 1-3-4-2 Ignition Timing T.D.C. Adv.

SPARK PLUG	ENGINE	Taxable Hp.
Size 7/8" Gap .020"	Bore 3¼" Stroke 4½"	17

INTAKE VALVE TIMING	EXHAUST VALVE TIMING
Open 2° A.T.C. Close 38° A.B.C.	Open 45° B.B.C. Close 3° A.T.C.

VALVE CLEARANCE Hot Intake .005" Exhaust .005"

CARBURETOR	COOLING SYSTEM	OILING SYSTEM
Tillotson	Type Ther. Cap 3 gal.	Type Pres. Cap 5 qts.

PISTON RING: Width 1/8" Diam 3 1/4" Gap

CLUTCH Plate 8 7"/8 X 1"/8 GEAR RATIO 5.1 AXLE Semi F

BRAKES

Front	Rear	Hand
24 3"/8 X 1½" X 5"/32	24 3"/8 X 1½" X 5"/32	24 3"/8 X 1½" X 5"/32

Lighting	Headlights	Dash & Tail	Side Lamps
Single Contact	21 C.P.	2 C.P.	2 C.P.

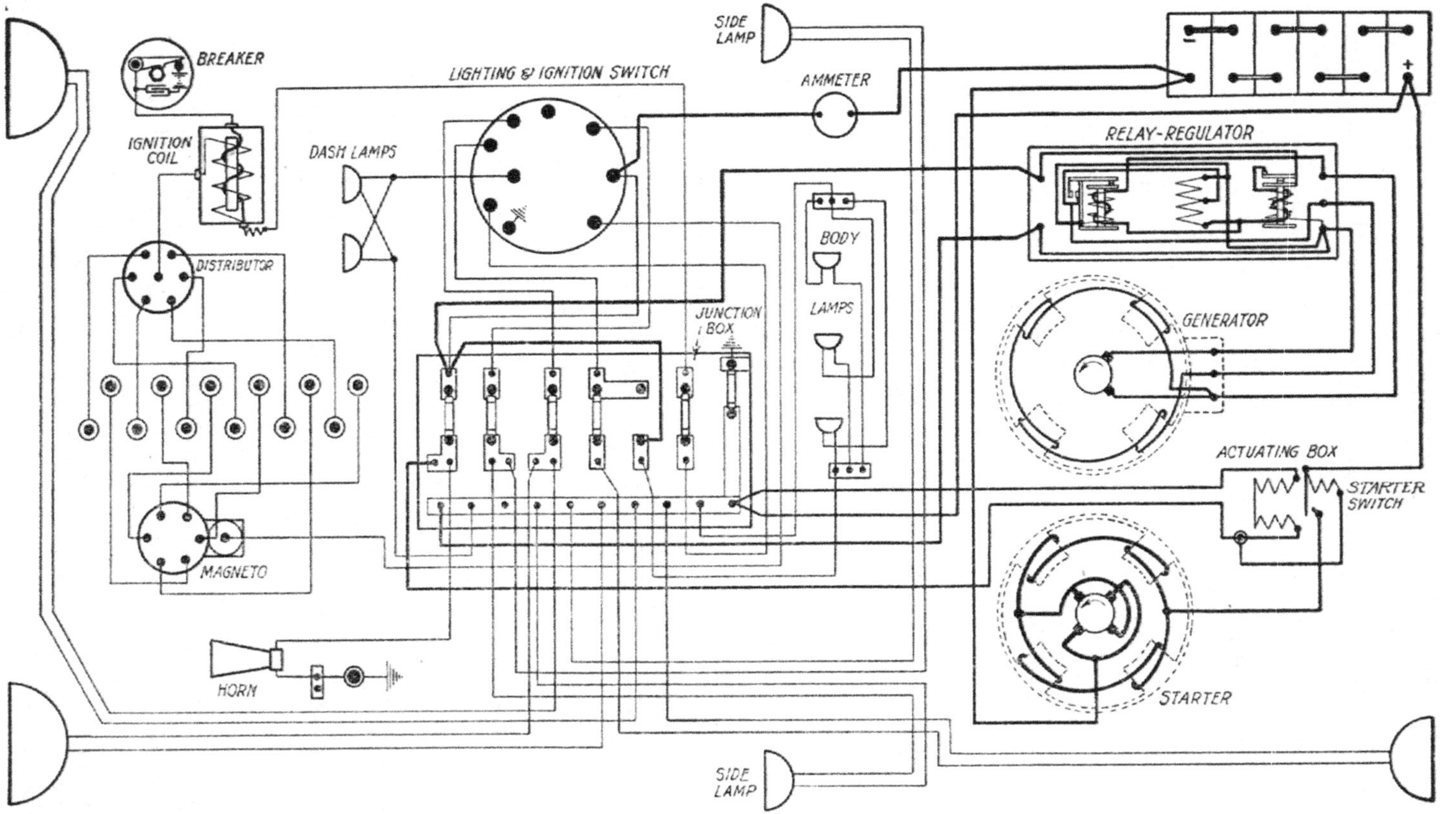

ROLLS-ROYCE WIRING DIAGRAM, 1923-1924 MODELS, BIJUR STARTING AND LIGHTING, BOSCH DUAL IGNITION SYSTEMS

R O L L S - R O Y C E Model All Year 1923-24

Bijur Starter & Generator | Bosch Dual Ignition

GENERATOR Model	Regulation	Max. Chg. rate
No. E X - 11	Relay	14 Amps.

Amps. Cut-in	Volts Warm Test	R.P.M.	Amps. Cut-in	Volts Cold Test	R.P.M.
0	- - -	600	7	- - -	750
13	- - -	800	14	- - -	900

RELAY Model	Closes	Opens	Contact Gap	Air Gap
K X L - R 125	11 MPH	9 MPH	.056	.050

STARTING Motor Model	Drive	Rotation
No. E X - 21	Gear	Anti-Clockwise

BATTERY	Type	Capacity	Bat. to Frame Con.
Exide	6-XXC-13-1	12 Volts 100 Amps.	None

IGNITION Coil Model	DISTRIBUTOR Model	Breaker Contact Gap
R R - Special	R R - Special	.014

FIRING ORDER	Ignition Timing	SPARK PLUG
1-4-2-6-3-5	Mag. 6° before TDC	Size 7/8 Gap .030

LIGHTING	Headlamps	Dash & Tail	Side Lamps
Double Contact	12 V 21 C.P.	12 V 4 C.P.	12 V 4 C.P

ENGINE	Cylinder	Valve tappet clearance
Bore $4\frac{1}{2}$ Stroke $4\frac{3}{4}$	No. 6 L Head	Inlet .002 Exhaust .003

CARBURETOR	COOLING	PISTON RING	OILING SYSTEM
Own	Pump	4.5" X .086"	Cap 6 qt. Type F&S

CLUTCH	REAR AXLE	Gear Ratio	TRANSMISSION
R-Cone - $14\frac{1}{2}$"	Full Flt.	3.25	Own 4 sp.

BRAKES	Position	Brake lining width & length
Rear wheel	Trans. or wheel W	SPECIAL COPPER LINERS

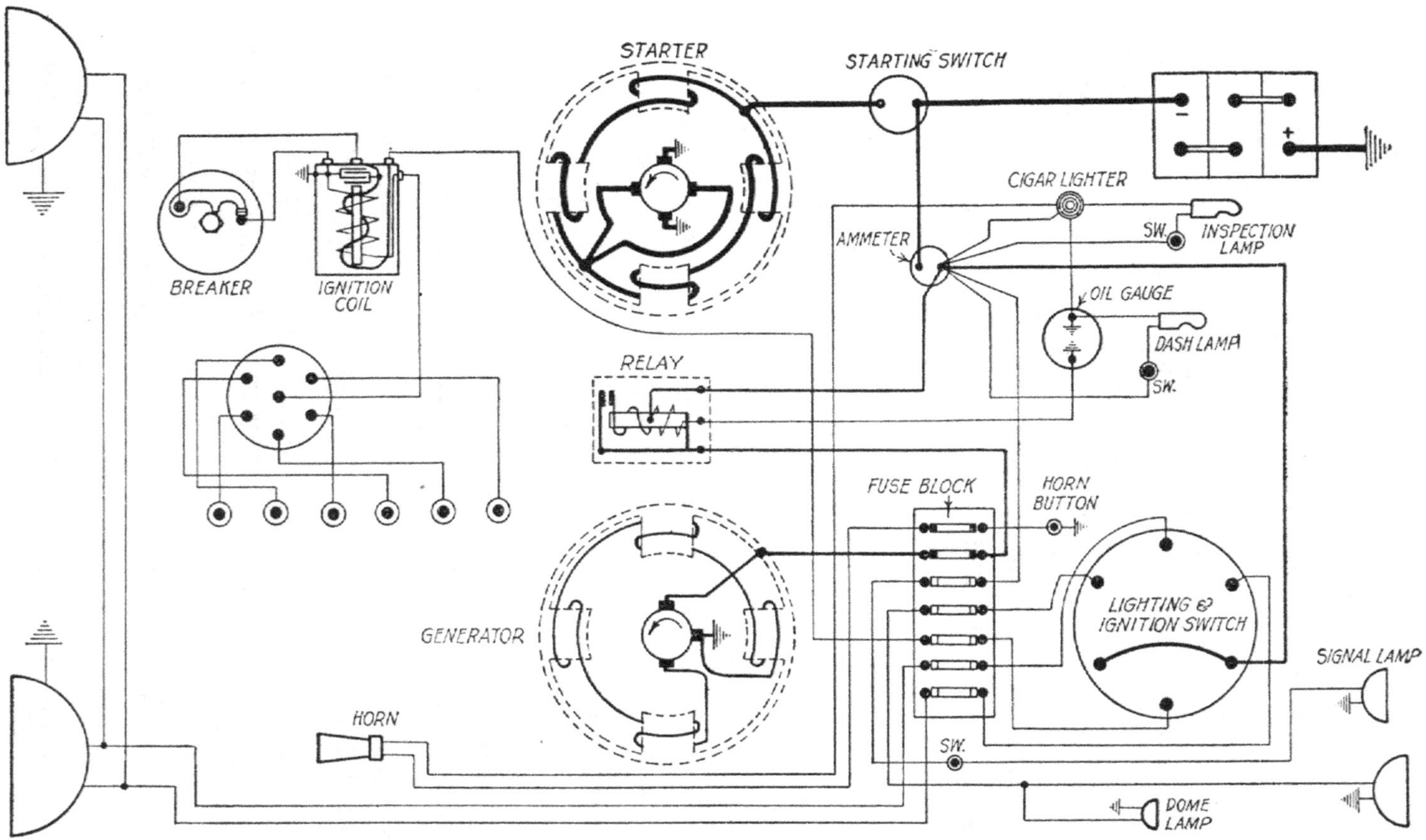

R & V KNIGHT WIRING DIAGRAM, 1923, MODEL H, AUTO-LITE EQUIPMENT

R. & V. KNIGHT Model H Year 1923

Auto-Lite Starter & Generator Auto-Lite Ignition

GENERATOR Model	Regulation	Max. Chg. rate
G J - 4020	Third Brush	16 Amps.

Amps. Cut-in	Volts Warm Test	R.P.M.	Amps. Cut-in	Volts Cold Test	R.P.M.
2.0	6.5	650	0.0	- - -	500
12.6	8.0	1600	16	8.3	1700

RELAY Model	Closes	Opens	Contact Gap	Air Gap
CB-1072	9 MPH	7 MPH	.030	.015

STARTING Motor Model	Drive	Rotation
ML-4091	Bendix	Anti-Clockwise

BATTERY Willard Type	Capacity	Bat. to Frame Con.
SJR-6	6 Volts 152 Amps.	Positive

IGNITION Coil Model	DISTRIBUTOR Model	Breaker Contact Gap
I G	I G	.020-.025

FIRING ORDER	Ignition Timing	SPARK PLUG
1-4-2-6-3-5	TDC at Retard	Size 7/8 Gap .025

LIGHTING	Headlamps	Dash & Tail	Side Lamps
Single Contact	21 C.P.	2 C.P.	4 C.P

ENGINE	Cylinder	Valve tappet clearance
Bore 3½ Stroke 4½ No. 6	K Head	Inlet - - Exhaust - -

CARBURETOR	COOLING	PISTON RING	OILING SYSTEM
Stromberg 1¼	Thermo	3½ X 5/16	Cap. 9 qt Type Force

CLUTCH	REAR AXLE	Gear Ratio	TRANSMISSION
Plate 10	Semi	5.30	Brown-Lipe

BRAKES	Position	Brake lining width & length
Rear wheel	Trans. or wheel W	Service 2½ X 43¼ Hand 2½ X 43¼

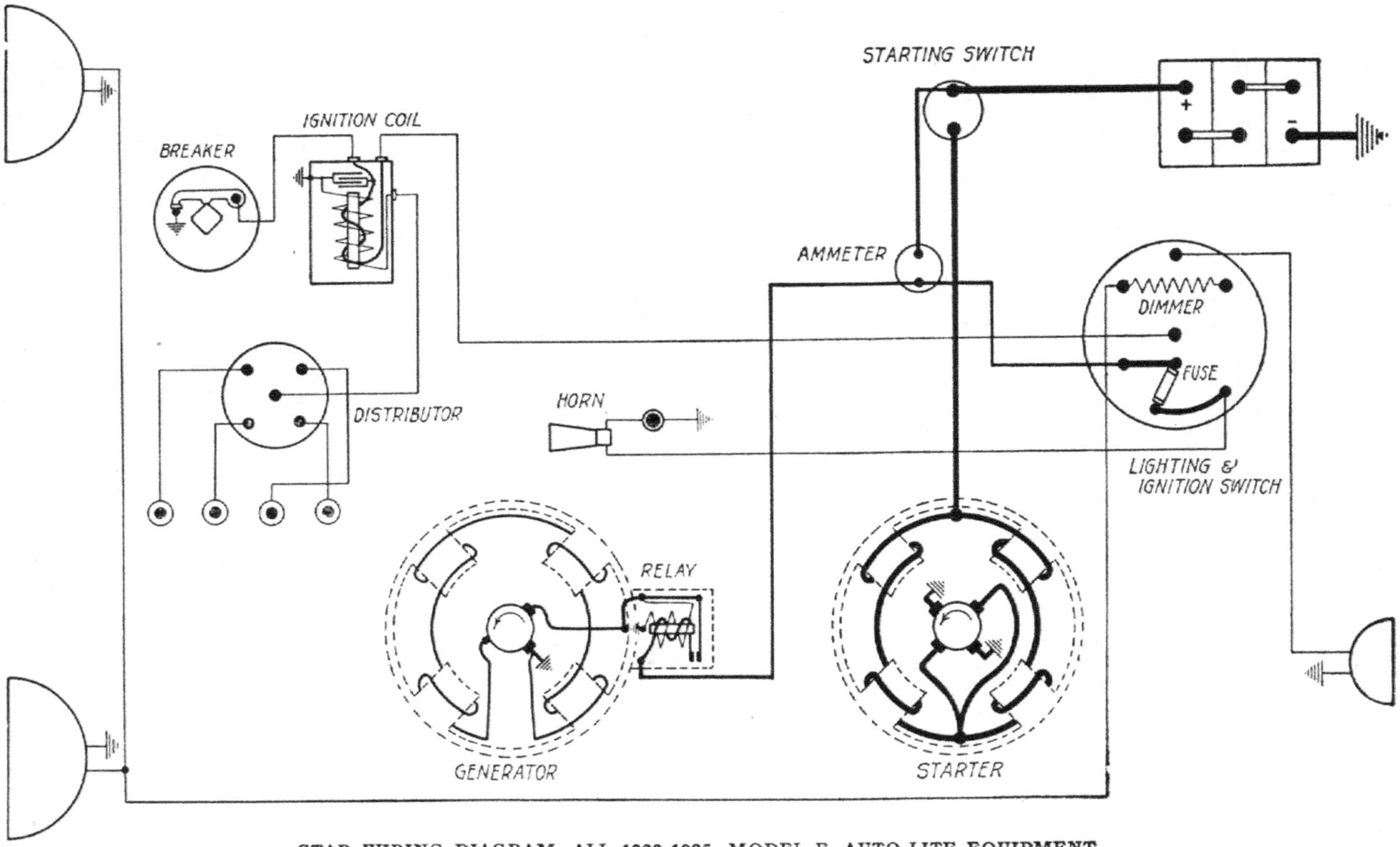

STAR WIRING DIAGRAM, ALL 1923-1925, MODEL F, AUTO-LITE EQUIPMENT

S T A R Model **All** Year **1923-24**

Auto-Lite Starter & Generator **Auto-Lite** Ignition

GENERATOR Model	Regulation	Max. Chg. rate
G J	Third Brush	16 Amps.

Amps. Cut-in	Volts Warm Test	R.P.M.	Amps. Cut-in	Volts Cold Test	R.P.M.
0	6.5	500	10-11	8.0	3300
16	7.5	1800	16	8.3	1700

RELAY Model	Closes	Opens	Contact Gap	Air Gap
G J	9 MPH	7 MPH	.025 - .035	.012

STARTING Motor Model	Drive	Rotation
M G	Bendix	Anti-Clockwise

BATTERY **U S L** Type Capacity Bat. to Frame Con.

C D N - 311 6 Volts 84 Amps. Negative

IGNITION Coil Model	DISTRIBUTOR Model	Breaker Contact Gap
I G	I G	.023

FIRING ORDER	Ignition Timing	SPARK PLUG
1-3-4-2	TDC at Retard	Size 7/8" Gap .030

LIGHTING	Headlamps	Dash & Tail	Side Lamps
Single Contact	22 C.P.	2 C.P.	- - - C.P.

ENGINE Cylinder Valve tappet clearance

Bore $3\frac{1}{8}$ Stroke $4\frac{1}{4}$ No. 4 L Head Inlet .008 Exhaust .008

CARBURETOR	COOLING	PISTON RING	OILING SYSTEM
Tillotson 1"	Pump	3 1/8 X 3/16	Cap. 5 qt. Type F&S

CLUTCH	REAR AXLE	Gear Ratio	TRANSMISSION
Plate 10"	Semi	4.75	Own

BRAKES Position Brake lining width & length

Rear wheel Trans. or wheel W. Service $1\frac{1}{2}$ X $27\frac{1}{2}$ Hand $1\frac{1}{4}$ X 27

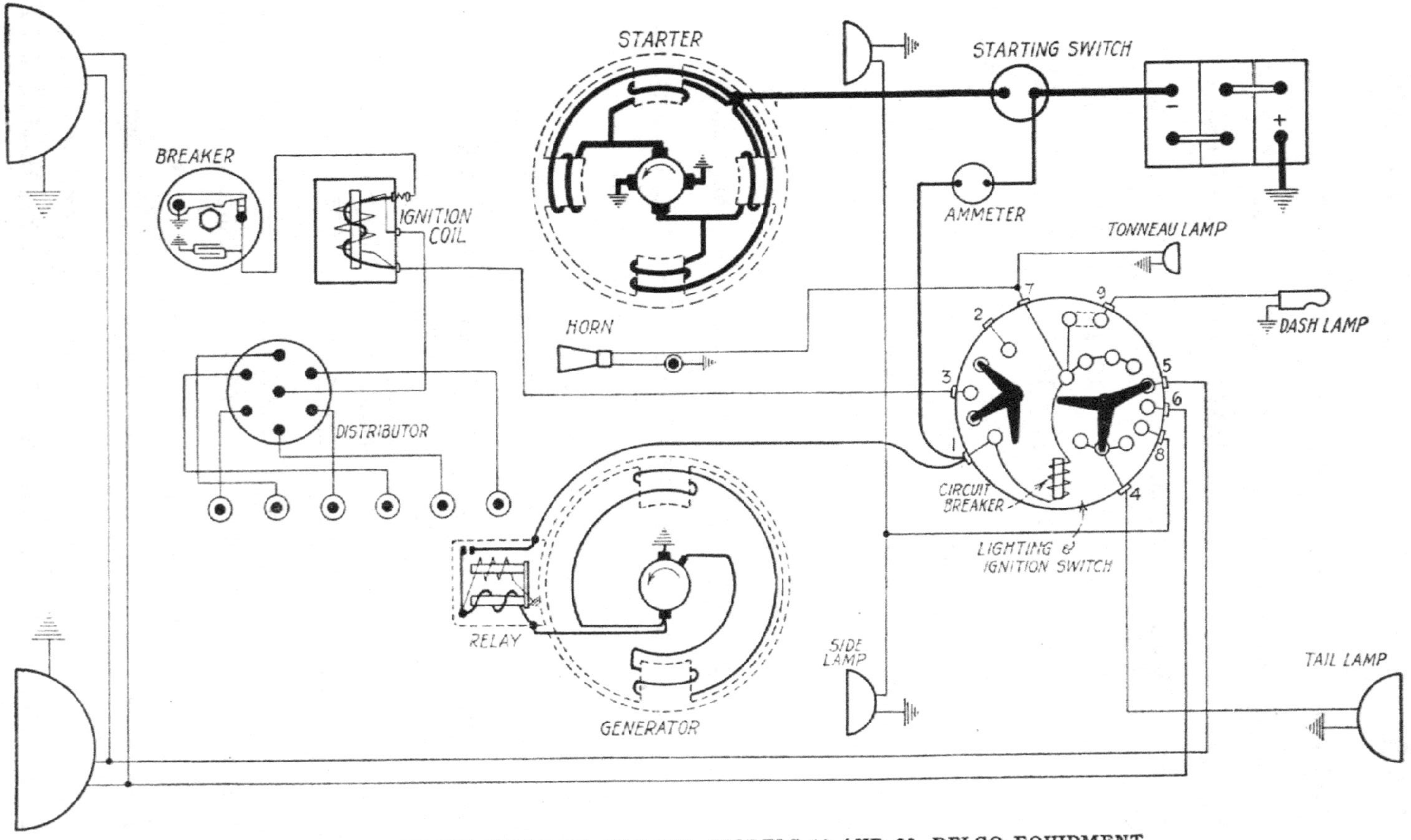

STEPHENS WIRING DIAGRAM, 1923-1924, MODELS 10 AND 20, DELCO EQUIPMENT

STEPHENS Models 10 & 20 Year 1923-24

Delco Starter & Generator Delco Ignition

GENERATOR Model	Regulation	Max. Chg. rate
No. 258	Third Brush	18 Amps.

Amps. Cut-in	Volts Warm Test	R.P.M.	Amps. Cut-in	Volts Cold Test	R.P.M.
4	—	800	9	—	1000
14-16	—	1600	17	—	2000

RELAY Model	Closes	Opens	Contact Gap	Air Gap
No. 5754	7 Volts	3 Amps.	Fixed	.015

STARTING Motor Model	Drive	Rotation
No. 262	Bendix	Anti-Clockwise

BATTERY U.S.L. Type	Capacity	Bat. to Frame Con.
313 - X	6 Volts 116 Amps.	Positive

IGNITION Coil Model	DISTRIBUTOR Model	Breaker Contact Gap
No. 2178	No. 5257	.020" - .027"

FIRING ORDER	Ignition Timing	SPARK PLUG
1-4-2-6-3-5	T.D.C. at Retard	Size 7/8" Gap .025"

LIGHTING	Headlamps	Dash & Tail	Side Lamps
Single Contact	21 C.P.	2 C.P.	2 C.P.

ENGINE — Cylinder — Valve tappet clearance

Bore $3\frac{1}{4}$ Stroke $4\frac{1}{2}$ No. 6 I Head Inlet .003 Exhaust .005

CARBURETOR	COOLING	PISTON RING	OILING SYSTEM
Stromberg 1 1/4"	Thermo	3 1/4 X 3/16	Cap. 7 qt. Type Force

CLUTCH	REAR AXLE	Gear Ratio	TRANSMISSION
Plate 10"	Semi	5.30	Mechanics

BRAKES	Position	Brake lining width & length
Rear wheel	Trans. or wheel T	Service 2 X $43\frac{1}{4}$ Hand $2\frac{1}{2}$ X $24\frac{1}{2}$

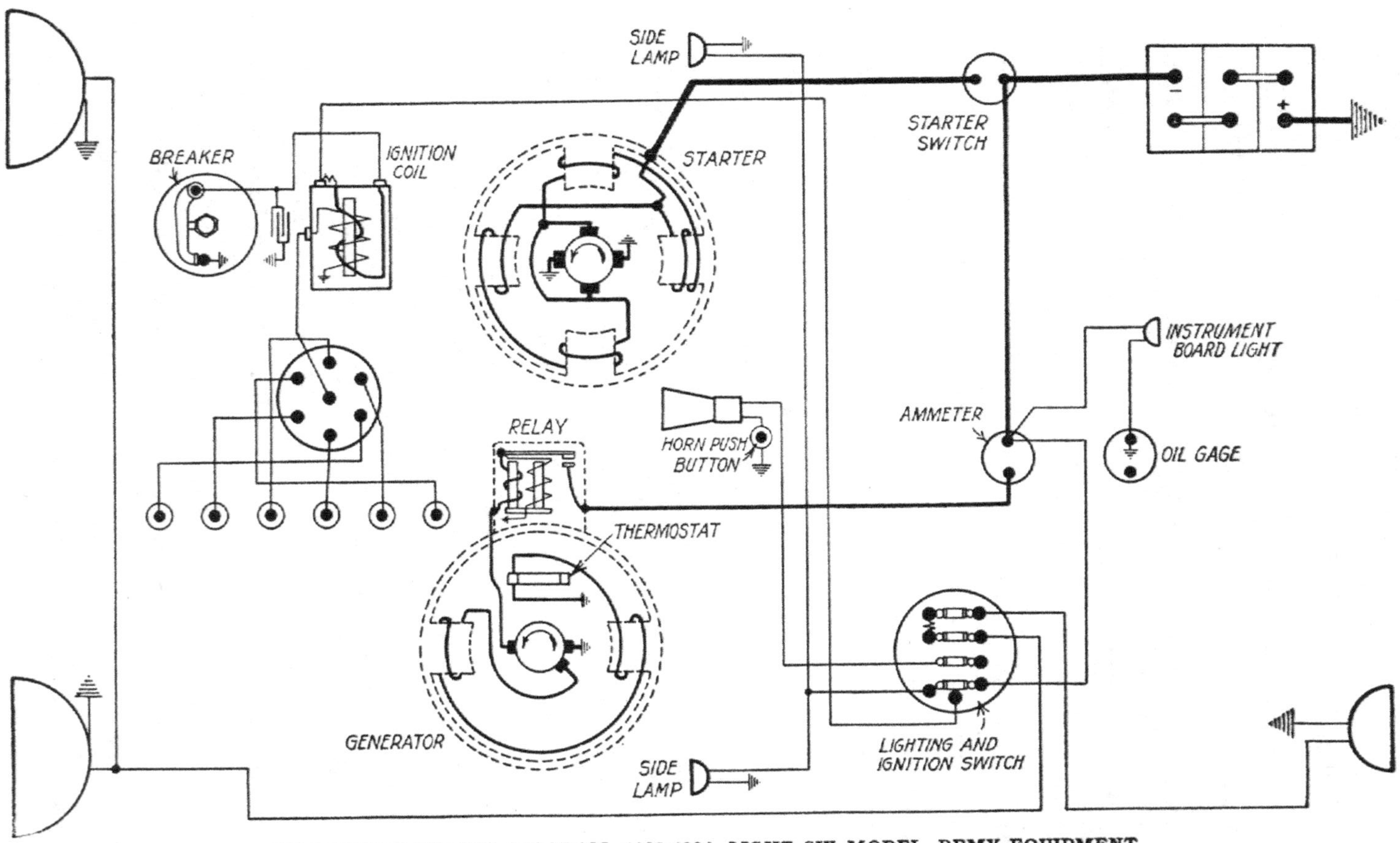

STUDEBAKER WIRING DIAGRAM, 1923-1924, LIGHT SIX MODEL, REMY EQUIPMENT

S T U D E B A K E R Model Light Six Year 1923-24

Remy Starter & Generator Remy Ignition

GENERATOR Model	Regulation	Max. Chg. rate
No. 917 - A	Third Brush	18 - 20 Amps.

Amps. Cut-in	Volts Warm Test	R.P.M.	Amps. Cut-in	Volts Cold Test	R.P.M.
6-9	7.5	1450	7	7.1	850
9-12	7.5	1800	18-20	8.3-8.8	1800

RELAY Model	Closes	Opens	Contact Gap	Air Gap
No. 264 - B	8 M P H	6 M P H	.012-.015	.012

STARTING Motor Model	Drive	Rotation
No. 720 - C	Bendix	Clockwise

BATTERY Willard Type Capacity Bat. to Frame Con.

S J R N - 3 6 Volts 90 Amps. Positive

IGNITION Coil Model	DISTRIBUTOR Model	Breaker Contact Gap
No. 284 - A	626 - A	.020 - .025

FIRING ORDER	Ignition Timing	SPARK PLUG
1-4-2-6-3-5	AD-SP at indicator	Size 7/8R Gap .025

LIGHTING	Headlamps	Dash & Tail	Side Lamps
Single Contact	21 C.P.	2 C.P.	4 C.P.

ENGINE Cylinder Valve tappet clearance

Bore $3\frac{1}{8}$ Stroke $4\frac{1}{2}$ No. 6 L Head Inlet .005 Exhaust .005

CARBURETOR	COOLING	PISTON RING	OILING SYSTEM
Stromberg 1"	Pump	3 1/8 X 3/16	Cap. 6qt. Type F. & S

CLUTCH	REAR AXLE	Gear Ratio	TRANSMISSION
Disc. 9 3/16	Semi	4.55	Own

BRAKES Position Brake lining width & length

Rear wheel Trans. or wheel W. Service 2 X 20 Hand $1\frac{3}{4}$ X 18

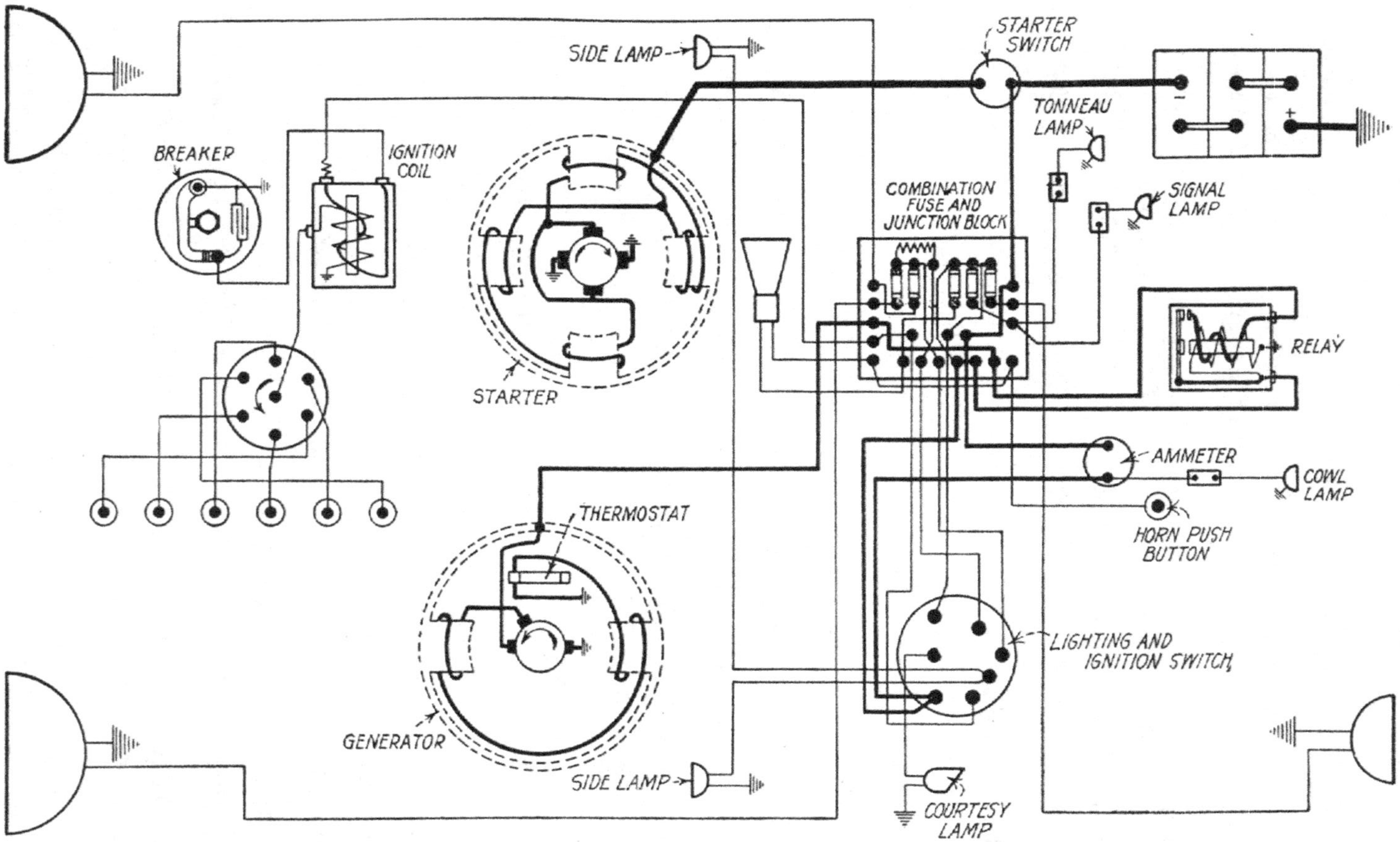

STUDEBAKER WIRING DIAGRAM, 1923-1924, SPECIAL SIX AND BIG SIX MODELS, REMY EQUIPMENT

S T U D E B A K E R Model Special Six Year 1923-24

Remy Starter & Generator Remy Ignition

GENERATOR Model	Regulation	Max. Chg. rate
No. 917 - E	Third Brush	18 - 20 Amps.

Amps. Cut-in	Volts Warm Test	R.P.M.	Amps. Cut-in	Volts Cold Test	R.P.M.
6-9	7.5	1450	7	7.1	850
9-12	7.5	1800	18-20	8.5	1800

RELAY Model	Closes	Opens	Contact Gap	Air Gap
No. 263 - F	10 M P H	8 M P H	.012 - .015	.012

STARTING Motor Model	Drive	Rotation
No. 723 - A	Chain	Clockwise

BATTERY Willard Type S J R - 4 — Capacity 6 Volts 110 Amps. — Bat. to Frame Con. Positive

IGNITION Coil Model	DISTRIBUTOR Model	Breaker Contact Gap
No. 284 - A	No. 626 - F	.020 - .025

FIRING ORDER	Ignition Timing	SPARK PLUG
1-5-3-6-2-4	T.D.C. at retard	Size 7/8R Gap .025

LIGHTING	Headlamps	Dash & Tail	Side Lamps
Single Contact	21 C.P.	2 C.P.	4 C.P.

ENGINE — Cylinder — Valve tappet clearance

Bore $3\frac{1}{2}$ Stroke 5 No. 6 L Head Inlet .005 Exhaust .005

CARBURETOR	COOLING	PISTON RING	OILING SYSTEM
Stromberg 1/4"	Pump	3 1/2 X 3/16	Cap. 6qt. Type F & S

CLUTCH	REAR AXLE	Gear Ratio	TRANSMISSION
Plate 12 3/4	Semi	4.33	Own

BRAKES	Position	Brake lining width & length
Rear wheel	Trans. or wheel W	Service 2 X $21\frac{1}{2}$ Hand $1\frac{3}{4}$ X $19\frac{1}{2}$

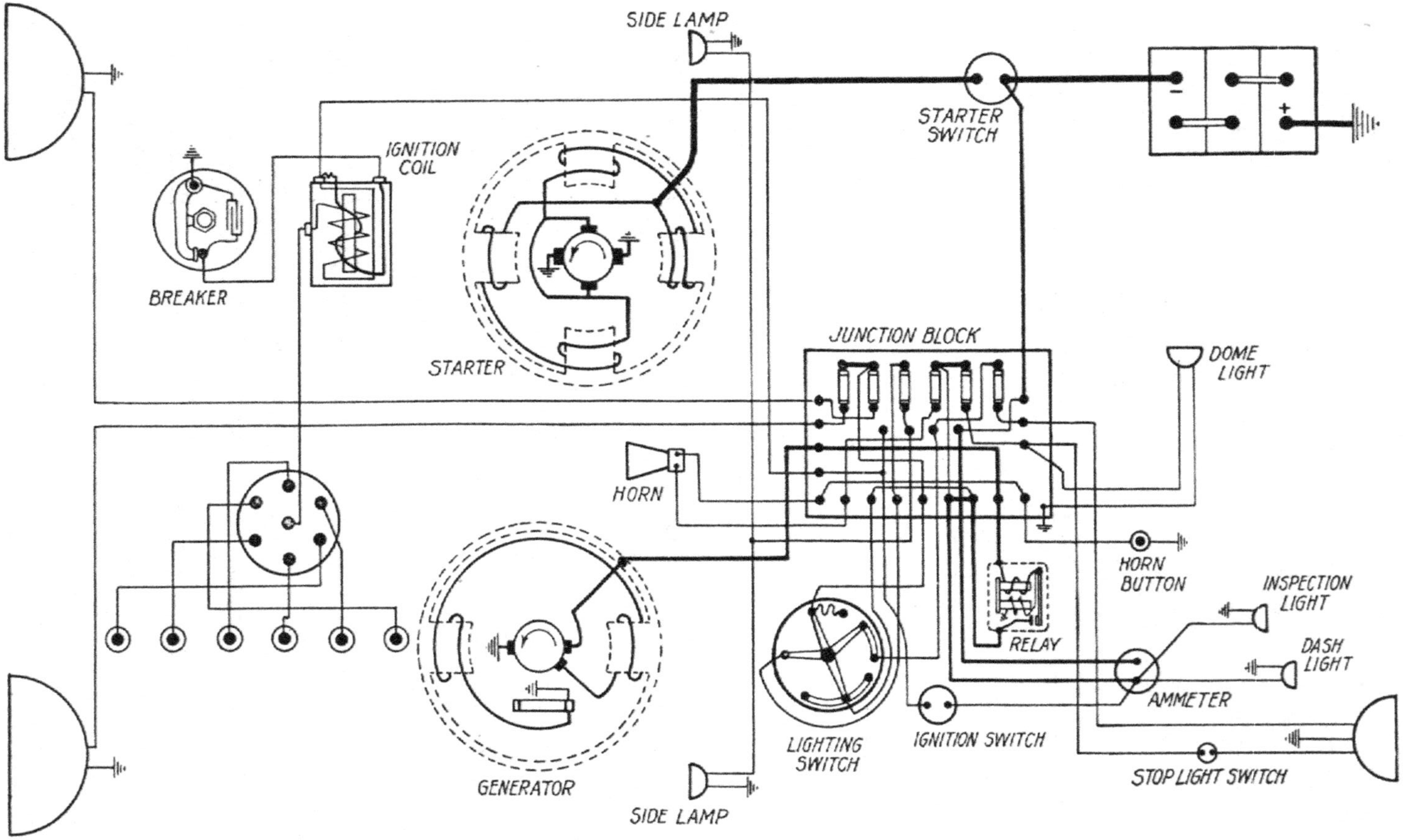

STUDEBAKER WIRING DIAGRAM, 1925, BIG AND SPECIAL SIX, MODELS EP AND EQ, REMY SYSTEM

S T U D E B A K E R Model E P & E Q Year 1925

Wagner Starter & Generator Wagner Ignition

Regulation	Max. Chg. rate and speed
3rd Brush	16 Amps. 1500

RELAY Air Gap	Contact Gap	Cut-in R.P.M.
.025"	.035"	580

BATTERY Willard Type S J R R - 4 Volts 6 Amps 110

Bat. to Frame Con. Pos. CONTACT BREAKER Gap .020"

Firing Order 1-5-3-6-2-4 Ignition Timing "UPDC 1-6" FL.Ret.

SPARK PLUG	ENGINE		Taxable Hp.
Size 7/8" Gap .025	Bore Big 6 - 3 7/8"	Stroke 5"	36
	Spec. 6 - 3½"	5"	29

INTAKE VALVE TIMING	EXHAUST VALVE TIMING
Open 7°-30' A.T.C. Close 37°-30' A.B.C.	Open 47° B.B.C. Close 11° A.T.C.

VALVE CLEARANCE Hot Intake .005" Exhaust .005"

CARBURETOR	COOLING SYSTEM	OILING SYSTEM
Big 6 - Ball & Ball Spec. 6 - Stromberg	Type P Cap 4.68 gal.	Type Pres. Cap 8 qts.

PISTON RING: Width 3/16" Diam. Same as Bore Gap

CLUTCH Plate GEAR RATIO Spec. 6 - 4-36; Big 6 - 3-69 AXLE Semi F

BRAKES

Front	Rear	Hand
		Trans.

Lighting	Headlights	Dash & Tail	Side Lamps
Single Contact	21 C.P.	2 C.P.	2 C.P.

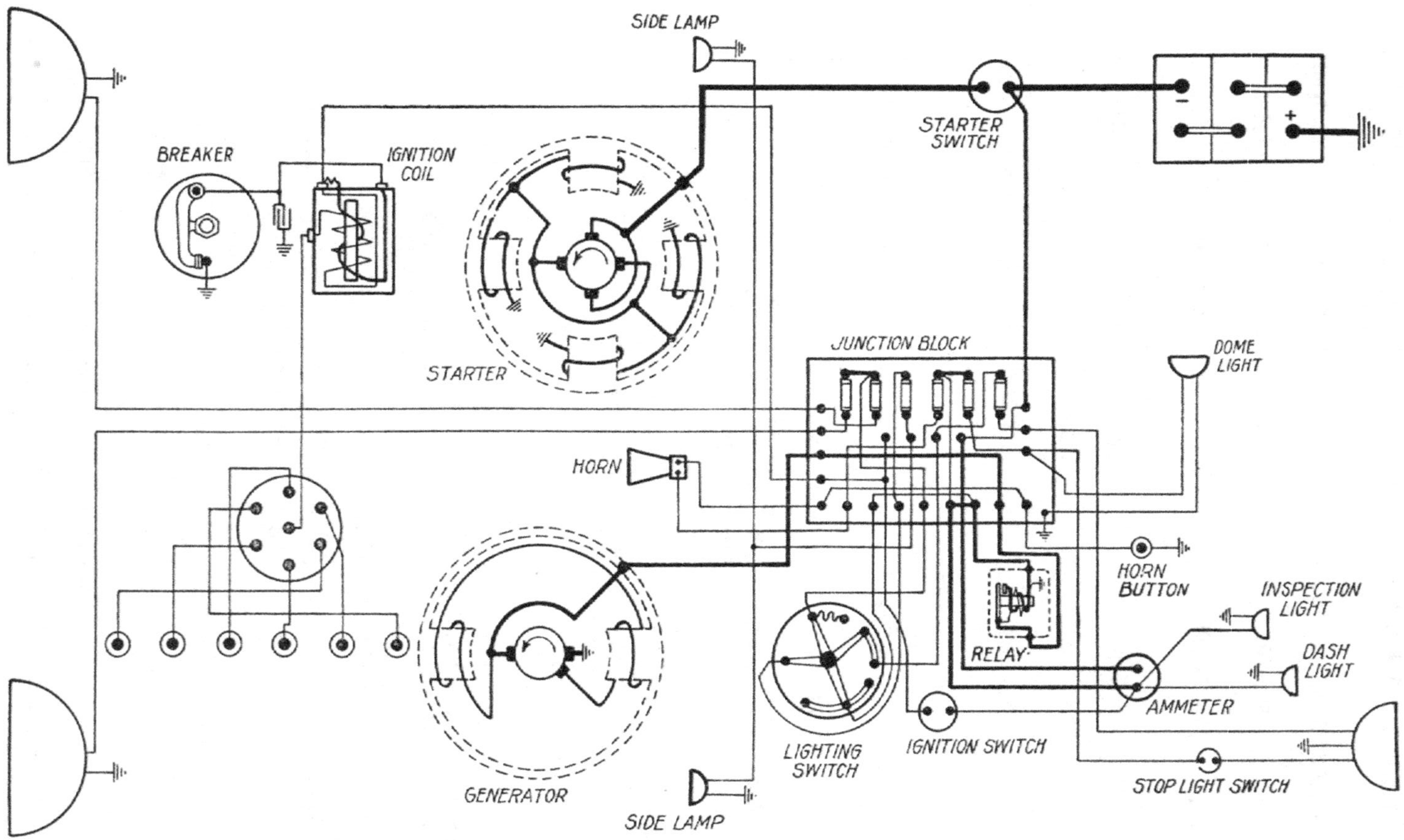

STUDEBAKER WIRING DIAGRAM, 1925, BIG AND SPECIAL SIX, MODELS EP AND EQ, WAGNER SYSTEM

S T U D E B A K E R Model E P & E Q Year 1925

Remy Starter & Generator Remy Ignition

Regulation: 3rd Brush

Max. Chg. rate and speed: 18-20 Amps. 1800

RELAY Air Gap	Contact Gap	Cut-in R.P.M.
.012"-.015"	.012"-.015"	8-10 M.P.H.

BATTERY Willard Type S J R N - 4 Volts 6 Amps 120

Bat. to Frame Con. Pos. CONTACT BREAKER Gap .020"

Firing Order 1-5-3-6-2-4 Ignition Timing "U.P.D.C. 1-6" FL.Ret.

SPARK PLUG	ENGINE		Taxable Hp.
Size 7/8" Gap .025	Bore: Big 6 - 3 7/8"; Spec. - 3½"	Stroke: 5"; 5"	36; 29

INTAKE VALVE TIMING: Open 7°30' A.T.C. Close 37°30' A.B.C.

EXHAUST VALVE TIMING: Open 47° B.B.C. Close 11° A.T.C.

VALVE CLEARANCE Hot Intake .005" Exhaust .005"

CARBURETOR	COOLING SYSTEM	OILING SYSTEM
Big 6 - Ball & Ball Spec. 6 - Stromberg	Type P Cap 4.68 gal.	Type Pres. Cap 8 qts.

PISTON RING: Width 3/16" Diam. Same as Bore Gap

CLUTCH Plate GEAR RATIO Spec. 6 4-36; Big 6 3-69 AXLE Semi F

BRAKES

Front	Rear	Hand

Lighting	Headlights	Dash & Tail	Side Lamps
Single Contact	21 C.P.	2 C.P.	2 C.P.

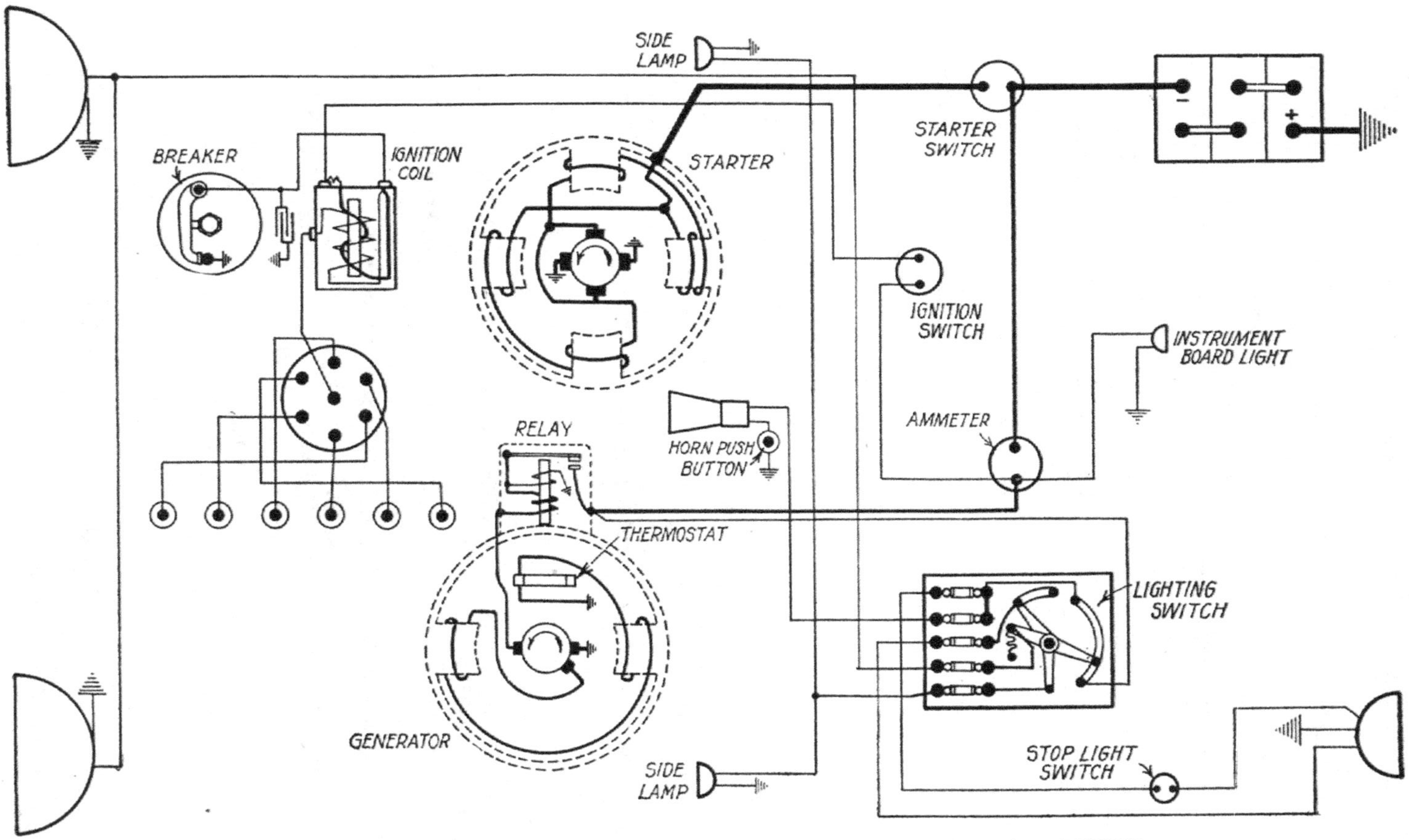

STUDEBAKER WIRING DIAGRAM, 1925, STANDARD SIX, MODEL ER, REMY SYSTEM

STUDEBAKER Model E R Year 1925

Remy Starter & Generator Remy Ignition

Regulation	Max. Chg. rate and speed
3rd Brush	18-20 Amps. 1800

RELAY Air Gap	Contact Gap	Cut-in R.P.M.
.012"-.015"	.012"-.015"	8-10 M.P.H.

BATTERY Willard Type S J R N - 3 Volts 6 Amps 90

Bat. to Frame Con. Pos. CONTACT BREAKER Gap .020"

Firing Order 1-4-2-6-3-5 Ignition Timing "UPDC 1-6" FL.Ret.

SPARK PLUG	ENGINE	Taxable Hp.
Size 7/8" Gap .025"	Bore 3 3/8" Stroke 4 1/2"	27

INTAKE VALVE TIMING	EXHAUST VALVE TIMING
Open 5° A.T.C. Close 35° A.B.C.	Open 40° B.B.C. Close 10° A.T.C.

VALVE CLEARANCE Hot Intake .005" Exhaust .005"

CARBURETOR	COOLING SYSTEM	OILING SYSTEM
Stromberg	Type P Cap 4 gals.	Type Pres. Cap 6 qts.

PISTON RING: Width 3/16" Diam. 3 3/8" Gap

CLUTCH Plate GEAR RATIO 4.18 AXLE Semi F

BRAKES

Front	Rear	Hand

Lighting	Headlights	Dash & Tail	Side Lamps
Single Contact	21 C.P.	2 C.P.	2 C.P.

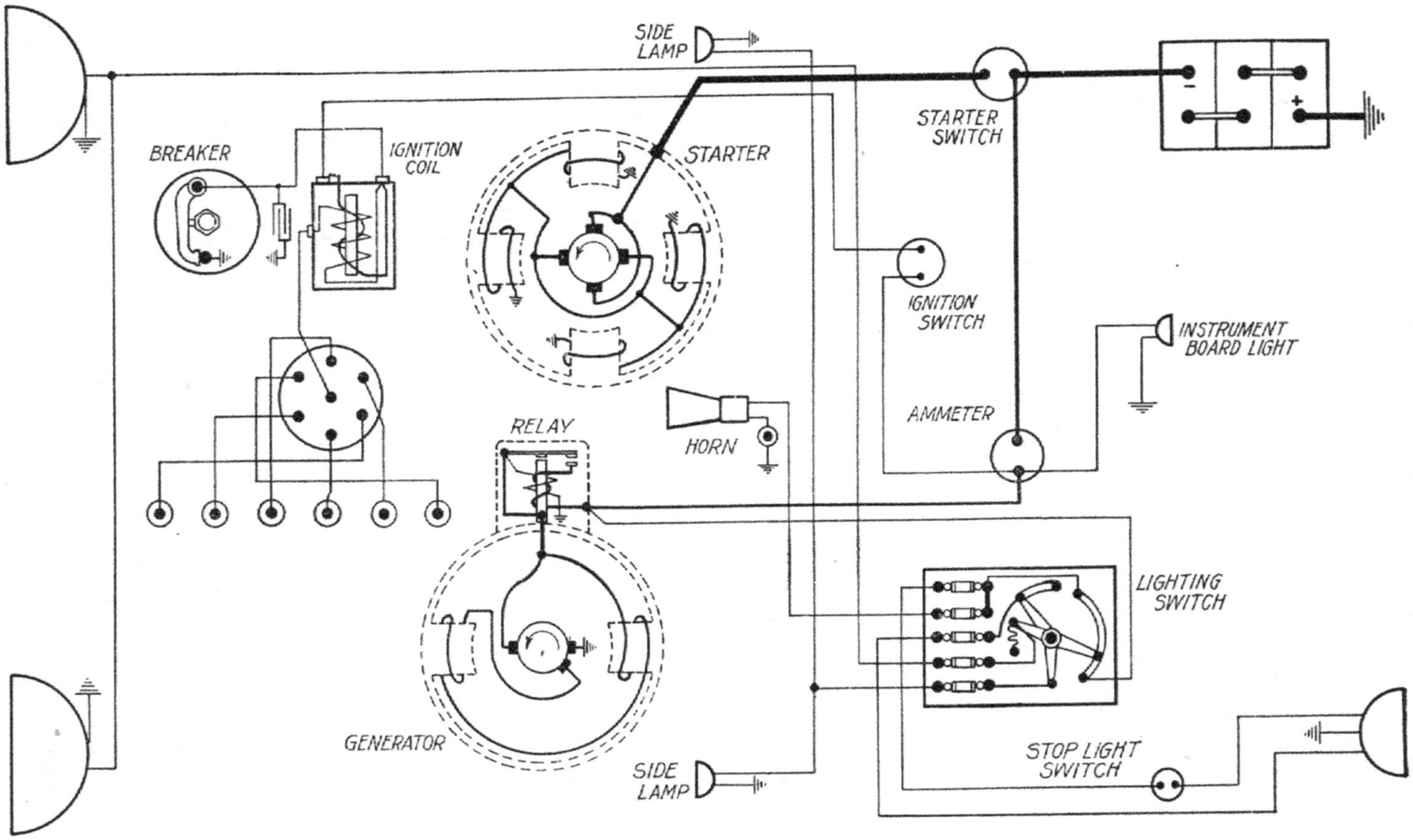

STUDEBAKER WIRING DIAGRAM, 1925, STANDARD SIX, MODEL ER, WAGNER SYSTEM

S T U D E B A K E R Model E R Year 1925

Wagner Starter & Generator Wagner Ignition

Regulation	Max. Chg. rate and speed
3rd Brush	15 Amps. 1500

RELAY Air Gap	Contact Gap	Cut-in R.P.M.
.025"	.035"	580

BATTERY Willard Type S J R R - 3 Volts 6 Amps 90

Bat. to Frame Con. Pos. CONTACT BREAKER Gap .020"

Firing Order 1-4-2-6-3-5 Ignition Timing "UPDC 1-6" FL. Ret.

SPARK PLUG	ENGINE	Taxable Hp.
Size 7/8" Gap .025"	Bore 3 3/8" Stroke 4 1/2"	27

INTAKE VALVE TIMING	EXHAUST VALVE TIMING
Open 5° A.T.C. Close 35° A.B.C.	Open 40° B.B.C. Close 10° A.T.C.

VALVE CLEARANCE Hot Intake .005" Exhaust .005"

CARBURETOR	COOLING SYSTEM	OILING SYSTEM
Stromberg	Type P Cap 4.2 Gal.	Type Pres. Cap 6 qts.

PISTON RING: Width 3/16" Diam. 3 3/8" Gap

CLUTCH Plate GEAR RATIO 4.18 AXLE Semi F

BRAKES

Front	Rear	Hand

Lighting	Headlights	Dash & Tail	Side Lamps
Single Contact	21 C.P.	2 C.P.	2 C.P.

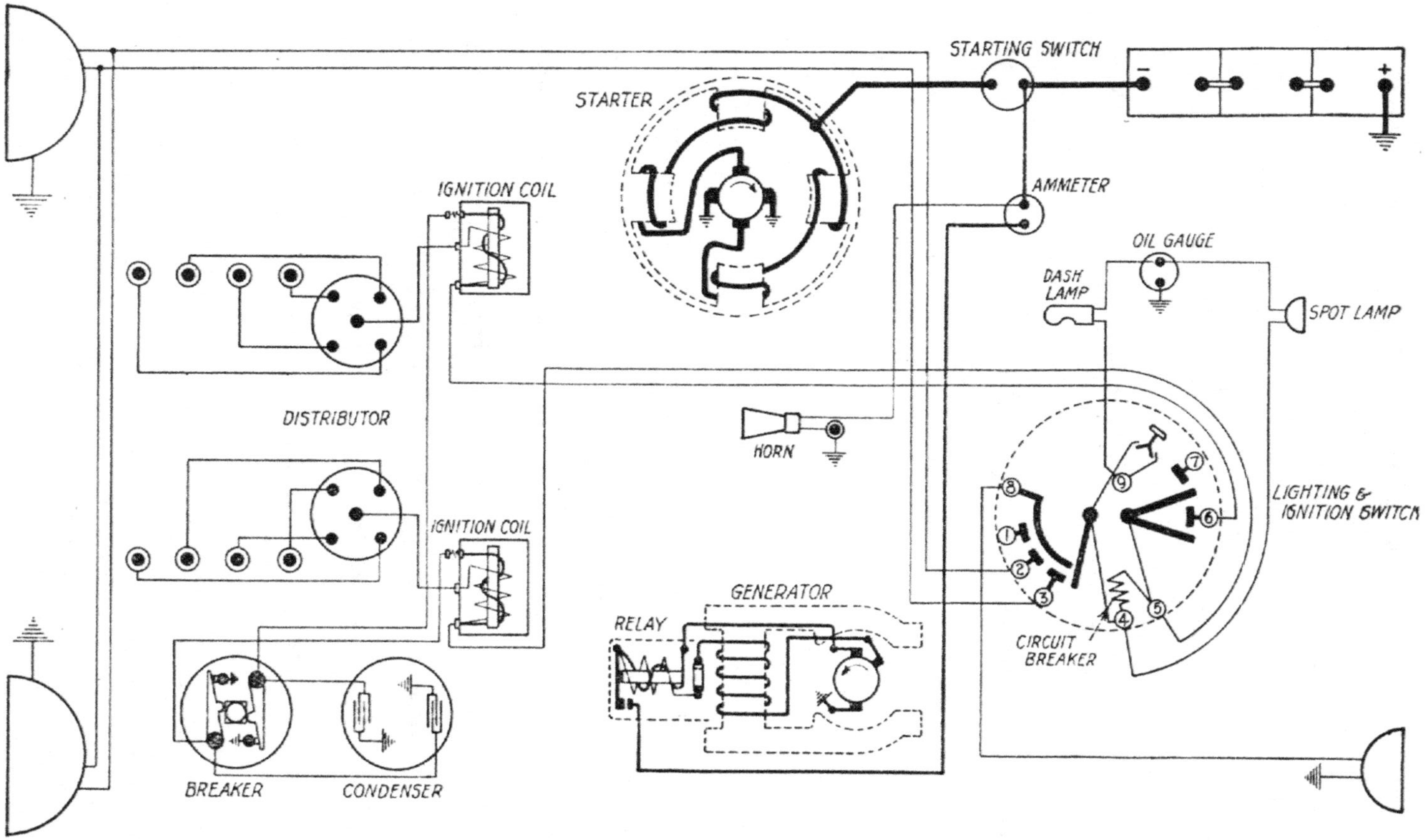

STUTZ WIRING DIAGRAM, 1923, SPEEDWAY 4 MODEL, REMY STARTING AND LIGHTING, DELCO IGNITION SYSTEMS

S T U T Z Model Sp. 4 Year 1923-24

Remy Starter & Generator Delco Ignition

GENERATOR Model	Regulation	Max. Chg. rate
O F	Third Brush	17 Amps.

Amps. Cut-in	Volts Warm Test	R.P.M.	Amps. Cut-in	Volts Cold Test	R.P.M.
0		275	7	7.15	475
12	7.65	675	17	8.2	1450

RELAY Model	Closes	Opens	Contact Gap	Air Gap
No. 264 - C	9 M P H	7 M P H	.013"-.015"	.013"

STARTING Motor Model	Drive	Rotation
No. 741 - B	Gear	Clockwise

BATTERY Willard Type Capacity Bat. to Frame Con.

S J R N - 5 6 Volts 125 Amps. Positive

IGNITION Coil Model	DISTRIBUTOR Model	Breaker Contact Gap
No. 2159	No. 5209	.018

FIRING ORDER	Ignition Timing	SPARK PLUG
1 - 3 - 4 - 2	1½" past T D C	Size 7/8" Gap .025"

LIGHTING	Headlamps	Dash & Tail	Side Lamps
Single Contact	21 C.P.	2 C.P.	4 C.P.

ENGINE Cylinder Valve tappet clearance

Bore $4\frac{3}{8}$ Stroke 6 No. 4 T Head Inlet .004 Exhaust .006

CARBURETOR	COOLING	PISTON RING	OILING SYSTEM
Stromberg 2½	Pump	4 3/8 X 1/4	Cap. 10 qt. Type F&S

CLUTCH	REAR AXLE	Gear Ratio	TRANSMISSION
Disc 9 3/16	Semi	3.75 3.50	Own

BRAKES Position Brake lining width & length

Rear wheel Trans. or wheel W Service 2¼ X 40¼ Hand 1¼ X 33

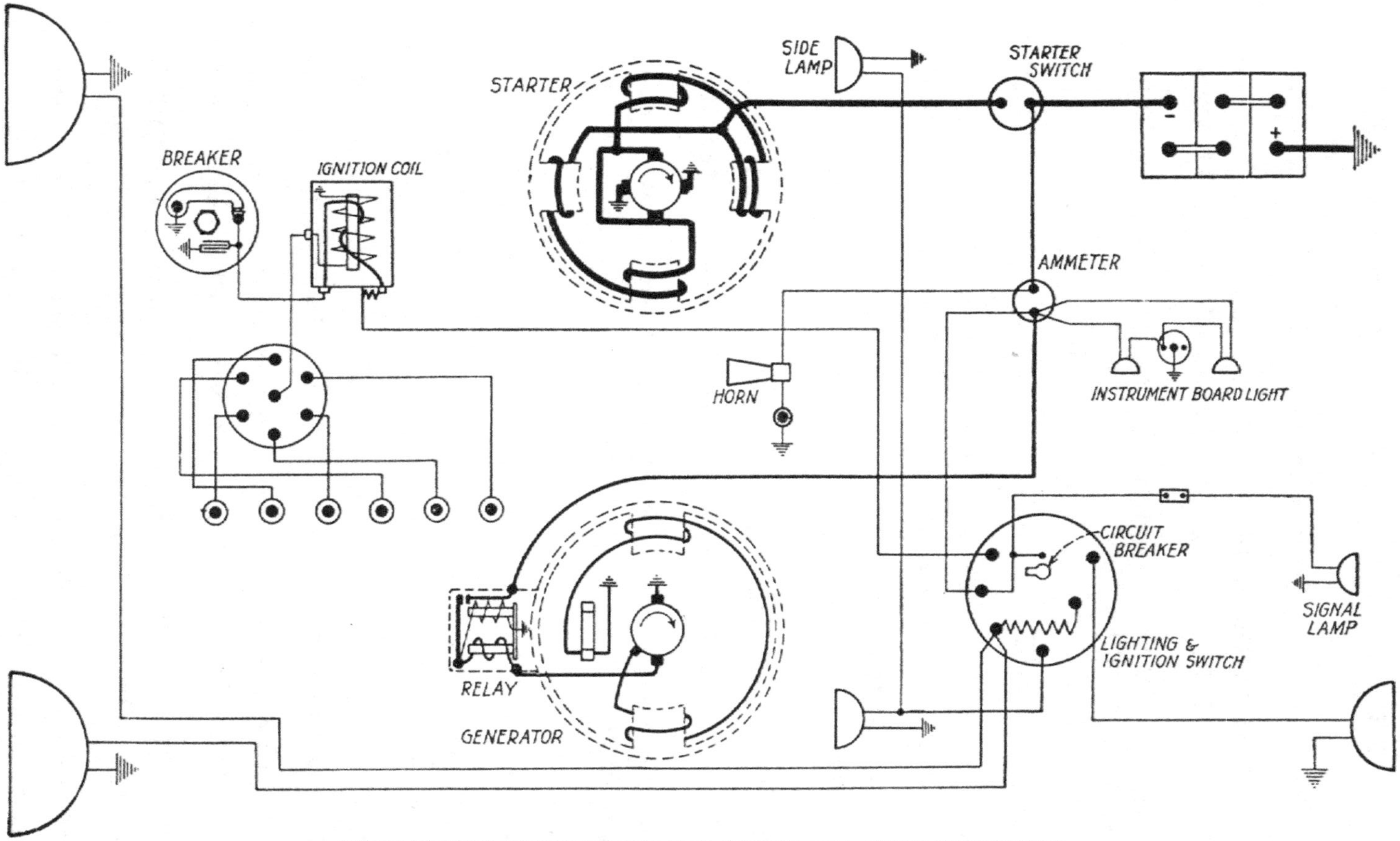

STUTZ WIRING DIAGRAM, 1923-1925, MODEL 6-90, REMY EQUIPMENT

S T U T Z . Model 6 - 90 Year 1923-24

Remy Starter & Generator Remy Ignition

GENERATOR Model	Regulation	Max. Chg. rate
No. 917 - J	Third Brush	18 Amps.

Amps. Cut-in	Volts Warm Test	R.P.M.	Amps. Cut-in	Volts Cold Test	R.P.M.
7	7.3	850	0	7.0	675
12	8.5	1800	18	8.4	1900

RELAY Model	Closes	Opens	Contact Gap	Air Gap
No. 265 - A	10 M P H	8 M P H	.014"	.020"

STARTING Motor Model	Drive	Rotation
No. 720 - J	Bendix	Clockwise

BATTERY Willard Type S R - 5 Capacity 6 Volts 120 Amps. Bat. to Frame Con. Positive

IGNITION Coil Model	DISTRIBUTOR Model	Breaker Contact Gap
No. 284 - P	No. 626 - H	.015"

FIRING ORDER	Ignition Timing	SPARK PLUG
1-4-2-6-3-5	T D C at Retard	Size 7/8" Gap .022"

LIGHTING	Headlamps	Dash & Tail	Side Lamps
Single Contact	21 C.P.	2 C.P.	

ENGINE Bore $3\frac{3}{8}$ Stroke 5 Cylinder No. 6 I Head Valve tappet clearance Inlet .003 Exhaust .005

CARBURETOR	COOLING	PISTON RING	OILING SYSTEM
Stromberg 1 1/2	Pump	3 3/8 X 5/16	Cap 8qt. Type Force

CLUTCH	REAR AXLE	Gear Ratio	TRANSMISSION
Plate 10"	Semi	5.09	Warner-Gear

BRAKES	Position	Brake lining width & length
Rear wheel	Trans. or wheel W	Service 2 X $43\frac{1}{4}$ Hand $1\frac{3}{4}$ X 44

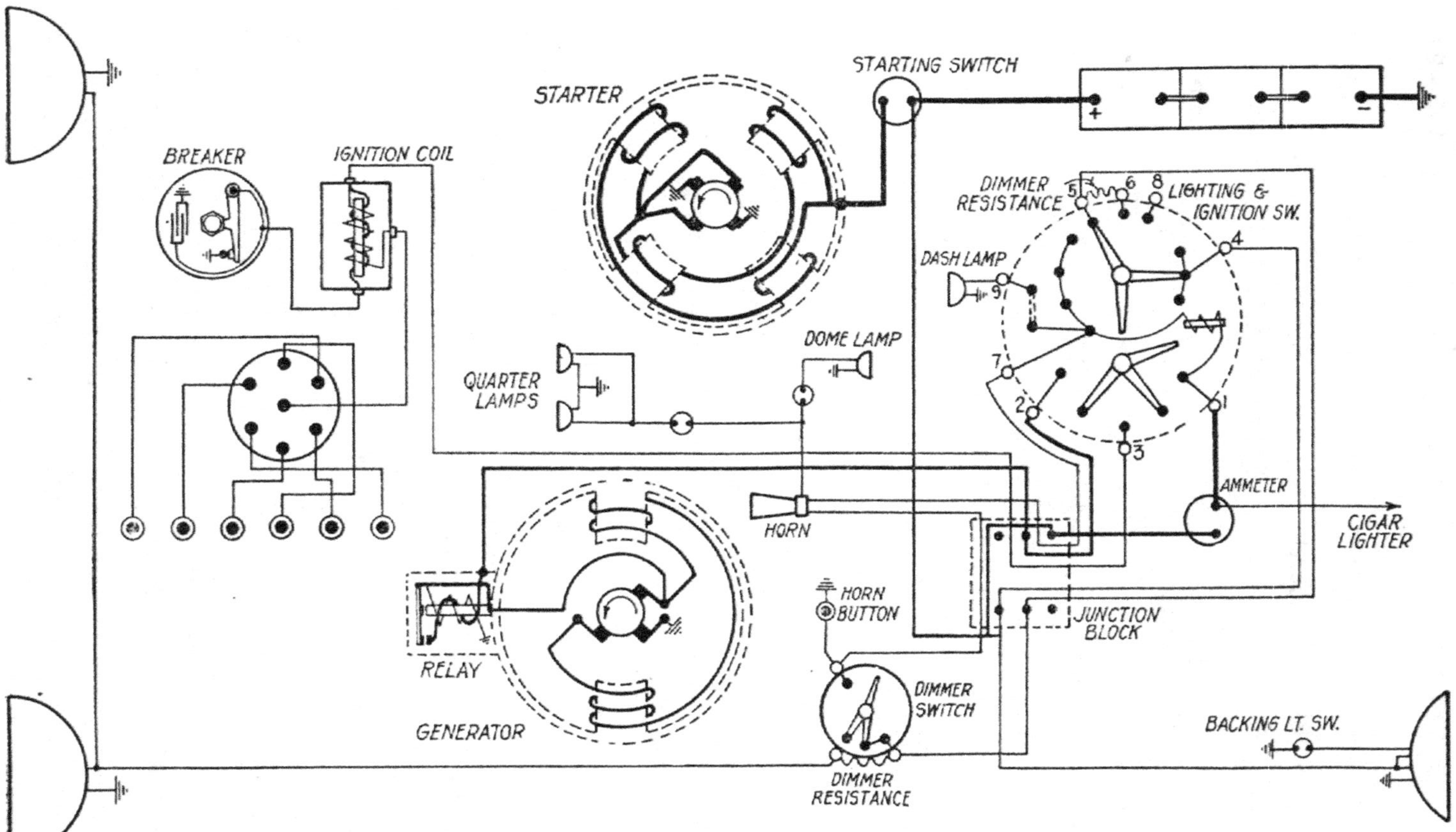

WILLS SAINTE CLAIRE "SIX" WIRING DIAGRAM, 1925, W-6, DELCO SYSTEM

WILLS St CLAIRE 6 Model W 6 Year 1925

Delco Starter & Generator Delco Ignition

Regulation: 3rd Brush

Max. Chg. rate and speed: 13-15 Amps. 1600

RELAY Air Gap	Contact Gap	Cut-in R.P.M.
.015"-.020"	.030"	

BATTERY Willard Type S I R - 5 Volts 6 Amps 145

Bat. to Frame Con. Neg. CONTACT BREAKER Gap .020"

Firing Order 1-5-3-6-2-4 Ignition Timing 7° B.T.C. on FL. adv

SPARK PLUG		ENGINE		Taxable Hp.
Size Met.	Gap .025"	Bore 3¼"	Stroke 5½"	25

INTAKE VALVE TIMING		EXHAUST VALVE TIMING	
Open	Close	Open	Close

VALVE CLEARANCE Hot Intake .006" Exhaust .012"

CARBURETOR	COOLING SYSTEM		OILING SYSTEM	
Schebler	Type Ther.	Cap 8 gal.	Type Pres.	Cap 10 qts.

PISTON RING: Width 1557" Diam 3¼" Gap .002" - .010"

CLUTCH Disc GEAR RATIO AXLE Semi F.

BRAKES

Front	Rear	Hand
35 7/8"x2¼"x3/16"	35 7/8"x2¼"x3/16"	12 5/8"x6"x5/16"

Lighting	Headlights	Dash & Tail	Side Lamps
Single Contact	21 C.P.	2 C.P.	2 C.P.

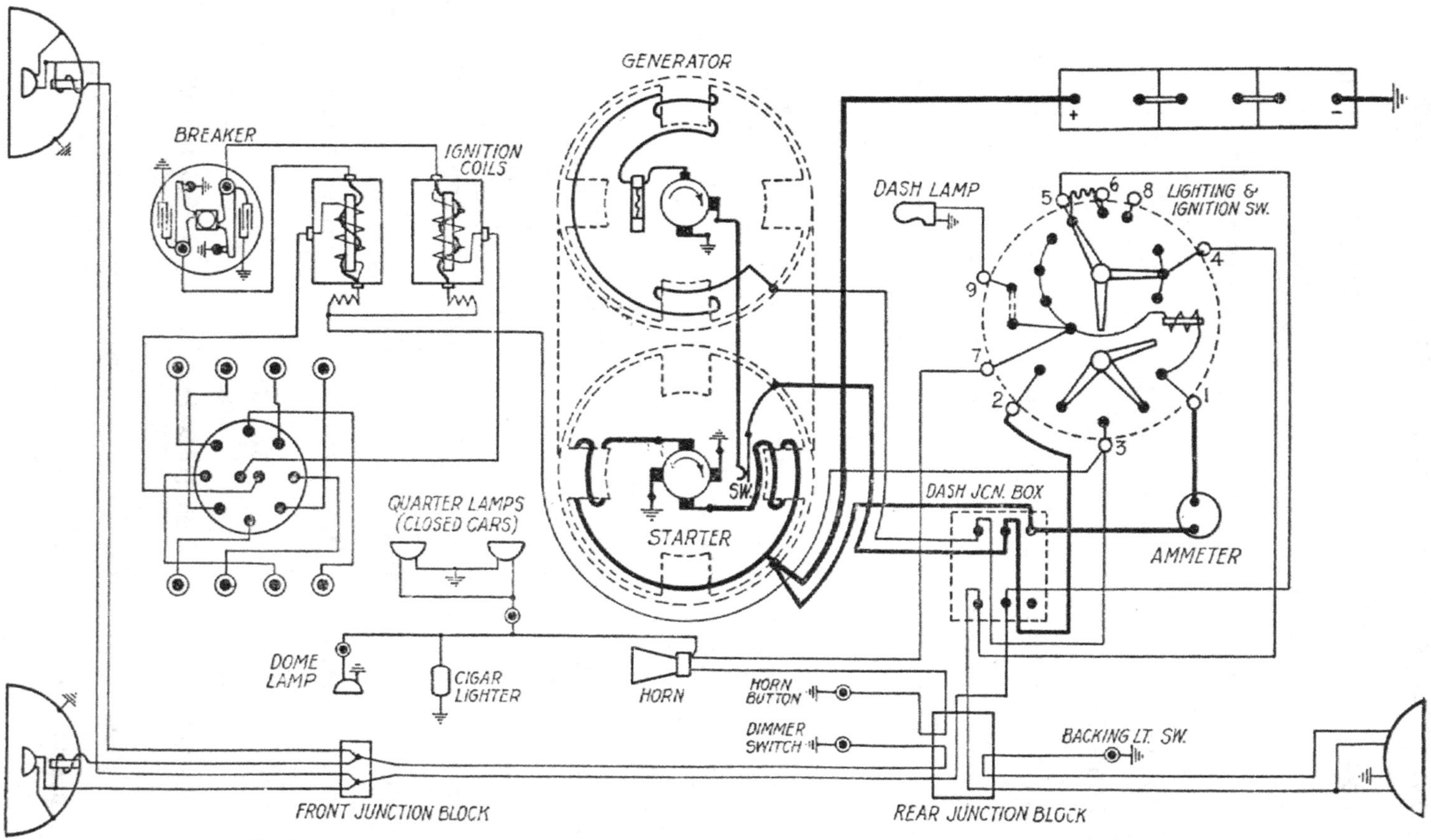

WILLS SAINTE CLAIRE "EIGHT" WIRING DIAGRAM, 1925, B-68 AND C-68, DELCO SYSTEM

WILLS St CLAIRE 8 Model B-68 C-68 Year 1925

Delco Starter & Generator Delco Ignition

Regulation	Max. Chg. rate and speed
3rd Brush	18-20 Amps. 2400

RELAY Air Gap	Contact Gap	Cut-in R.P.M.
None		

BATTERY Willard Type S I R - 5 Volts 6 Amps 140

Bat. to Frame Con. Neg. CONTACT BREAKER Gap .020"

Firing Order 1-8-2-7-4-5-3-6 Ignition Timing 7° A.T.C. Ret.

SPARK PLUG		ENGINE		Taxable Hp.
Size	Gap .025"	Bore 3¼"	Stroke 4"	34

INTAKE VALVE TIMING		EXHAUST VALVE TIMING	
Open 10° A.T.C.	Close 35° A.B.C.	Open 40° B.B.C.	Close 5° A.T.C.

VALVE CLEARANCE Hot Intake .006" Exhaust .012"

CARBURETOR	COOLING SYSTEM		OILING SYSTEM	
Schebler	Type Ther.	Cap 7½ gal.	Type Pres.	Cap 7 qts.

PISTON RING: Width 5/32" Diam. 3¼" Gap .002" - .010"

CLUTCH Disc GEAR RATIO AXLE Semi F.

BRAKES

Front	Rear	Hand
35 7/8"x2¼"x3/16"	35 7/8"x2¼"x3/16"	12 5/8"x6"x5/16"

Lighting		Headlights	Dash & Tail	Side Lamps
Single	Contact	21 C.P.	2 C.P.	2 C.P.

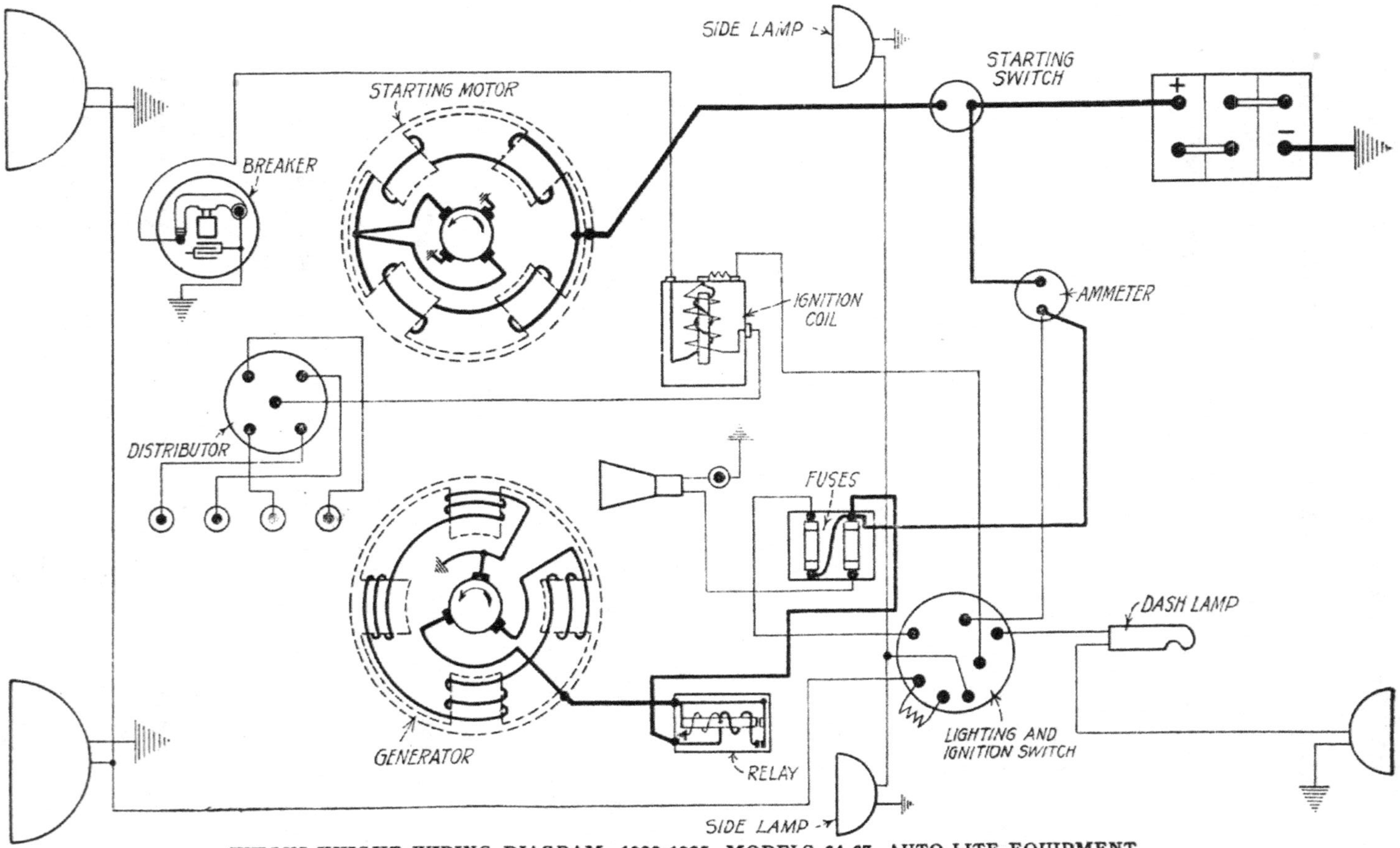

WILLYS-KNIGHT WIRING DIAGRAM, 1923-1925, MODELS 64-67, AUTO-LITE EQUIPMENT

W I L L Y S-K N I G H T Model 64-67 Year 1923-24

Auto-Lite Starter & Generator Auto-Lite Ignition

GENERATOR Model			Regulation	Max. Chg. rate	
G J			Third Brush	16 Amps.	
Amps. Cut-in	Volts Warm Test	R.P.M.	Amps. Cut-in	Volts Cold Test	R.P.M.
0	6.5	500	10-11	8.0	3300
16	7.5	1800	16	8.3	1800

RELAY Model	Closes	Opens	Contact Gap	Air Gap
G J	7-8M	5-7M	.025-.035	.012

STARTING Motor Model	Drive	Rotation
M H	Bendix	Anti-Clockwise

BATTERY U S L Type Capacity Bat. to Frame Con.

HDN-317X 6 Volts 166 Amps. Negative

IGNITION Coil Model	DISTRIBUTOR Model	Breaker Contact Gap
I G	I G	.018

FIRING ORDER	Ignition Timing	SPARK PLUG
1-3-4-2	12° past TDC	Size 7/8 Gap .025

LIGHTING Headlamps Dash & Tail Side Lamps

Single Contact 6-8V-21 C.P. 3-4V-2 C.P. - - - C.P.

ENGINE Cylinder Valve tappet clearance

Bore $3\frac{5}{8}$ Stroke $4\frac{1}{2}$ No. 4 X Head Inlet - - Exhaust - -

CARBURETOR	COOLING	PISTON RING	OILING SYSTEM
Tillotson $1\frac{1}{4}$	Thermo	$3\frac{5}{8}$ x $\frac{3}{16}$	Cap. 6 qt Type Force

CLUTCH	REAR AXLE	Gear Ratio	TRANSMISSION
Disc 10"	3/4 Flt.	4.44	Own

BRAKES Position Brake lining width & length

Rear wheel Trans. or wheel W Service $1\frac{3}{4}$ X $39\frac{1}{2}$ Hand $1\frac{1}{2}$ X $39\frac{1}{2}$

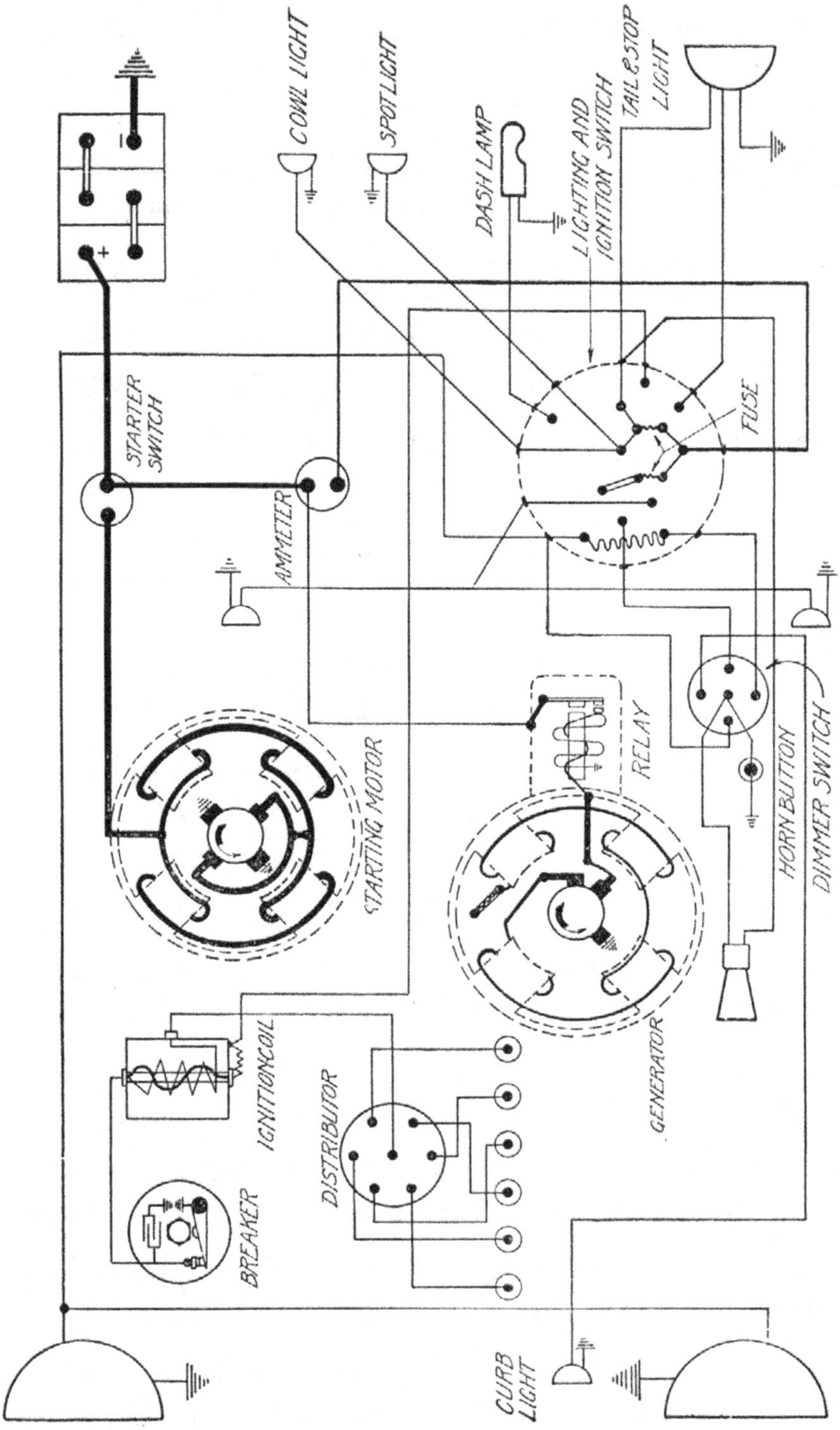

WILLYS-KNIGHT WIRING DIAGRAM, 1925, MODEL 66, DEJON SYSTEM

WILLYS KNIGHT Model 66 Year 1925

Dejon Starter & Generator Dejon Ignition

Regulation	Max. Chg. rate and speed
3rd Brush	15.5 Amps. 1600

RELAY Air Gap	Contact Gap	Cut-in R.P.M.
.010"	.025"-.035"	480

BATTERY U.S.L. Type 3-HVX-8X Volts 6 Amps 170

Bat. to Frame Con. Neg. CONTACT BREAKER Gap .018"

Firing Order 1-5-3-6-2-4 Ignition Timing 8° B.T.C. Adv.

SPARK PLUG	ENGINE	Taxable Hp.
Size 7/8" Gap .025"	Bore 3 1/4" Stroke 4 3/4"	25

INTAKE VALVE TIMING	EXHAUST VALVE TIMING
Open ... Close ...	Open ... Close ...

VALVE CLEARANCE ... Intake ... Exhaust ...

CARBURETOR	COOLING SYSTEM	OILING SYSTEM
Tillotson	Type Ther. Cap 5½ gal.	Type Pres. Cap 5 qts.

PISTON RING: Width 3/16" Diam. 3 1/4" Gap ...

CLUTCH Disc GEAR RATIO 5.1 AXLE 3/4 F

BRAKES

Front	Rear	Hand

Lighting	Headlights	Dash & Tail	Side Lamps
Single Contact	21 C.P.	3 C.P.	3 C.P.

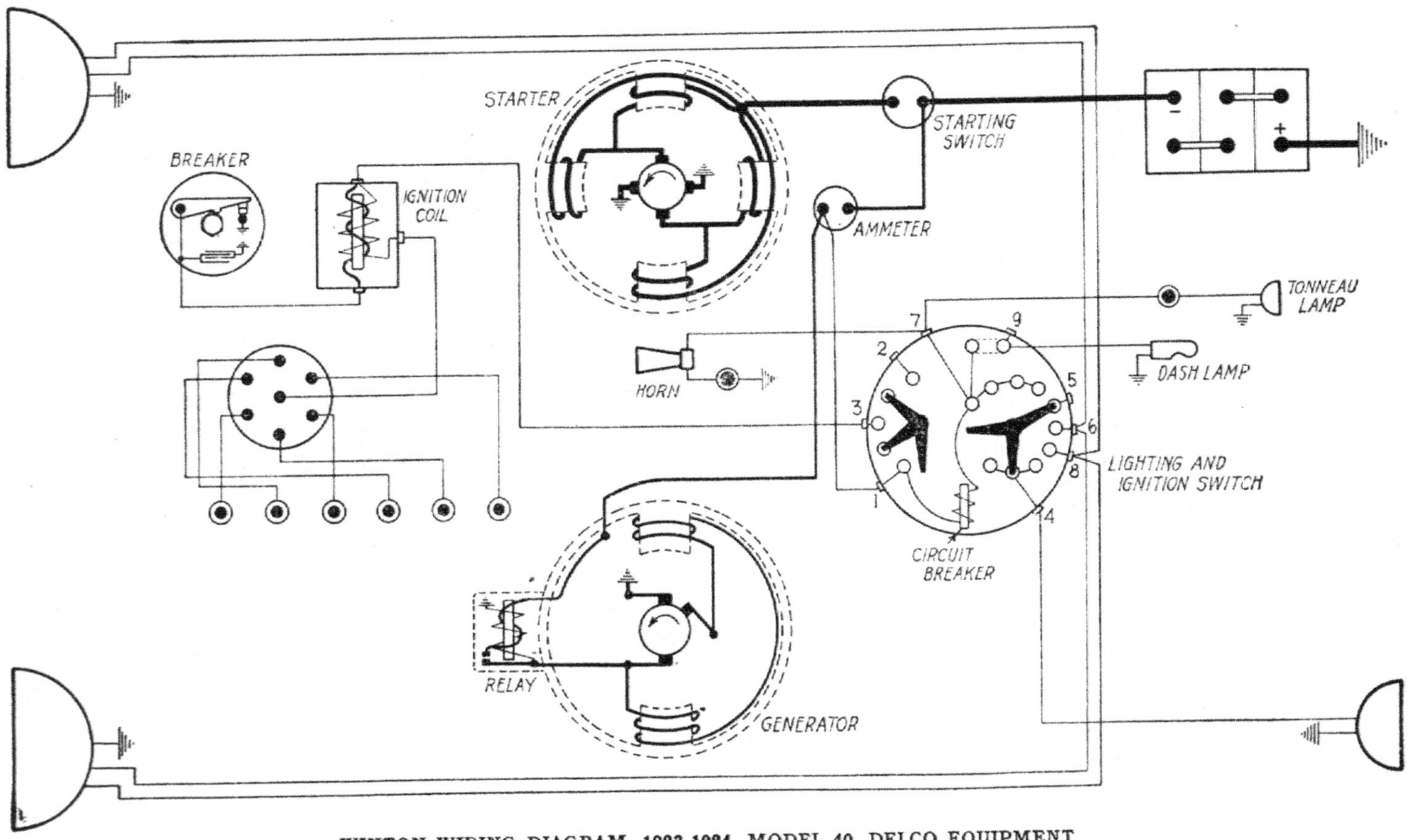

WINTON WIRING DIAGRAM, 1923-1924, MODEL 40, DELCO EQUIPMENT

WINTON Model 40 Year 1923-24

Delco Starter & Generator Delco Ignition

GENERATOR Model	Regulation	Max. Chg. rate
Nos. 210 & 243	Third Brush	18 Amps.

Amps. Cut-in	Volts Warm Test	R.P.M.	Amps. Cut-in	Volts Cold Test	R.P.M.
5	- - -	700	11	- - -	1000
16	- - -	1600	15	- - -	1800

RELAY Model	Closes	Opens	Contact Gap	Air Gap
No. 243	10 M P H	8 M P H	Fixed	.030"

STARTING Motor Model	Drive	Rotation
Nos. 209 & 244	Bendix	Anti-Clockwise

BATTERY Willard Type SJRN-5 Capacity 6 Volts 125 Amps. Bat. to Frame Con. Positive

IGNITION Coil Model	DISTRIBUTOR Model	Breaker Contact Gap
No. 2178	No. 5208	.020" - .027"

FIRING ORDER	Ignition Timing	SPARK PLUG
1-5-3-6-2-4	T.D.C. at Retard	Size 7/8" Gap .030

LIGHTING	Headlamps	Dash & Tail	Side Lamps
Single Contact	21 C.P.	2 C.P.	4 C.P.

ENGINE Bore 3 3/4 Stroke 5 1/4 Cylinder No. 6 L Head Valve tappet clearance Inlet .003" Exhaust .005"

CARBURETOR	COOLING	PISTON RING	OILING SYSTEM
Rayfield 1 1/2"	Pump	3 3/4 X 3/16	Cap. 8qt. Type Force

CLUTCH	REAR AXLE	Gear Ratio	TRANSMISSION
Disc 11 1/2	Semi	4.58	Brown-Lipe

BRAKES Rear wheel Position Trans. or wheel W. Brake lining width & length Service 2½ X 54 Hand 2½ X 49½

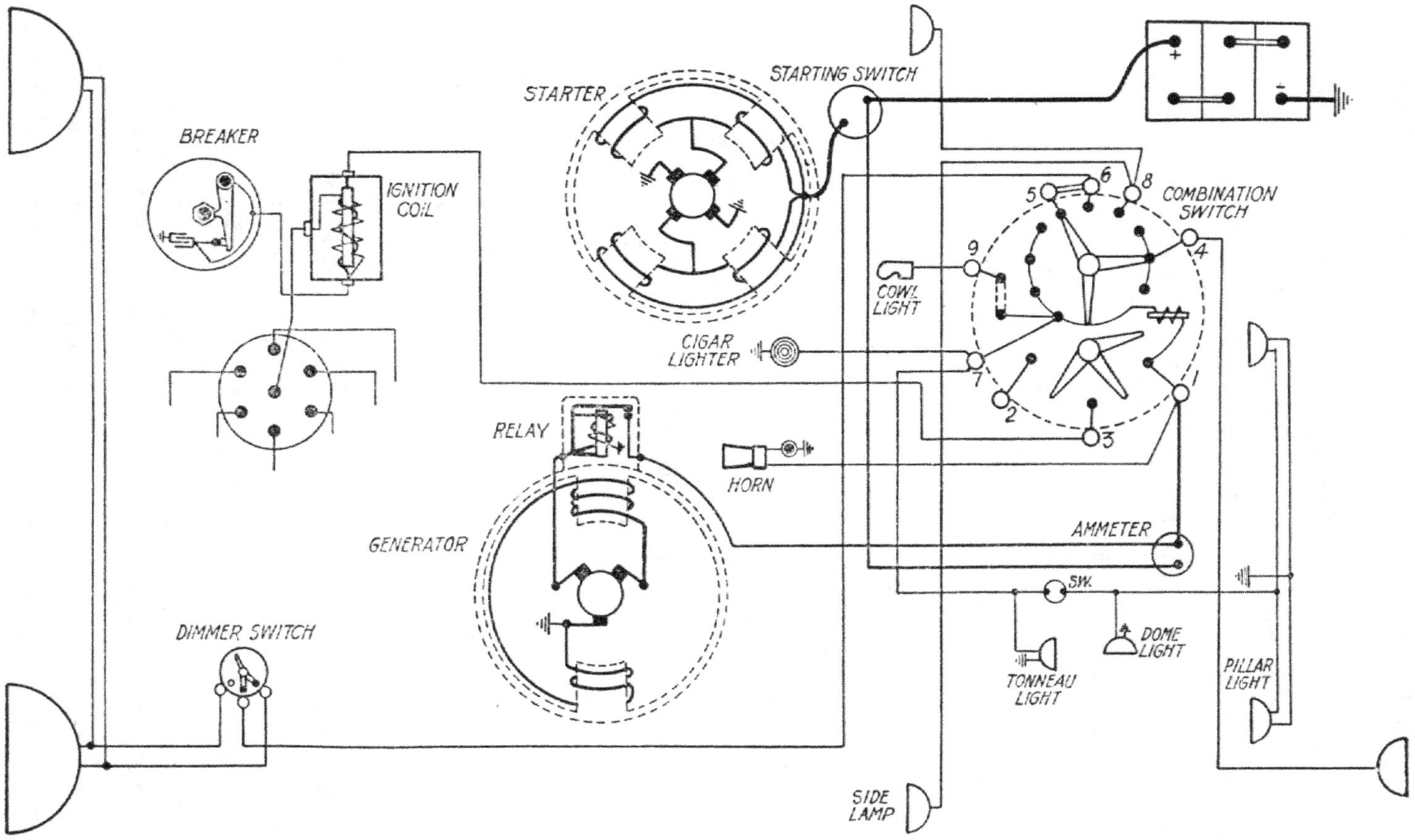

BUICK WIRING DIAGRAM, 1926, MASTER SIX

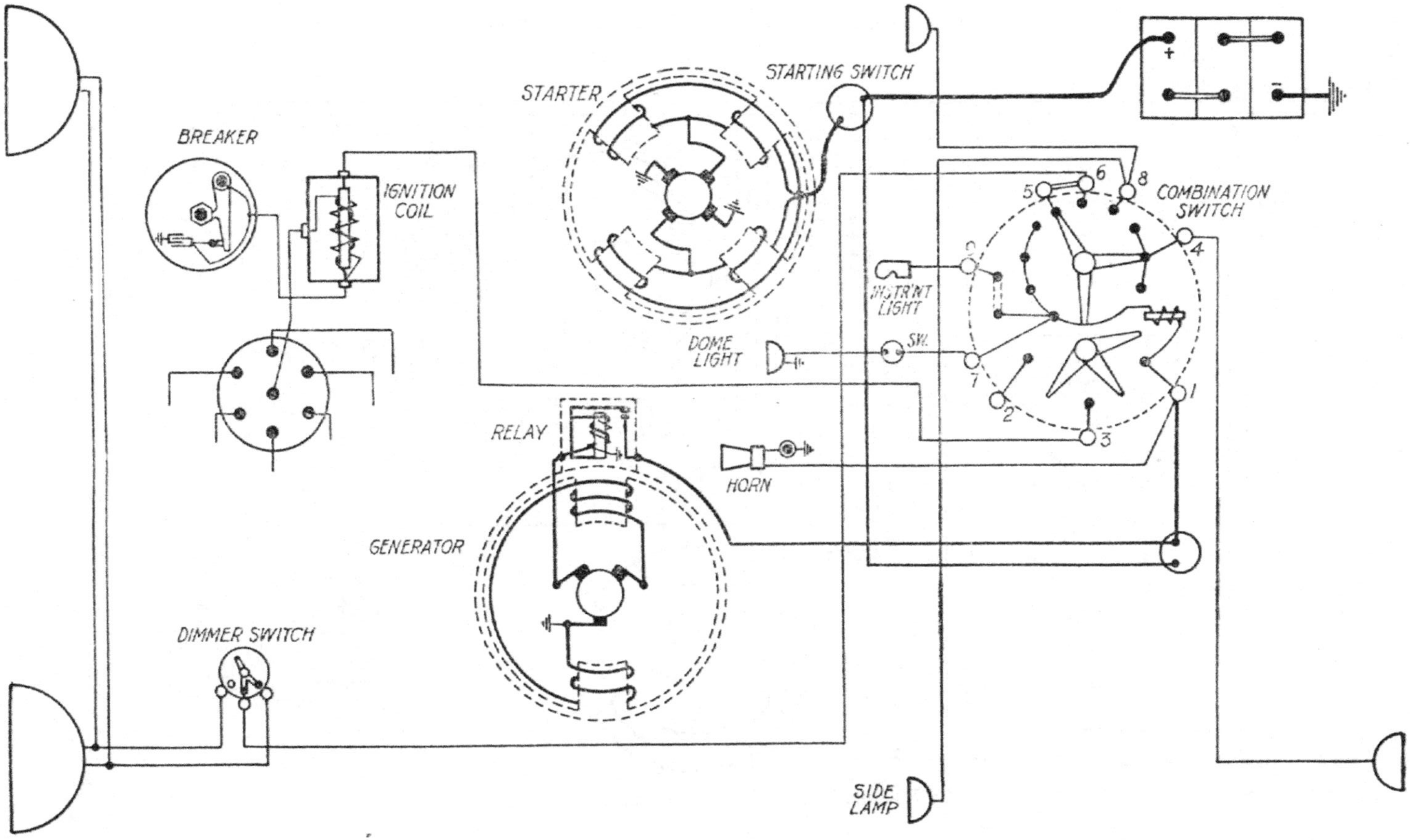

BUICK WIRING DIAGRAM, 1926, STANDARD SIX

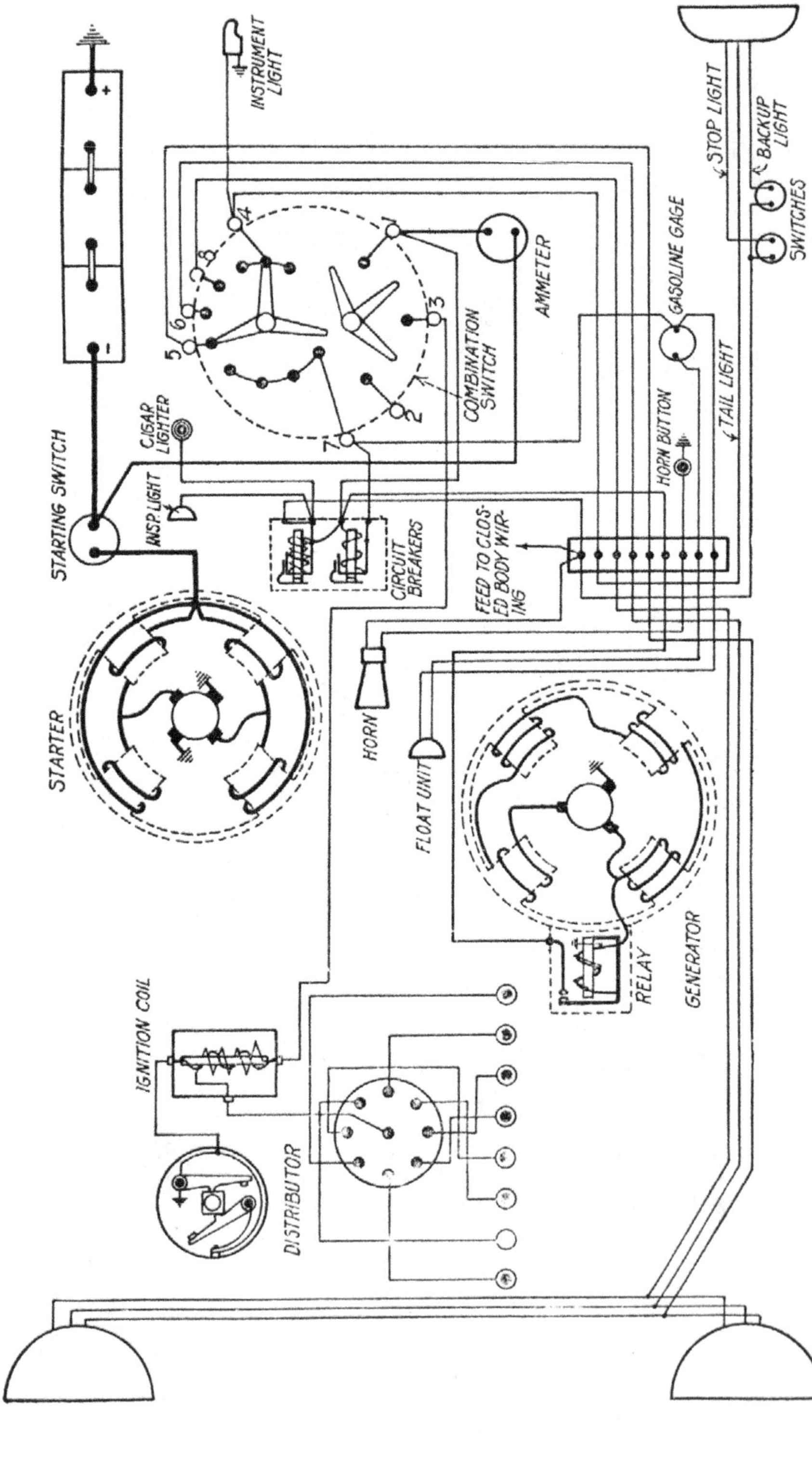

CADILLAC WIRING DIAGRAM, 1926, DELCO SYSTEM

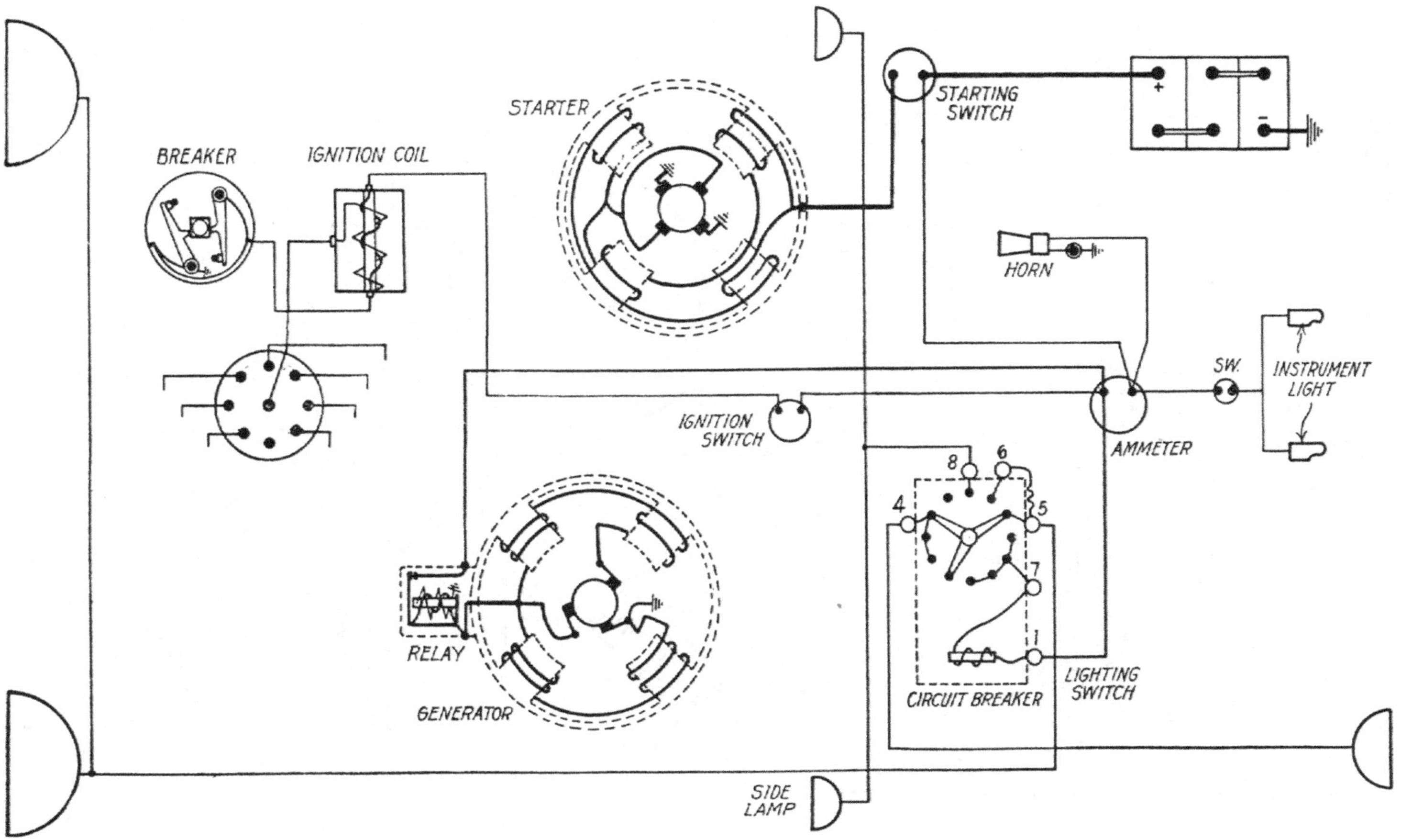

DIANA WIRING DIAGRAM, 1926

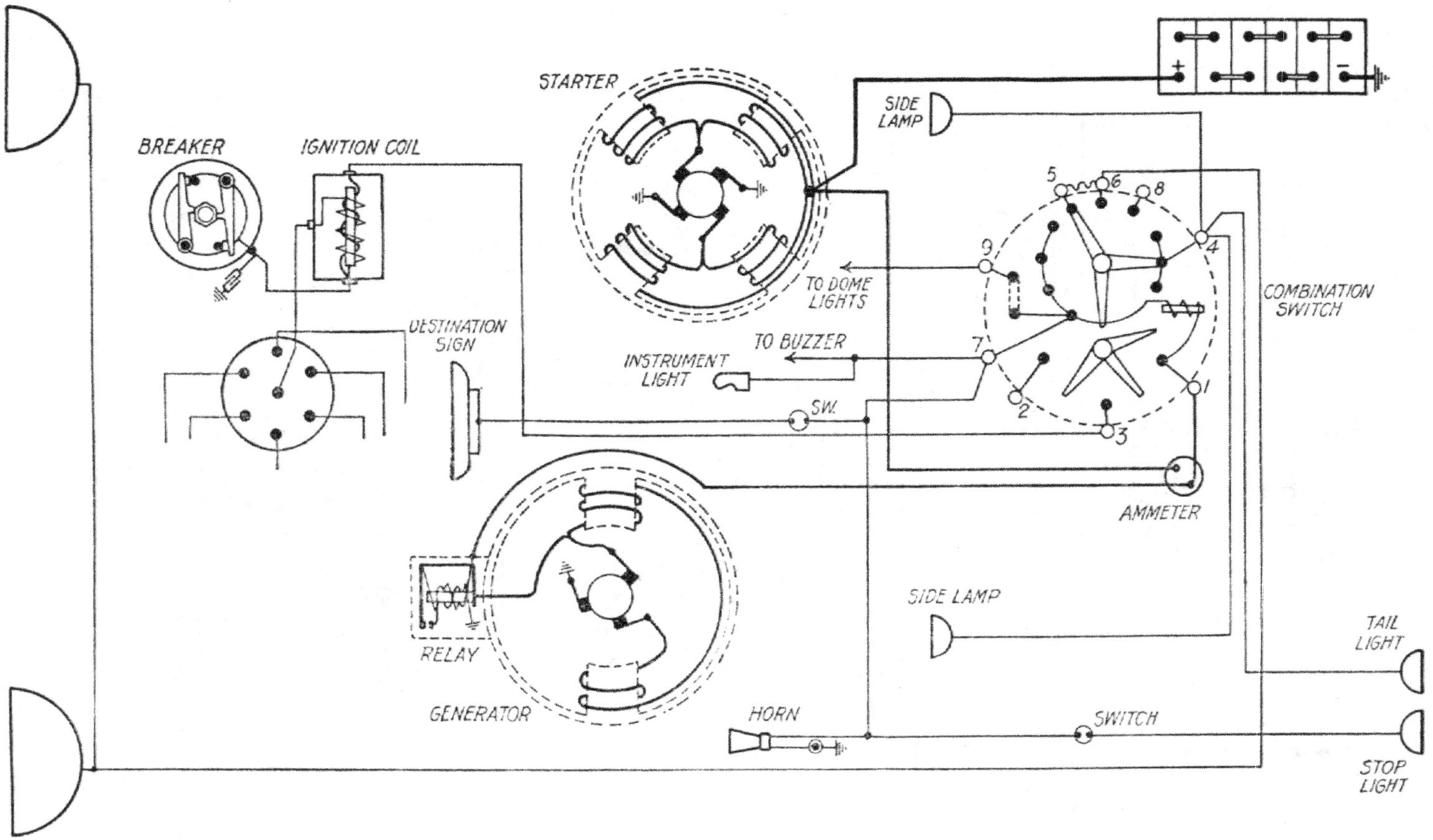

HALL SCOTT WIRING DIAGRAM, 1926, FAGEOL BUS

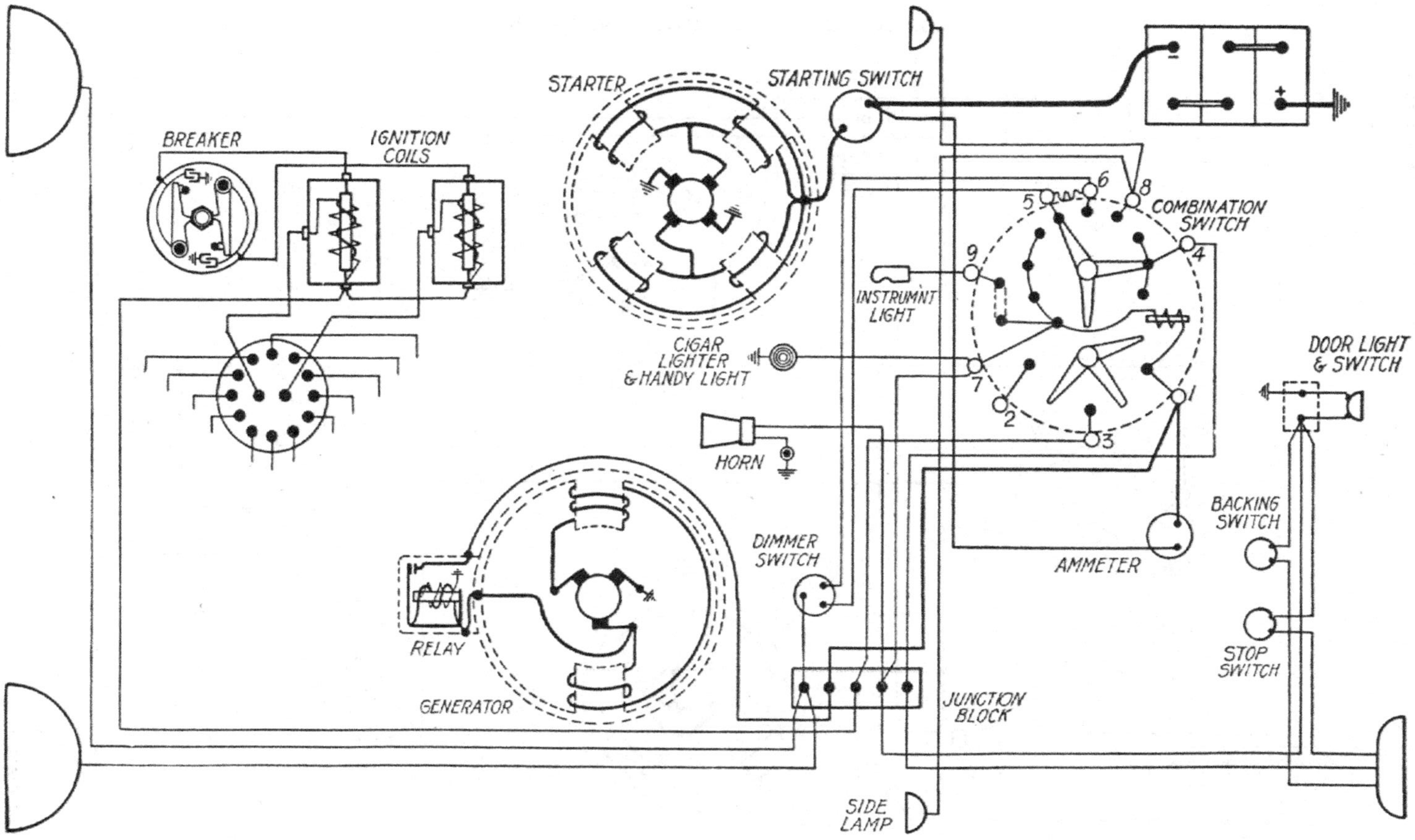

MARMON WIRING DIAGRAM, 1926

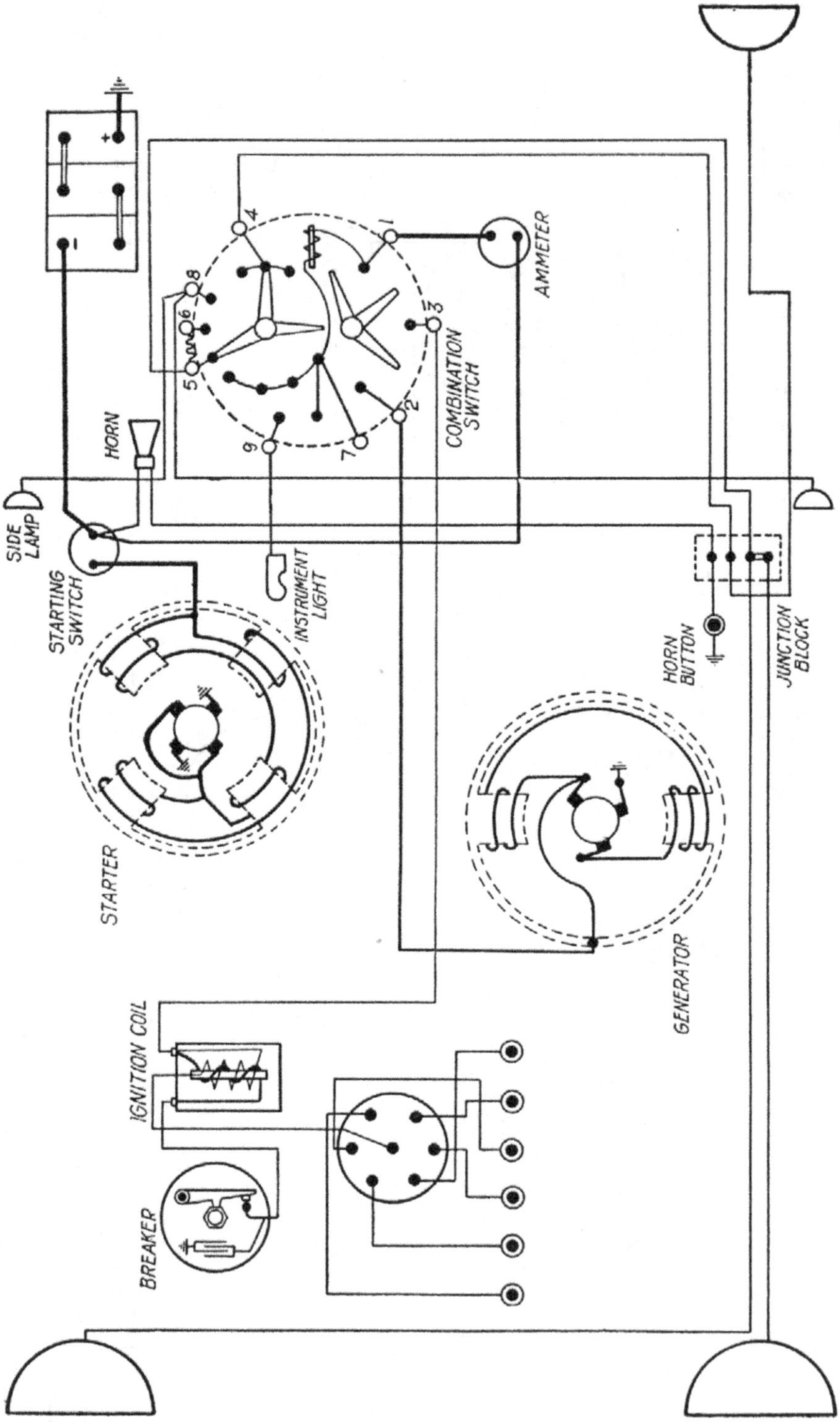

NASH WIRING DIAGRAM, 1926, SPECIAL SIX

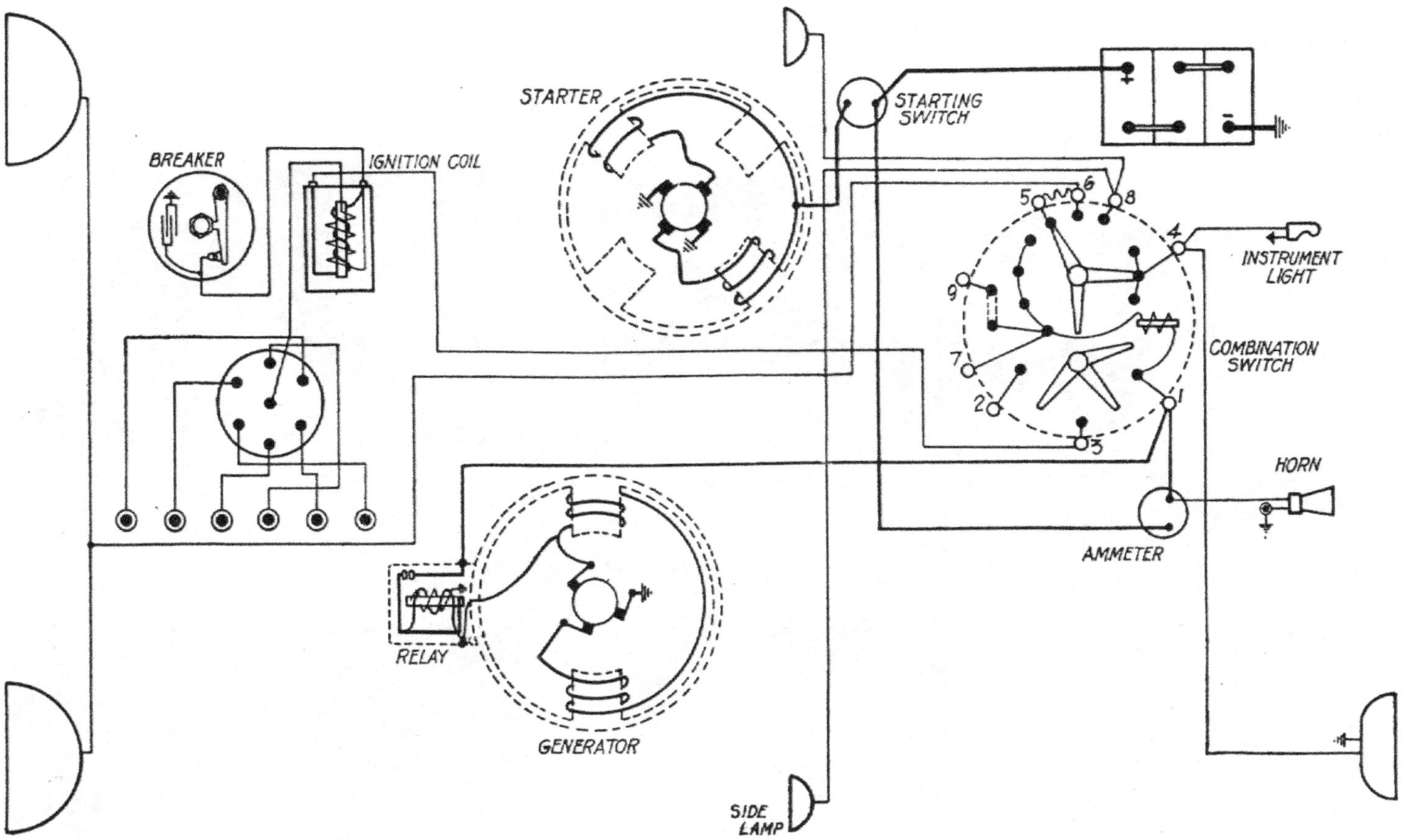

OLDSMOBILE WIRING DIAGRAM, 1926, DELCO SYSTEM

THE BETTER BUICK, MODEL 50, SEVEN-PASSENGER, MASTER-SIX SEDAN, $1995

SUMMARY OF ELECTRICAL PRINCIPLES

QUESTIONS AND ANSWERS

The importance of a knowledge of the fundamental principles of electricity and of its characteristics to the man who wishes to familiarize himself with the electrical apparatus on the automobile to the point where he can readily diagnose and remedy its ills has already been dwelt upon. To bring these out more clearly and make them easier to memorize, they are repeated here in the form of a brief résumé in questions and answers.

Q. What is electrical pressure? To what may it be compared?

A. Electrical pressure is electromotive force, usually termed e.m.f., or voltage, also potential, and may be likened to water under pressure in a pipe or to compressed air in a container.

Q. Of what does this electrical pressure consist, and how is it measured?

A. It is represented by the difference of potential between two points in a circuit, and it is measured in volts.

Q. What does the unit volt represent?

A. The volt is the amount of e.m.f. required to force a current of one ampere through a resistance of one ohm.

Q. What is the ampere?

A. It is the unit of current flow.

Q. What is the ohm?

A. The unit of resistance represented by a length of wire that will pass one ampere under a pressure of one volt.

Q. In what unit is the volume of current flow measured?

A. In the coulomb, which is the equivalent of one ampere per second.

Q. What is resistance, and of what may it consist?

A. Any element which tends to retard the flow of the current is resistance. It may consist of the wire of the circuit itself; the windings of different apparatus in the circuit, such as an induction coil or a motor; the filament of a lamp; a switch; or the like.

Q. Can undesirable resistance be interposed in a circuit in any ways other than those already mentioned?

A. Yes, by the use of wire too small to carry the amount of current required by the apparatus.

Q. What is the effect of using wires too small for the current?

A. The wires waste a great deal of the current in heat and, if much too small for the purpose, are likely to become overheated to a point at which they will burn the insulation off or to actually become fused by the current.

Q. What determines the voltage in an electrical circuit?

A. The potential, or voltage, of the source of supply, such as a storage battery, in which case the voltage will be constant less the drop caused by the resistance of the circuit; or, in the case of a lighting generator, it will depend upon the design of the latter (winding, etc.) and the speed at which it is running.

Q. How may the voltage be varied?

A. In the case of a battery, by varying the number of cells, each cell of a storage battery giving approximately 2 volts. In a generator, by varying the windings of the field and the armature and by increasing or decreasing the speed at which it runs. On a circuit having a higher voltage than desired, by the insertion of an amount of resistance calculated to give the drop required.

Q. Can lamps of a certain voltage be burned on a circuit having a higher voltage?

A. Not if inserted directly in such a circuit. For example, the standard 6-volt lamp cannot be used directly on a 6- or a 12-cell storage-battery circuit as employed for the lighting and starting systems of many cars. The filament would immediately burn out, as its thickness is calculated to a nicety to become incandescent when current of the voltage for which it is designed is passed through it, and anything in excess of this voltage will fuse the wire.

Q. How can lamps of lower voltage be used on such circuits without the employment of a wasteful resistance to cut the voltage down?

A. By cutting down the number of cells employed for the lighting, as, for example, where 12 cells are used to operate the starting motor, the battery is divided into four groups of 3 cells each for the lighting, these groups delivering current at 6 volts, while the complete battery has a potential of 24 volts. This is termed putting the battery into *series-multiple* connection, which is explained further under the head of "Circuits".

Q. When the voltage is lower than that required by the lamp, what happens?

A. The lamp filament will give only a dull red glow with a voltage drop of but 20 per cent or less of the total, since there is insufficient potential to cause the current to bring the filament wire to incandescence.

Q. Is the insertion of any apparatus, such as lamps, a motor, etc., in a circuit having a voltage higher than that for which they are designed likely to damage them?

A. Yes, it will burn them out if, for instance, 110-volt lamps or motors are connected to 220-volt current, or 6-volt lamps put on a 12-volt circuit.

Q. Does the opposite also hold true?

A. No. The apparatus will merely fail to function properly if put on a circuit of a voltage lower than that for which it is designed.

OHM'S LAW

Q. What is Ohm's law?

A. It is the basis of all computations concerning the flow of an electric current. It is stated as *current equals voltage divided by resistance* and may be transposed to find any of the three factors, as, *resistance equals voltage divided by current,* so that, given any two of the factors, the third may be readily determined.

Q. How is the power equivalent of an electric current expressed?

A. Power equals current times voltage, the product being watts, as one volt times one ampere equals one watt.

Q. How many watts are there in a horsepower?

A. 746. Electrical horsepower, however, is usually figured in kilowatts, or units of one thousand watts, generally abbreviated to KW.

Q. Given a 6-volt storage battery fully charged and a circuit including a starting motor, the total resistance of which (idle) is .1 ohm, how much current will pass through the motor?

A. As current equals voltage divided by resistance, we have $6 \div .1 = 60$ amperes.

Q. If, instead of a heavy stranded cable between the battery and motor, we substitute a fine wire having a resistance of 10 ohms, how much current will pass?

A. Only .6 ampere.

Q. What would happen if a very small wire were employed to connect the starting motor with the battery?

A. Not sufficient current would reach the motor to operate it, and the wire would probably be fused by the heating effect of the heavy current.

Q. If one horsepower be required to turn the engine over at 100 r.p.m., and the car is equipped with a 6-volt battery, how many amperes will be necessary to start?

A. As power divided by voltage equals current, $746 \div 6 = 124\frac{1}{3}$ amperes.

Q. How is the power equivalent usually expressed?

A. Power equals current times voltage.

Q. As the voltage is one of the chief determining factors, what effect does doubling it have?

A. Reduces by one-half the amount of current required, exactly the same as doubling the pressure of a steam boiler reduces correspondingly the volume of steam necessary to perform the same amount of work.

Q. If the voltage be cut in half, what will be necessary to perform the same amount of work?

A. The number of amperes, or amount of current, must be doubled.

MAGNETISM

Q. What is magnetism?

A. It actually is electricity in another form and is evidenced by the attraction or repulsion that one magnet exerts on another, or that any piece of magnetized metal has for objects of steel or iron.

Q. How is this relation between magnetism and electricity shown?

A. By the fact that they are interchangeable. By passing a current of electricity through a coil surrounding an iron or steel bar, it becomes magnetic; upon moving a magnetized piece of metal close to a coil of wire, a current of electricity is induced in the wire.

Q. What is meant by the polarity of a magnet?

A. Upon being magnetized, a bar of steel will attract other pieces of metal (iron or steel) indiscriminately, but upon being brought close to another magnet, it will display an attraction at one end and a repulsion at the other for the second magnet. In other words, the magnetic attraction at both ends is not the same. These ends are termed the poles, one north and the other south, by analogy with the compass which is merely a magnetized needle having a natural tendency to point north and south.

Q. What other characteristics do the poles of a magnet display?

A. They show that the force of the magnet is practically concentrated at these poles, as the magnetic attraction is very much less at any other part of the bar.

Q. What is the law of magnetic attraction and repulsion?

A. Like poles repel one another and unlike poles attract. In other words, if a bar magnet be suspended, and the north pole of a second magnet be held close to the north pole of the suspended magnet, the latter will swing away; if the south pole of the second magnet be approached to the north pole of the suspended magnet, the latter will swing toward the former until they touch.

Q. How does the force of this attraction or repulsion vary?

A. Inversely as the square of the distance, i.e., separating the poles by twice the distance reduces the force acting between them to one-fourth its value. For example, if two magnets exhibit a strong attraction for each other at a distance of one-half inch, the attraction will be four times stronger when they are separated by only one-fourth inch.

Q. What are the chief magnetic substances?

A. Iron and steel.

Q. What is meant by the magnetic field?

A. The space immediately surrounding the poles and at which the magnetic force is most plainly apparent, as shown by the experiments with filings which graphically illustrate the field of influence of the magnet, and from which the term in question originates.

Q. What is a magnetic circuit?

A. The path followed by the magnetic flux, or flow, from one pole to the other.

Q. What analogy is there between the poles of a magnet and the flow of a current in an electric circuit?

A. The current is said to flow from the positive, or north, pole of a battery or generator, to the negative, or south, pole to complete the circuit, exactly as the lines of force in a magnet flow to complete the magnetic circuit.

Q. How can the polarity of a current flowing in a wire be determined by a simple experiment?

A. Hold a small pocket compass close to the wire. If the needle of the compass is attracted at its north pole to the wire, the current flowing in the latter is negative (south pole), as unlike poles attract, and *vice versa.* This will be true only when a direct current is flowing in the wire, since an alternating current, as the term indicates, alternates in polarity with every cycle.

Q. What are lines of force?

A. The invisible flow of magnetic influence from the north to the south pole of a magnet or about any conductor carrying an electric current.

Q. What is a solenoid?

A. A hollow coil of wire through which a current may be passed to produce a magnetic field.

Q. What is the difference between a permanent magnet and an electromagnet?

A. When a piece of hard steel has been magnetized, either by being rubbed on another magnet or by being placed in a solenoid through which a current is passed, the steel retains a large percentage of its magnetism when removed from this magnetic field and is said to be a permanent magnet. An electro magnet consists of a soft iron or steel core on which a coil of wire is wound. When a current passes through the wire, the coil becomes strongly magnetic, but when the current ceases, the magnetism does likewise.

Q. When a bar of iron is placed partly in the coil of a solenoid through the winding of which a current is passed, what takes place?

A. The bar is strongly attracted to the center of the coil and held there.

Q. How is this principle taken advantage of in electric starting and lighting systems on the automobile?

A. It is employed for the operation of electromagnetic switches for the starting motor, and it is also the principle upon which the electromagnetic gear shift depends for its operation.

Q. What effect has the insertion of an iron core in a solenoid?

A. It greatly increases the flow of magnetism through the solenoid, with the same amount of current passing through the winding of the latter.

Q. What effect has reversing the direction in which the current is passed through the winding of a solenoid?

A. It reverses the polarity of the latter so that if the core were a bar of hard steel, it would be drawn into the opening of the solenoid with the current in one direction, and expelled from it when the current was reversed.

Q. What bearing have the principles of magnetic attraction and repulsion and of magnetic polarity on electric generator and motor operation?

A. They are the fundamental principles upon which the operation of all electric generators and motors are based.

INDUCTION

Q. What is the principle of electric induction?

A. If a circuit carrying an electric current be opened and closed quickly in the case of direct current, and a coil of wire be held close to this circuit, a current will be induced in the coil. If the latter be wound on an iron core, the induced current will be very much stronger, and if both the active circuit and the coil are on the same magnetic core, the maximum inductive effect will be produced. The latter is, in effect, a transformer, and if an alternating current be sent through the first circuit, or coil, there is no need to make and break the circuit as where the current is direct.

Q. Why will a transformer not operate on direct current without making and breaking the circuit constantly?

A. It is necessary to magnetize and demagnetize the core, or, where there is no core, to produce a magnetic field and then destroy it, in order to produce an inductive effect.

Q. Why will it operate on alternating current without making and breaking the circuit?

A. Because the alternating current intermittently rises to its maximum in one direction, then drops to zero and rises to its maximum in the opposite direction, that is, the direction or the polarity of the current changes with every cycle. The transformer core is accordingly magnetized to full strength with a certain polarity, is then demagnetized and again remagnetized with the opposite polarity, and it is this rise and fall in the strength of the magnetic field from zero to maximum, first in one direction and then in the other, that causes the inductive effect.

Q. What is a cycle?

A. It consists of one alternation from zero to maximum in one direction, back to zero and then to the maximum in the opposite direction, and back again to zero. The ordinary house-lighting supply current is 60 cycles, i.e., it alternates 60 times per second, or 3600 times per minute. It is owing to this extreme rapidity in alternation that no flickering is apparent in an incandescent lamp fed by alternating current.

Q. Where alternating current is not available, how can a transformer be operated?

A. By making and breaking the circuit at a high rate of speed, as with a vibrator used on automobile induction coils.

Q. In general, why is no vibrator necessary on a coil when fed with current from the magneto?

A. Because the magneto supplies an alternating current.

Q. On the so-called dual system of ignition, the same coil without any vibrator is used with both the battery and magneto as a source of current. How is this effected?

A. The circuit breaker, or interruptor, of the magneto takes the place of the vibrator when the battery is used. The current from the magneto alternates twice per revolution and the circuit is broken at the breaker points the same number of times.

Q. What relation does the induced current bear to the current from the source of supply?

A. This depends upon the transformer and the purpose for which it is intended. On the automobile where it is desired to raise the current to a high voltage to enable it to bridge the gap of the spark plugs, the transformer is known as a *step-up* type, i.e., it takes current at a low voltage and transforms it to one of high voltage, or tension. The original, or primary, current passes through a winding of a comparatively small number of turns of coarse wire on a core of soft iron wires. Directly over this winding is a second one consisting of a great number of turns of very fine wire. This is known as the *secondary winding*, and the current induced in it is termed a *secondary current*. The voltage of this secondary current depends upon the voltage of the source of supply and the proportion that the number of turns in the secondary winding bears to that of the primary winding.

Q. Is the transformer used in any other form or type on the automobile?

A. In the so-called true high-tension type of magneto, the transformer is made integral with the armature, the fine wire, or secondary winding, being placed directly over the coarser winding that serves to generate the current. The step-up is the only type of transformer used on the automobile.

CONDUCTORS

Q. Do materials differ greatly in their ability to conduct electricity, and which are the most efficient in this respect?

A. They vary all the way from absolute insulators to those metals which will pass the electric current with the minimum resistance, such as silver, copper, and aluminum.

Q. Do the characteristics of a material affect its current-conducting ability?

A. Very greatly. The harder copper is, the poorer its conductivity, and this is likewise the case with steel.

Q. Name the different materials in the order of their current-conducting ability.

A. Silver in pure state, soft copper, brass, aluminum, iron, steel, carbon, German silver, etc.; also water, depending upon how alkaline or acid it is.

Q. Is German silver a good conductor?

A. No. It is known as *high-resistance* conductor and is accordingly used chiefly for winding resistances and not for the wires of a circuit.

Q. What are some good insulators?

A. Wood, glass, resin, paraffin wax, silk, cotton, asbestos, rubber, and similar mineral or vegetable substances.

Q. Are they always equally good insulators, regardless of their condition?

A. They are efficient as insulators only when dry. The presence of moisture on any of them affords a path for the current to cross them.

Q. What effect on the ability of the conductor to carry a current has the amount of material used?

A. The resistance is increased with a decrease in size and is also increased directly as the length of the conductor.

Q. Where, for mechanical or other reasons, it is not practical to use copper or aluminum, how can an equally efficient conductor of some other material be provided?

A. By increasing the amount of material employed in the same proportion that its conductivity bears to that of copper. For example, assuming that steel is only one-thirtieth as good a conductor as copper, thirty times as much of it must be employed to give the same conductivity.

Q. Give an example of this?

A. The single-wire system of connecting the starting and lighting outfit on an automobile. A small copper cable forms one side of the circuit, while the entire chassis forms the other. The ordinary trolley-road circuit is another, the small overhead wire forming one side of the circuit, and the rails on which the car runs, the other.

Q. Name some of the materials which are employed for their high resistance to the current.

A. German silver, iron wire, cast iron in the form of grids of small cross-section, and carbon. Very fine copper wire is also employed where the resistance desired is not very great, and space considerations permit its employment.

Q. What is meant by the "specific conductivity" of a material?

A. Its ability to conduct the current as compared with that of pure silver which has a specific conductivity of one.

Q. Does this ability of a conductor to convey the current vary particularly with a great increase in voltage?

A. Yes. The so-called high-tension current which has been stepped-up in a transformer from the 6-volt potential of the 3-cell storage battery to many thousand volts for ignition purposes will cross surfaces and penetrate materials that are perfect insulators to the low-tension current. For example, the high-tension current will leak across a moist wooden surface or it will sometimes puncture the one-fourth inch of rubber and cotton insulation of the secondary cable.

Q. What is one of the chief effects of transforming a current at a low voltage to one of high potential?

A. It enables the current to leap an air gap, the width of which is proportioned to the voltage itself. The greater the voltage the greater the width of the gap it will jump. This is the principle on which the spark plug is based.

HIGH-TENSION CURRENTS

Q. When a current of 2 amperes at 6 volts, such as would be consumed by the ordinary ignition coil from a storage battery, is transformed to a high potential, is the amount of current still the same? In other words, can 2 amperes at 6 volts be transformed or stepped-up to 2 amperes at 10,000 volts?

A. No. The current decreases as the voltage increases. For example, to make the comparison more clear, consider a current of 10 amperes at 100 volts. This is passed through a step-up transformer, of which the ignition coil is a type, and is given a potential of 1000 volts. The current, however, would then be 1 ampere, that is, the current decreases in the same proportion that the voltage is increased. The opposite is also true. By passing this current of 1 ampere at 1000 volts through a step-down transformer, it may be converted into a current of 100 amperes at 10 volts. It will be noted that the product of volts times amperes in any of the above instances cited, or of any possible combinations that can be made, is always the same. In other words, a certain amount of energy is sent through the transformer, and the same amount, barring losses due to the transformation process itself, is taken out.

Q. Is there any mechanical analogue of this process of transforming a current up or down to impress upon it a greater or lesser potential?

A. There is nothing in mechanics that corresponds exactly to this peculiar property of electricity. The resulting change in the form in which the energy is applicable as a result, however, may readily be compared with mechanical standards. For example, we may have in a very small boiler, a pressure of 1000 pounds to the square inch, but a volume of only one cubic foot of steam. This small amount at its high pressure represents the equivalent in energy of 10 cubic feet of steam at a pressure of 100 pounds.

Q. What is the object of stepping the current up to such high voltages?

A. On the automobile, simply to enable it to jump the gap of the spark plug and fire the charge. In ordinary commercial service, to permit of sending it long distances with a minimum expenditure for copper wire and a minimum loss in the amount of energy transmitted.

CIRCUITS

Q. What is meant by an electric circuit?

A. The path by which the electrical energy, or current, is said to flow from and return to its source.

Q. Is a circuit absolutely necessary in order to permit of utilizing electricity?

A. Unless there is a circuit or complete path for the current, it does not flow.

Q. Must a circuit be comprised completely of wires leading from, and returning to, the source, such as the battery or generator?

A. No, it is not necessary that wire be used for both sides of the circuit. One side or the other may be composed of a *ground*, such as the tracks of a trolley system, the overhead wire constituting the other side of the circuit, or in the case of a single-wire lighting and starting system in which one cable is employed to conduct the current from the battery to the starting motor and lights. and the chassis itself forms the ground return for both.

Q. How many forms of circuits are there in general use?

A. Three: the series, the multiple, and the series-multiple. In the first, all apparatus in the circuit is in series. That is, all the current from the source must pass through each instrument or light in turn to complete the circuit. In the multiple type of circuit, every instrument or light on it is independent of all the others. Lights may be turned on or off, motors started or stopped, without interfering in any way with any of the others. As its name indicates, the series-multiple is a combination of the two forms of circuits. For example, in using incandescent lamps to cut down the current for charging a storage battery from the lighting mains, the lamps themselves are in multiple, but the whole bank of lamps is in series with the storage battery. See illustration on charging storage battery direct from lighting mains.

Q. Which of these forms of circuit is in most general use on the automobile?

A. All three will be found on practically every car equipped with a starting and a lighting system. For instance, the starting motor is operated in series with the battery, while the lamps are wired in multiple for the side and head lights, and the speedometer and tail light are wired in series as a branch of the multiple-lighting circuit, thus giving a series-multiple circuit. The ignition distributor, coil, and battery are in series.

Q. What is meant by a grounded circuit?

A. This is ordinarily used to indicate that through lack of insulation at some part of the wire, or similar injury, the circuit has been shortened, owing to this bare wire touching a *ground,* thus permitting the current to return to its source without passing through whatever instruments there may be on the circuit. A *grounded circuit,* however, is also one in which one side consists of a ground return instead of having two wires. This is frequently distinguished by being termed a *ground-return* circuit.

Q. What is a short=circuit?

A. As the term indicates, a completion of the circuit short of the point or apparatus which the current is intended to reach. The example just cited is a short circuit as well as a ground, sometimes termed a *grounded short-circuit.* In other words, the abrasion of the insulation of one of the conductors has permitted the current to escape by a convenient path of return which, being of less resistance than the one it is intended to take, prevents any current from reaching the apparatus in the circuit. A ground is practically always a short-circuit, but the reverse is not always true, that is, a short-circuit need not necessarily be a ground, as in a double-wire circuit, but the two conductors may come together at a point where the insulation is worn, or winding of a coil may break down and cause a short-circuit.

Q. What are some typical examples of grounded circuits on the automobile?

A. Both the primary and secondary sides of the ignition circuit and the starting and lighting circuits of the so-called single-wire systems in which the chassis is always used as a ground return for all the circuits employed.

HYDRAULIC ANALOGUE

Q. What is a hydraulic analogue, and what bearing has it on an electrical system?

A. It is a comparison of the electrical system with a hydraulic or water-pressure system and serves to make clear the resemblance or analogy that exists between the principles upon which both operate.

Q. What type of hydraulic system is similar to an electrical system consisting of a generator, external circuits, and lamps, motors, or the like, as a load?

A. A constant-pressure system in which the pumps keep the water in the pipes under a certain amount of pressure corresponding to the demand. When the demand increases, the supply does likewise and *vice versa.* (In the case of the pumping system, this is not automatic, but is controlled by the attendant.)

Q. To what does the pressure of such a pumping system correspond in the electrical system?

A. To the voltage, or electromotive, force.

Q. Can there be voltage, or potential, in an electrical system without a flow of current?

A. Yes, exactly as in the pumping system in which there is always a constant pressure on the water in the pipes whether the water is escaping through any of the outlets or not. In other words, there may be pressure but no flow. The same thing is true of the generator. If it be turning at its normal speed and is wound to produce current at 100 volts, there will be a potential of 100 volts across its terminals, even though there are no lamps or motors switched on in the external circuit.

Q. How does the resistance of the pipe lines in the water system compare with the resistance of the wires in a circuit to the electric current?

A. It is nearly the same. It varies inversely as the size of the pipe and directly as its length. The smaller the pipe the greater the resistance per foot; the longer the pipe the greater the total resistance. In the same way, the resistance to the electric current increases with the decrease in the size of the wire and increases with the length of the wire, the chief difference being that bends or turns in the wire do not add to the electrical resistance, whereas bends in the pipe impose greatly added resistance to the flow of water.

Q. What comparison may be made between the speed at which the generator and the pumps run?

A. The greater the speed, the greater the pressure in the case of the pumps, and of the voltage in the case of the generator. Below a certain speed, usually termed the normal speed, there is a sharp falling off in the pressure in both. Neither can be operated safely at an excessive speed.

Q. What is the cause of the increase in voltage with increasing speed in the case of the generator?

A. Voltage, or electromotive force, is generated by the coils, or windings, of the armature cutting the magnetic lines of force of the field of the generator. The greater the number of times that these coils pass through the lines of force per minute, the greater the voltage will be.

Q. How does fall in pressure correspond to voltage drop?

A. To reach the end of the piping system, the water must overcome the resistance of the latter to its passage, and the friction involved robs it of some of its pressure in overcoming this resistance. Consequently, there is less pressure at the outlet a mile away from the pumps than there is at the pumps themselves. The same thing is true of the electric circuit. The current must force its way through the wires by reason of its voltage or pressure and, in so doing, some of the voltage is lost in overcoming the resistance of the wires, joints, switches, and the like. In both cases allowance for this loss is made by increasing the pressure at the source by an amount equivalent to the loss in transmission. For example, in electric street-railway work the motors are wound to operate on current at 500 volts, while the generators in the powerhouse produce current at 550 to 600 volts, the difference being known as the *voltage drop*.

Q. Is this an important matter on the automobile where the circuits are so short?

A. It is of considerable importance, particularly in connection with the starting motor circuit. The circuits are very short, but the initial voltage is likewise very low, so that the percentage available for voltage drop is correspondingly limited. For example, a drop of one volt in a 110- to 115-volt lighting circuit is negligible, being less than 1 per cent, but a drop of one volt in a 6-volt circuit represents almost 17 per cent and would accordingly be prohibitive. As poor connections, dirty switch contacts, dirty commutator, and worn brushes are apt to increase the resistance to a point where the voltage drop is in excess of this, the importance of properly maintaining these parts of the system may be appreciated.

Q. How does the flow of water correspond to the flow of current?

A. In both cases, the amount is proportionate to the resistance of the outlet and to the pressure back of the current, whether water or electricity. In other words, the volume of water that will flow depends upon the size of the outlet (the smaller the outlet the greater the resistance to the flow) times the pressure back of it. In the same way the number of amperes that will flow when the circuit is closed depends upon the voltage of the circuit divided by the resistance (Ohm's law). For example, the ordinary 16 c.p. carbon-filament lamp for a 110-volt circuit has a resistance of 220 ohms, which, divided by 110, gives ½ ampere as the current that will flow when the lamp is switched into the circuit.

Q. Can the piping system properly be compared with an electric circuit?

A. In practically every way except that of the return required for the latter. For example, the opening of a series of outlets in the piping system reduces the pressure in proportion to the number opened; so in connecting a number of different pieces of apparatus in series in an electric circuit, the voltage through each will decrease as another is added. It may also be compared with a parallel or multiple circuit in that the opening of one outlet does not prevent drawing water from another. A break in a main corresponds to a short-circuit or a ground in that no water can then be drawn from any outlet beyond the break. The comparisons between the piping system and the circuit are not exact, owing to the lack of any necessity for a return in the case of the water piping, but they serve to make clearer some of the fundamentals of the electric circuit.

GENERATOR PRINCIPLES

Q. What makes it possible to generate a current of electricity by mechanical means?

A. The fact that electricity and magnetism are different manifestations of the same force and that, given one, the other may be produced. Also the fact that they are readily interchangeable, i.e., one may be readily converted into the other.

Q. On what fundamental principle does the generation of electricity in this manner depend?

A. That of induction.

Q. How is it utilized?

A. By revolving a coil of wire in the field of a magnet.

Q. What occurs when this is done?

A. An e.m.f. is generated in the coil.

Q. Describe the simplest form of generator.

A. Such a generator consists of a horseshoe magnet between the poles of which a coil of wire is revolved.

Q. What governs the strength of the e.m.f. or potential, thus generated?

A. The speed with which the conductor or wire revolves, or is said to "cut the lines of force" of the magnetic field.

Q. How can this potential be further increased?

A. By winding the coil of wire on an iron core, as the iron becomes strongly magnetic and greatly increases the inductive effect.

Q. What is this simplest form of generator consisting of horseshoe magnets for the field and of a single winding on an iron core termed, and for what is it employed on the automobile?

A. It is known as a *magneto* and is generally employed for producing the current needed for ignition purposes.

Q. Can such a generator be directly employed for charging a storage battery or for lighting lamps?

A. No, it cannot be used for charging purposes, since it generates an alternating current. Moreover, owing to the small number of poles (two), its single winding, and the high speed at which it is driven, it produces very little current but a high e.m.f., as this is desirable for ignition. It cannot be used for lighting purposes for the same reason, i.e., the simple winding produces an alternating current with a very perceptible interval between the alternations, or cycles, so that a lamp would flicker very badly. As its e.m.f., or voltage, is proportionate to its speed and as there is no method of controlling it, the lamp would be burned out as soon as the magnetc was speeded up.

Q. What are the essentials of this simple form of generator?

A. The field consisting of the horseshoe magnets, and the armature consisting of a soft iron or steel core, usually in the form of an H, in the slots of which, the single winding of comparatively coarse wire is wound.

Q. Why is the field of a magneto usually referred to as a "permanent field?"

A. Because it consists of so-called permanent magnets. Naturally, they are not permanent in the real sense of the word, but their magnetism is constant while it lasts and it decreases only very gradually under the influence of heat and vibration.

Q. Why does heat affect the magnetism of the field of a generator of this type?

A. Because a piece of hard steel that is strongly magnetic when cold loses its magnetism altogether when raised to a sufficiently high temperature. In other words, if heated to a bright red and then cooled, it is no longer a magnet, and the steel must be remagnetized. Constant vibration has the same effect, but it is much slower.

Q. Is there any other way of increasing the voltage of such a generator besides running its armature at a higher speed?

A. Yes, by increasing the number of turns of wire in the winding, which has the same effect as revolving a single coil at a higher speed.

Q. How is the current produced by a simple form of generator, such as the magneto, conducted to an outside circuit?

A. Ordinarily, this would be done by means of slip rings, i.e., plain bands of copper mounted on the armature shaft with narrow copper brushes bearing on these rings, as is the case with large alternating-current generators. But as the ignition system of the automobile is a grounded circuit, one end of the armature winding of the magneto is connected directly to the core of the armature, and the other is led to a small V-shaped ring or to an insulated stud on the end of the shaft against which either a copper or a carbon brush is held by a small spring.

Q. What is the cause of the alternating cycle of the magneto, and at what points in the revolution of the armature does it occur?

A. In revolving in the field of the magnets, the armature passes successively from the field of influence of a north pole to one of opposite polarity, so that the direction of the e.m.f. is reversed.

When the armature is in a horizontal position in the field, the e.m.f. curve is at zero; as it turns, the edges of the armature core pass the ends of the pole pieces of the field, and the e.m.f. rises sharply to a maximum as the central line of the core passes the ends of the poles, when it is said to be cutting the maximum number of lines of force. It drops off again quickly from this point and again reaches zero, when the armature is in a vertical position. As its ends come under the influence of opposite poles, the curve again rises, but is now in the opposite direction, or of opposite polarity. In other words, it passes from zero to maximum and back again in every half revolution, or 180 degrees.

Q. How can a generator be made to produce a direct, instead of an alternating current?

A. The current is always alternating as generated in the armature, but it may be conducted to the outside circuit as a unidirectional, or so-called direct, current by the addition of a commutator.

Q. Can such a current be produced by the addition of a commutator to the simple single-coil winding already mentioned in connection with the magneto?

A. Yes, but as the commutator would have but two parts, the e.m.f., while passing in one direction, would be strongly pulsating.

Q. What is a commutator and how does it convert the alternating current produced in the armature to a direct current in the outside circuit?

A. It consists of a number of segments of copper, one for each coil terminal of the armature, i.e., two for each complete coil of the winding. These segments are insulated from one another with mica, and brushes bear at opposite points of the conducting hub thus formed by the segments. As the terminals of the armature coils are connected to segments that are opposite one another (in the simplest forms of winding), and as the brushes, also opposite one another, are set at points so that they pass from one segment to another when the rate of cutting is at a minimum in the armature winding, their relation to the latter is changed each half revolution. In other words, at the point in the revolution where the polarity of the e.m.f. generated reverses, the relation of the brushes to the winding is also reversed, so that the direction of the e.m.f. is accordingly always the same.

Q. How is the pulsating nature of the direct current thus generated overcome?

A. By adding coils and commutator bars to the armature so that new coils come into action before the e.m.f. produced in those just preceding them under the brushes has an opportunity to drop much below the peak or maximum. Thus, only the peak of the wave is utilized, and the e.m.f. of a direct current consists of a series of these wave peaks overlapping one another.

Q. Are permanent magnets used for the fields of all generators?

A. No, only for those of magnetos. In other types, an electromagnetic field is used.

Q. What are the advantages of the permanent field for use in connection with the magneto?

A. It is always at its maximum strength, so that the magneto generates a powerful e.m.f., even though turned over very slowly. Regardless of the speed of the armature, the strength of the field remains the same, so that no controlling devices are necessary to prevent the armature from burning out, owing to excessive speed.

Q. What is an electromagnetic field and how is it produced?

A. It is based on the fact that when a current of electricity is sent through a winding surrounding an iron core, the core becomes strongly magnetic. It accordingly consists of windings on the fields of the generator, in addition to those on the armature. Depending upon the particular type of generator in question, either all or only part of the current produced in the armature is sent through the windings of the field. The latter is then said to be *self-excited* in that it depends upon no outside source.

Q. Is the self-excited field characteristic of all generators except the magneto?

A. Yes, of all direct-current generators. Large alternating-current generators are said to be separately excited, a smaller direct-current generator being employed solely for the purpose of rendering the fields of the larger machine magnetic.

Q. What is a series-wound generator, and why is this type not used on the automobile?

A. It is one in which the entire current generated in the armature is passed through the field windings. It does not generate until a high speed is reached. Its voltage varies sharply as its speed, and it may have its polarity reversed by the battery if its speed drops below a certain point, consequently, it is not fitted for charging storage batteries. (In fact, the series-wound generator is practically obsolete, except for some special uses.)

Q. What are shunt- and compound-wound generators?

A. In the former, the windings of the armature and of the fields are in multiple, or shunt, so that only a certain amount of current, depending upon the difference in the resistance of the outside circuit and that of the fields, passes through the windings of the latter. As the load and consequently the resistance of the outside circuit increases, more current passes through the shunt, and the fields become more strongly magnetic, thus increasing the output so that the generator is, to a certain extent, self-regulating.

In the compound type, there is, in addition to the main shunt winding on the fields, an auxiliary winding of heavier wire (lower resistance) which is connected in series with the armature. As in a series-wound generator, the amount of current exciting the fields is directly proportional to the speed, more current in proportion passes through the compound winding than through the shunt winding as the load is increased, and the generator is self-regulating to a much greater degree. The compound-wound type of generator is in practically universal use on the automobile as well as for general power purposes.

Q. What is meant by the term "self-regulating" as used in the preceding paragraphs?

A. The generator automatically produces more current in response to the demand occasioned by an increase in the load, without any change in its driving speed.

Q. How is this accomplished?

A. The amount of current produced by the generator depends upon the strength of its magnetic field in which the armature revolves. The magnetism of this field represents the so-called lines of force. The greater the number of lines, or the more powerful they are per unit of pole-piece surface, the greater the volume of current that will be generated. In practical usage, this is referred to as the *magnetic flux*, or flow, through the armature. By increasing or decreasing the amount of this magnetic flux through the armature, the current output can be controlled within close limits.

Q. What is meant by the "load" on a generator?

A. The lamps, motors, storage battery, or similar apparatus to which it is supplying current.

Q. As the speed of the generator itself does not increase, how does it provide for an increase in the load?

A. By absorbing more power from its driving unit. For example, if a generator be operating with only ten 100-watt lamps in the circuit, it is requiring approximately one and one-half horsepower to drive it. Now, if another group of ten lamps of the same size be switched on, the amount of power demanded by the generator of its engine will be doubled. This may be very readily demonstrated in a rough way by fitting a handcrank to any small automobile generator and turning the machine over with one lamp in the outside circuit. It will be found very easy to spin the generator very rapidly by hand, as practically no resistance is felt. Now connect in the circuit a discharged storage battery, and the additional power required to turn the machine will at once be very perceptible.

Q. What are the brushes and what purpose do they serve?

A. They are strips of copper or carbon (the latter is now almost universally used), which serve to conduct the current generated in the armature to the outside circuit and to the field windings by bearing on the revolving commutator. Except where an additional brush is employed for regulating purposes, there is usually one brush for each pole of the field, i.e., a bipolar generator is fitted with two brushes, a four-pole with four brushes. The brushes are held against the commutator by springs. Soft copper embedded in carbon is also employed, especially for low-voltage generators, such as the lighting generator on the automobile.

ELECTRIC MOTORS

Q. Is there any difference in principle between the electric generator and the electric motor?

A. Fundamentally, they are the same, as is evidenced by the fact that either is reversible, that is, an electric generator, when supplied with current from an outside source (of the proper voltage, of course), will operate as a motor, and a motor, when driven by an outside source of power, will generate an electric current. They are naturally not interchangeable in practice, owing to differences in design and winding. The generator is wound to produce the maximum amount of current at a certain voltage with a given horsepower, while the motor is designed to produce the maximum amount of power with the minimum current.

Q. What is the operative principle of the electric motor?

A. That of magnetic attraction and repulsion.

Q. How is it applied?

A. As in the generator, both the fields and the armature of the motor consist of electromagnets. The brushes and the commutator serve the same purpose of reversing the direction of the current through the armature coils every time a different pair of commutator segments passes under the former. As has already been explained, reversing the direction of current flow through the winding of an electromagnet reverses the polarity of the magnet itself. To simplify the illustration, take a bipolar motor with a two-pole armature having but a single winding. When the current is switched on, the armature is at a 45-degree angle, so that its poles are just under the poles of the field. As the commutator causes the current to flow through the armature winding in a reverse direction to that of the fields, unlike poles will be created. They will attract each other, and the armature will revolve a small part of a revolution, until it is directly in the strongest part of the field of the influence of the field magnets. Just as this point is reached, however, the brushes pass on to new segments of the commutator, and the direction of the current in the armature coils is instantly reversed. The polarity of the armature core is also reversed, so that there are now like poles opposed to one another, and they repel, causing the armature to complete another part of its revolution, when the former conditions are again established and the armature is again attracted. In a bipolar motor with a simple two-pole armature, there would be two phases of attraction and repulsion per revolution. In larger motors this is multiplied by the number of poles in the field and the number of coils on the armature.

Q. As an electric motor in running fulfills all the conditions necessary for the generation of an e.m.f., what becomes of this voltage?

A. It constitutes what is termed a *counter e.m.f.* and serves the useful purpose of increasing the resistance of the motor when in operation, thus reducing the amount of current necessary to drive it. For example, when the motor is standing idle, the resistance of its windings is low. It is for this reason that large direct-current motors (one h.p. or over) cannot be started without the aid of an outside resistance to cut down the starting current, otherwise the armature would be burned out. As the armature speeds up, the counter e.m.f. generated opposes that of the driving current and accordingly increases the resistance. The heating of the windings in operation further serves to increase the resistance, as the resistance of most metals increases with a rise in their temperature.

Q. How many types of motors are there, and what type is most generally used for automobile starting?

A. As they correspond exactly to generators, there are the same number of types, i.e., series, shunt, and compound wound. The series type is almost universally employed on the automobile and is also very largely used on trolley cars.

Q. If the series-wound generator is of so little practical application, how is it that the series-wound motor is found so advantageous?

A. The same characteristics that are a disadvantage in the generator are correspondingly valuable in a motor, which explains why generators and motors are not interchangeable in practice, as already mentioned. A series-wound machine is essentially a variable-speed machine, and this is not desirable in a generator, while it is in a motor. The series type of motor has a very heavy starting torque, or pull, which increases as the speed of the motor decreases. This is exactly what is wanted to overcome the inertia of the gasoline engine. Its current consumption falls off very quickly as its speed increases, and it has a very liberal overload capacity, being capable of carrying loads up to five times the normal, for short periods.

Q. As the speed of the series motor decreases in proportion to the load, what happens when the load is suddenly relieved as in the starting of the gasoline motor?

A. The electric motor tends to race, or *run away*.

Q. How is this prevented on the automobile?

A. The method employed differs in different systems, but, as a rule, the starting of the gasoline engine automatically opens the starting motor circuit, or means are provided for greatly reducing the amount of current it receives the moment the load is removed.

Q. Are either shunt= or compound=wound motors used on automobiles?

Q. What is a motor=generator, and what is it employed for?

A. As its name indicates, it consists of two units, a motor and a generator, the former having an alternating current, and the latter a direct current. It is employed for converting an alternating current into a direct current, so that it may be utilized for charging storage batteries. The alternating-current supply is used simply for running a motor of that type to which is directly coupled a direct-current generator. There is no electrical connection between the two machines.

Q. Are motor=generators ever used on automobiles?

A. No, but the combination of a direct-current generator and a starting motor in one machine, as in the single-unit systems, is frequently so-called through error. This single unit is variously termed a *dynamotor* and a *genemotor* to distinguish it from a motor-generator.

Q. How are the two radically different purposes for which the generator and the motor must be designed combined in one machine?

A. By putting independent windings on the fields and the armature, and, in some instances, by employing two commutators at different ends of the armature shaft.

BATTERIES

Q. What other method is there of producing an electric current besides that of driving a dynamo?

A. The use of batteries known as primary and secondary cells.

Q. What is the difference between these two types?

A. In the primary cell, the current is generated by means of the chemical reaction taking place between electrodes of different materials in an acid or alkaline solution, one electrode being dissolved in the solution as the chemical action continues.

The secondary cell is the storage battery. This does not generate a current of electricity as in the case of the primary cell, nor does it actually store electricity as its name would indicate. The passing of a current through its elements brings about a chemical conversion of the latter, which is reversed when the current flows out of the cell.

IGNITION INSTRUCTIONS

Q. How many different systems of ignition are in use on the automobile today?

A. Generally speaking, only one, known as the high-tension system. The low-tension system used in earlier days has been obsolete for a number of years. The single classification, however, may be subdivided into several others which are known by their distinguishing features, the first being determined by the source of current supply, as magneto- and battery-ignition systems. These two classes may be divided further according to the type of magneto employed, such as the duplex, the dual, and the double-spark types. All battery systems are fundamentally the same, only differing in the type of circuit breaker and distributor employed—whether direct driven from the engine or combined with the lighting generator, and in the type of controlling switches and auxiliary devices.

Q. Why is the system generally used termed "high-tension" system?

A. Because the low-voltage current of the battery or magneto must be passed through a certain type of step-up coil, known as an inductance coil, to give it a sufficiently high voltage to cause it to jump the air gap in the spark plug. A voltage of 12000 to 30000 volts is required to do this since the compressed gases in the combustion chamber increase the resistance offered to the current at the spark gap. In designing the system a sufficient margin must be allowed over the required voltage to care for any resistances which may enter the circuit, such as a loose connection, a partially broken cable or a dirty joint, insuring a spark under any condition.

Q. Of what does a high-tension system consist?

A. The essential parts of a high-tension ignition system are: (1) a source of current, such as a dry battery, the storage battery of the lighting and starting system, the direct-current generator of the latter, or a magneto; (2) a step-up transformer or induction coil, the primary winding of which is in circuit with the source of current supply; (3) a contact breaker or interrupter to open this circuit periodically, i.e., once every other revolution for each cylinder of a four-cycle engine; (4) a distributor in circuit with the secondary winding of the coil and provided with as many contacts as there are cylinders; (5) a spark plug for each cylinder; (6) primary and secondary cables for the respective connections, and a controlling switch to open and close the supply circuit or to change from one supply circuit to another, where both a battery and a magneto are employed.

Q. How do these essentials vary in different systems?

A. Where a battery is depended upon for the current supply, the interrupter and the distributor are usually combined in an independent device which is driven from the camshaft of the engine.

In the case of a magneto, both the interrupter and the distributor are integral with it. This does not apply to the Ford magneto which has a separate low-tension timer and uses no distributor, as there is a vibrating coil for each cylinder.

In what are commonly known as modern battery systems, the timer and distributer may be either mounted separately, as first mentioned, or combined with the lighting generator.

CURRENT SUPPLY AND APPLICATION

Magnetos

Q. How many types of magnetos are there in general use?

A. Two general classes, the low-tension and the high-tension, and various special types, such as the dual, the double-spark, the duplex, and the inductor magnetos.

Q. What is the difference between low-tension and the high-tension magnetos?

A. The low-tension magneto has only a single winding on its armature the current being generated at low voltage and transformed by passing through an independent coil, whereas the high-tension magneto generates the current in one winding and steps it up through another, both on the same armature.

Q. What is a dual magneto and why is it so called?

A. It is a low-tension type, the interrupter and distributor of which are also employed in connection with a battery for starting. It is so called because these essentials are common to both the magneto and the battery sides of the system.

Q. What is a double-spark magneto?

A. One provided with two distributors designed to produce two sparks simultaneously at two different plugs in the same cylinder.

Q. What is a duplex magneto?

A. One designed to permit of passing the battery current through the armature of the magneto to facilitate starting, the magneto and battery both acting together to produce the spark at low speeds. To accomplish this a commutator is mounted on the armature shaft and the battery connected to it, the magneto being of the high-tension type. This commutator causes the battery current to alternate in direction with that produced by the magneto so that it is said to be "in phase" with the latter.

Q. What is an inductor magneto and how does it differ?

A. An inductor is employed instead of an armature, the windings being stationary. The inductor is simply a revolving piece of metal which alternately opens and closes the magnetic circuit. In the K-W inductor magneto this winding is of copper ribbon and is placed between the poles of the inductor; in the Dixie magneto it is a conventional induction coil placed in the hollow of the magnets above the inductor.

Q. Why can a magneto not be run in either direction equally well?

A. Owing to the contour of the cam which serves to open the contacts of the interrupter. This must be designed to operate the magneto either as a "right-hand" or a "left-hand" machine.

Q. How is a magneto timed?

A. Disconnect its drive from the engine. Turn engine over by hand until the piston of cylinder No. 1 is exactly at the upper dead center on the firing stroke. Turn the armature shaft of the magneto to a point where the contacts of the interrupter are just beginning to open; the brush of the distributor which is then making contact with the distributor segment should be connected to the spark plug of cylinder No. 1. The next brush should be connected to cylinder No. 2 or No. 3, according to whether the firing order is 1- 2- 4- 3 or 1- 3- 4- 2. The armature of the magneto should be coupled to its driving shaft in the position as determined for the first cylinder.

Q. If after timing a magneto in this manner, it is found that the spark-timing lever does not give sufficient advance or retard, what should be done?

A. Remove the cover from the distributor housing of the

magneto, the piston of cylinder No. 1 being at upper dead center of firing stroke and the interrupter contacts just about to open as directed for timing. Note the relative position of the segment and the distributor brush which should be making contact with it. If the segment has already passed the brush, the spark-timing lever being at the maximum advance position, remove the distributor gear from its shaft and from engagement with the pinion on the armature shaft. Move it back one tooth (against the direction of its rotation which is the opposite of that of the armature pinion) and remesh with pinion. If this does not bring it into contact with the brush, move back another tooth. Should the distributor segment not have reached the brush when in the position as given above, move the distributor gear forward, or in the direction of its rotation, one or two teeth, and remesh.

Q. How can the various types of magnetos be identified, as installed on car?

A. The two types in most general use may be distinguished at once by their external connections. The so-called dual type can be identified by its separate coil, or transformer, mounted on the face, or the front, of the dash, under the hood, and the connecting cables from the magneto to this coil. One of these connections is for the primary of the coil, and the other is for the secondary; the other ends of both coils are joined together and connected to a common ground wire, so that the coil has only three connections. The true high-tension type can at once be recognized by the fact that its only external wires are those connecting the distributor plate of the magneto directly with the spark plugs.

Q. Of these various types, which are most commonly used?

A. The dual type will be found on most low-priced cars (except the Ford), and the straight high-tension type on higher priced cars not using battery ignition.

Q. In searching for faults, is it easier to locate trouble in one type than in the other, and is the procedure different in each case?

A. Owing to its having but a single winding on its armature, and to the fact that all of its connections are external, the dual low-tension magneto is the simpler of the two; but the exposed location of its connections makes them more subject to default than those of the high-tension magneto. The procedure differs in that failure to operate, in the case of a dual magneto, may be caused by injury to, or the breaking of, some of these external connections; whereas, with the high-tension type, the cables are so short and so direct that the fault is likely to lie in the magneto itself.

Q. Where would be the most likely places to look for the cause of failure of a dual magneto to operate?

A. In about the order of their liability to occur, these would be as follows: broken or faulty connection at one of the coil terminals; primary or secondary cable, connecting magneto with coil, grounded through chafing or from being soaked with oil and water; ground between the dry cells and the metal battery box (this last is a not infrequent ground and is very annoying to locate, as it will occur one moment and disappear the next, owing to the vibration breaking the contact); dirt, oil, or excessive wear at the primary collector brush, and similar conditions at the brush which conveys the high-tension current from the coil to the distributor; failure of the coil through breakdown; failure of the condenser, causing the interrupter points to burn away rapidly; a break in the armature winding. All of these last are rare causes of trouble

Q. How are these faults best remedied?

A. Poor connections at the coil termimals can be overcome by baring about an inch and a half of the cable of insulation, twisting a $\frac{1}{4}$-inch loop in the end, and, after cleaning and wetting the braided end with soldering flux, dipping the loop into molten solder. The terminal nuts can be screwed down hard on these loops without opening or injuring the stranded wires, and they will make solid and permanent terminals. Where cables have become so oil soaked that the integrity of their insulation is suspected, they should be replaced, and the new wires properly supported. If either of the brushes is at fault, due to excess of oil and dirt, clean with gasoline, and true up the ends square with a fine file. Should the brushes have worn unevenly, or, in the case of the primary brush, taken on a hard, glazed surface, treatment with the fine file will remedy the trouble. When they have worn down to a point where the spring no longer holds them in good contact, new brushes and springs (attached) should be inserted.

Q. When examination shows no fault at any of these points, how can the coil be tested?

A. Disconnect the coil from the magneto, and connect to the battery terminal one wire from a spare battery of four dry cells or, if more convenient, to an ignition storage battery. Fasten the ground connection from the coil to some handy metal part of the chassis; lay the secondary cable from the coil on the chassis so that its bared end is not more than ¼ inch from the metal of the motor, or chassis. Then connect another length of wire to the opposite terminal of the testing battery. (The dry cells should be in series.) Scrape the end of this wire clean, and touch it rapidly to some part of the motor. A spark should occur every time it is touched, showing that the primary winding of the coil is uninjured, and, if the secondary is likewise uninjured, a spark should jump between the bared end of the secondary cable and the adjacent metal, every time the circuit is closed with the testing wire. The occurrence of these two sparks show the coil to be in proper working condition. If the spark occurs at the testing wire, but no high-tension spark takes place at the end of the secondary cable, it indicates that the secondary winding of the coil has broken down. Should the spark, taking place every time the testing wire is touched to a part of the motor, be very bright and hot, it is quite likely that the condenser has become punctured. Unless the failure of the secondary is due to a broken connection between the fine wire of the winding and the terminal, it must be sent to the maker for repairs. This is the case also when the condenser has been punctured. The magneto will continue to work with the condenser short-circuited, but this will cause a rapid burning away of the expensive platinum contact points of the interrupter.

Q. How is the magneto armature winding tested for a break?

A. Employ the test battery already mentioned. Touch one wire to the armature shaft, and the other to the collector end which is insulated from the shaft. A spark should result if the winding is intact. Failure to obtain a spark would indicate a break in the winding, and the armature should be returned to the maker for repairs. In making tests with a battery in this manner, always make sure that the connections from the battery have not pulled loose nor become broken, before finally accepting the lack of a spark, at the point where it should occur as conclusive evidence of a fault in the part being tested. Otherwise, a failure of the testing apparatus itself may be put down as a fault in the part being tested.

Q. Is a spark the best indication obtainable in making such tests?

A. With a fresh battery of dry cells, the self-induction of the coils tested, such as the winding of the armature or the primary of the coil, will give a bright spark that cannot be mistaken; but, if an audible indication be desired, the battery may be placed in a box and a common electric door bell, or buzzer, mounted on the box. The bell, or buzzer, must be connected in series with the testing wires, so that, when the circuit is completed, the current passes through it. Then the success of the test will be evidenced by the sounding of the bell, or buzzer, as long as the circuit is closed. In the case of the secondary winding of the coil, the spark is all that is necessary. Should a more visible signal be desired, the lamp-testing set, described in connection with trouble hunting on the starting and lighting system, may be employed. In this case, the test battery is dispensed with and the 110-volt lighting circuit used, but a 10-ampere fuse-block and fuses should be inserted in the testing circuit to guard against accidental short-circuits in handling the testing apparatus.

Q. As secondary cable is expensive, and the owner does not usually want it replaced unless absolutely necessary, how can it be tested for faults?

A. Connect the test battery to the coil, as previously described, but, instead of relying upon breaking the circuit by hand, insert a buzzer, or bell, in series, between the battery and the primary of the coil. This will give a vibrating contact, and will keep the coil working continuously. Connect the piece of cable to be tested to the secondary of the coil and support it well, clear of the ground, such as the motor or chassis. Take another piece of secondary cable and connect one end of it to the ground. Bare the other end, and pass this along the entire length of the cable being tested, very close to, or actually touching, the insulation of the cable under test. If there are any weak spots in the insulation, a spark will jump through it to the testing wire. In case a vibrating coil is at hand for making this test, it will not be necessary to insert the buzzer as mentioned.

Q. What is likely to result when there are weak spots in the insulation of the secondary cables?

A. The high-tension current will escape through them to the ground, because of the nearness or actual contact of the secondary cable with the motor cylinders or other metal parts. This leakage will be neither visible nor audible, unless the insulation is very bad, and the failure to fire will usually be attributed to the spark plug instead.

Q. Should primary cables be tested in the same manner.

A. It is not necessary; as, unless the insulation is actually worn off, there will be no escape of current, owing to its low voltage. Where solid instead of flexible primary wire is employed, as on some old cars, or where the owner has made replacements, a test for a break in the wire itself under the insulation may be made with the aid of the battery and buzzer alone.

Q. Are the causes of failure similar in a true high-tension magneto?

A. No. As the primary-generating winding and the secondary, or high-tension winding, are both on the armature of the magneto itself, and all connections, except the high-tension leads from the distributor to the spark plugs, are made internally, there are no outside cables, terminals, or coils to default. Unless the repair man has become proficient in testing electrical apparatus and has familiarized himself with the construction of the high-tension type, he will find it preferable to refer to the manufacturer any cases of trouble in this class of apparatus. In fact, the maker usually absolves himself from any responsibility in case a magneto of this type has been taken apart. Even where the repair man is capable of dismantling and testing this type of magneto, its repair would ordinarily be beyond his facilities, so that it is better to refer it to the maker at once.

Q. Are there any faults, peculiar to the "duplex" or to the "double-spark" types of magnetos, which are not encountered in the others?

A. In the case of the duplex type, there may be a failure to work of the battery connections or of the battery commutator on the armature shaft, i.e., the commutator which throws the battery current into phase with the armature current of the magneto when starting. Since the advent of the self-starter, this type of magneto has had no particular advantage to recommend it, and will be found only on older cars. As the only difference between the double-spark and the usual magneto is a duplication of the distributor to give two sparks in the cylinder instead of one, its treatment is the same.

There are simply two distributors to maintain instead of one. This type is also of limited application and will be found only on comparatively few cars of several years back.

Q. Why is it important that the magneto be accurately timed?

A. Unless the spark occurs at exactly the right moment, the motor will not operate efficiently. If it occurs too soon, the explosion will tend to retard the piston; if too late most of the power that would have been derived from the compression will be lost. As the lag, i.e., the time intervening between the moment the contact points are opened at the interrupter and the occurrence of the spark at the plug, is negligible in the magneto, it must be more accurately timed than the old battery and vibrator-coil system.

Q. Does the magneto ever fail to operate through lack of oil, and what attention should be given to its lubrication?

A. Prior to 1910, when some of the lower-priced magnetos were made with plain bearings, this naturally occurred, but, with the adoption of high-grade annular ball bearings for the magneto shaft, it is practically unknown. However, even the high-grade ball bearing will not operate without lubrication. If it runs dry, the balls are apt to rust and ruin the bearing. Most of these bearings are packed with vaseline, or similar light grease, when the machine is assembled and require no attention during the life of the average car. Where this has not been done, as on some of the older machines, a few drops of fine sewing-machine oil once a year will suffice. The pivot and roller of the contact-breaker arm also should be oiled once or twice a year with one drop of oil to each, using a toothpick.

Q. When taking a magneto apart, as where it is necessary to remagnetize the fields, why is it of the greatest importance to reassemble them in the same way?

A. Unless this is done, the polarity of the fields will be reversed and the machine will not generate properly. The maker usually identifies the polarity of the fields by marking the magnets so that it is easy to reassemble them in the proper manner. They are usually stamped with the polarity.

Q. When remagnetizing the field magnets of a magneto, why is it important that their polarity should not be changed in the process?

A. The effect would be exactly the same as if the remagnetizing were carried out properly, so far as their polarity were concerned, and then the magnets were put back the wrong way. The machine would not generate. For example, the marking on the side of the Eisemann magneto indicates the north pole of its fields. If in remagnetizing, this side of the field is made the south pole, and it is then correctly assembled, it will be evident that the polarity of the entire field has been reversed.

Q. In the case of the dual magneto, how can trouble, caused by the grounding of the dry cells against the metal battery box, be overcome?

A. By making a tight-fitting wooden box to hold the battery, this wooden box fitting inside the metal one.

Q. Does the care required by an inductor type of magneto vary from that necessary for other types?

A. No. So far as its outside connections go, it is the same. That is, in a dual type installation, using an external coil, or transformer, the causes of trouble and their remedies will be the same as in any dual system, as already given; and this also applies to the high-tension dual type.

Q. In the order of their occurrence, what are the commoner causes of failure of the magneto, due to wear of its parts?

A. In practically every case of failure of the magneto that is not otherwise apparent at a glance, an inspection should always be made of the contact points in the breaker box. Unless they open properly when the cam strikes the lifter, no current reaches the outside circuit and the coil is not energized. Next, inspect the contact of the collector brush; see that it is clean; that it is making good contact over its entire surface; and that its spring is holding it firmly in place. After this, inspect the distributor.

Q. How often should the contact points in the breaker box require attention?

A. This will depend upon the type of magneto, and when it was made. On some of the earlier types, prior to 1909, no condensers were used on many of the lower-priced magnetos, and the points required attention every 3,000 or 4,000 miles. Others will run two or three times this distance without requiring attention. If, when the points have been found in poor condition, they have not been properly trued up and adjusted, they are apt to require attention again much sooner, as any irregularities in their contact faces will cause them to burn away much more rapidly.

Q. As the platinum contacts of the magneto breaker box are expensive, how far down can they be allowed to wear before it is necessary to replace them?

A. This will depend upon the amount of adjustment provided. As long as there is sufficient platinum left to provide a true surface on each contact point, it will not be necessary to replace them if they can be adjusted so that the cam opens and closes them properly.

Q. Since the magneto-armature circuit is normally closed upon itself and only opens when a spark is required at one of the plugs, how is the magneto ignition shut off?

A. By short-circuiting the armature around the contact points. Instead of opening the entire ignition circuit, as in the case of a battery where it is necessary to save current, the generating part of the circuit is closed. This is true of all high-tension magnetos.

Q. When the contact points of the interrupter of the magneto fail to separate, what is the result?

A. The armature remains short-circuited upon itself throughout the revolution, and no current reaches the outer circuit, so that the engine will not fire on the magneto at all.

Q. What is the result when the contacts open but the gap is not wide enough?

A. Erratic missing, probably at all speeds, but more pronounced when the engine is running slowly. The increased kick given the movable contact arm by the cam, when the engine is running at a higher speed will cause it to fire more regularly.

Q. What happens when contact points are separated too far?

A. The missing is likely to be more noticeable at high speeds than at low speeds; as, when turning very fast, the looseness of the movable contact arm may prevent it from closing the circuit again in time for the next cylinder to fire.

Q. What is the proper distance for the setting of the contact points of the magneto interrupter, or breaker box?

A. Approximately $\frac{1}{64}$ inch, or the equivalent of an ordinary sheet of paper. When the points are properly adjusted, it should be possible to insert the paper between them and move it around without binding.

Q. How are the contact points usually adjusted?

A. Practically every magneto manufacturer now supplies an adjustment gage, which serves also as a screwdriver or small spanner, according to the construction of the interrupter. The thickness of the metal represents the distance that the contacts should open. In all except the old types produced several years ago, the interrupter may be removed without the use of any tools, and the adjustment made without the necessity of removing the magneto from its bedplate. The following instructions cover the Eisemann dual type in this respect, and are typical: "Insert from behind through the hole in the plate carrying the make-and-break mechanism, the metal adjuster which we supply with every magneto: by means of a flat wrench hold the nut on platinum screw (contact) and turn the platinum screw upward or downward until there is a gap of $\frac{1}{64}$ inch between the contacts, i.e., a piece of paper must pass between them without getting jammed", Fig. 143.

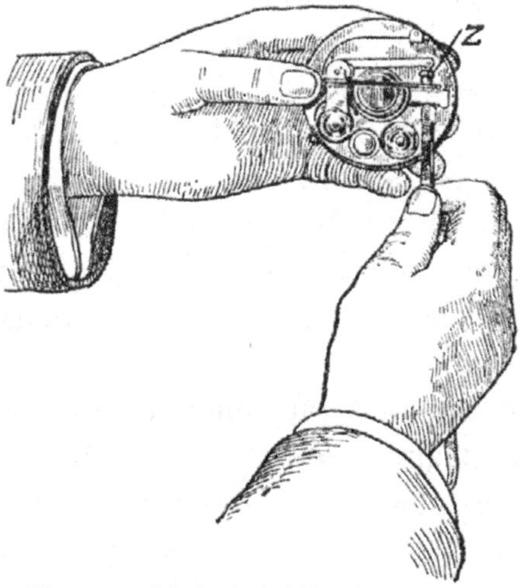

Fig. 143. Method of Adjusting Magneto Contact Points

Q. How often should the magnets of a magneto need remagnetizing?

A. No definite time nor average can be given for this. In some cases, the makers state specifically that the magnets will never need remagnetizing during the entire life of the machine, unless they are taken off and allowed to stand without a "keeper", i.e., a piece of soft iron placed across the pole pieces, the magnets themselves being placed in their usual relation as when on the machine. In others, it has been nothing unusual to require this once a year, or after 6,000 to 10,000 miles running.

Q. How can it be determined definitely whether the magnets actually need remagnetizing or not?

A. Many cases of weak or faulty ignition are attributed to loss of strength in the magnets, when they are actually due to something else. Before deciding that the magnets themselves are at fault, every part of the system should be gone over thoroughly. See that the contact points are clean and true, and that they are properly adjusted; that the low-tension collecting brush, or contact, is clean and is making good contact; also, that the distributor is clean. Inspect all terminals and connections. Test the condenser. Disconnect one of the spark-plug leads, keep it away from any metal part of the chassis and, with the engine running, note whether a spark occurs at the safety gap. In a high-tension magneto, this gap is ually located in the hollow of the magnets back of the distributor; in a low-tension type, having a separate coil, it is usually on the coil. Take the spark-plug lead which has been detached (engine running) and approach its metal terminal gradually to the spark-plug end, or to some metal part of the engine, noting the maximum distance that the spark will bridge. If this is one-quarter inch or more, the magnets are not at fault.

Q. When there is no question but that the magnets are the cause of faulty ignition, how can the magnets be remagnetized?

A. With the aid of a small electromagnet, as described in the chapter on this subject. Care must be taken to see that the polarity of the magnets is not reversed in the process, as this will render the magneto altogether inoperative.

Q. How is the ground connection made in a magneto?

A. One side of the winding of the armature in a low-tension type, and of both windings, i.e., the primary and secondary are grounded by being electrically connected directly to the core of the armature itself, in the high-tension type. The fastening of the magneto in its bedplate on the engine, completes this ground connection.

Q. Is it ever advisable to insert paper liners, or liners of any material, between the magneto and its bedplate?

A. If on an old car, it is necessary to resort to liners or shims to correct the alignment of the magneto with its driving shaft, nothing but thin sheets of brass or iron should ever be employed, as the use of any insulating material for this purpose would break the ground connection and prevent the magneto from functioning properly. But even on old cars, where the lack of alignment of the magneto with its shaft is such as to be plainly perceptible to the unaided eye, the Oldham coupling usually employed on the driving shaft, allows sufficient play to compensate for this. Any lack of alignment is more likely to be due to carelessness in replacing the magneto after it has been removed from the engine for attention, than to any fault in the placing of the bedplate and driving shaft.

Q. What will be the result of the magneto being lined up incorrectly?

A. The universal coupling is apt to wear badly, and the side pressure, which the lack of alignment causes, may not be taken care of entirely by the coupling, so that the pressure has to be taken in part by the magneto shaft and its bearing. This will result in undue wear of the bearing and, in time, may permit the armature core of the magneto to strike the pole pieces, or sides of the tunnel, so that it will jam. On many of the higher-priced cars, a flexible coupling, consisting of leather discs, is employed for the magneto drive.

Q. Which is more likely to occur, separation of the points by too great a distance or their coming together so that they do not open?

A. The result of wear will usually be to increase the distance between the points, but the writer has experienced the opposite on an old car where the cam apparently wore down faster than the points and prevented their opening. This was probably due to improper hardening of the cam itself or to the fact that the roller was harder than the cam. Many of the later models do not employ a roller in the movable contact arm, the cam striking directly against a projection on the arm itself. An example of the roller type, is the K-W interrupter.

Q. Is it possible for the adjustment of the contact points to wear to such a degree that the magneto will fire the engine at high speeds, but not at all at low speeds?

A. Instances of this nature on old magnetos have proved a puzzling cause of ignition failure. The extra kick of the cam at high speeds would separate the points, whereas at low speeds they remained together.

Q. Inspection having shown other parts of the ignition system to be in good order, how can it be determined whether the magneto itself is at fault?

A. Run the engine on the battery, and switch from battery to magneto while the engine is running at a good speed. If the magneto is not operating properly, the difference in running will be perceptible immediately; in case the contact points are not opening, the explosions will cease at once upon throwing the switch to "magneto".

Q. What is meant by "ignition timing"?

A. A four-cycle motor is so-called because there are four parts to each cycle: viz, suction, compression, firing, exhaust. It is evident that ignition is necessary only at one part of the cycle and that the spark must occur at a certain time with relation to the carrying out of that part of the cycle. Determining the point at which the spark is to occur in that part of the cycle is usually referred to as the "timing of the spark", or as the "ignition timing".

Q. Is it necessary that this be carried out with precision?

A. In the high-speed automobile motors of the present day, it must be extremely accurate with relation to the movement of the piston and must be exceedingly rapid in action; otherwise, the efficiency of the motor would suffer greatly. In the slower-running motors with fewer cylinders, of several years ago, neither of these factors was of such great importance; but the present-day high-speed multi-cylinder motors could never be operated satisfactorily on ignition apparatus of the type familiar on 1910 models, for example.

Q. Why would a variation or a lack of precision affect the running of the motor?

A. Speeds are now so high that the time factor is reduced to exceedingly small fractions of seconds. For example, take a high-speed four. It turns at 2500 r.p.m.: as there are two explosions per revolution in a four-cylinder motor, this would mean 5000 sparks per minute, or one spark every .012 second. This is cutting time pretty fine, but in a modern twelve-cylinder motor, it is finer still. Say the twelve runs at 3000 r.p.m.: in a twelve there would be six explosions per revolution, requiring 18,000 sparks per minute, or one for every .0033 second—thirty-three ten-thousandths of a second—an incredibly brief space of time in which to carry out a combined mechanical and electrical function.

Q. How is this extreme accuracy obtained?

A. By the use of precision machine work in the distributor, approaching that of a fine watch, and a great amount of "advance" to compensate for any lag in either the mechanical or electrical functioning of the apparatus.

Q. What is meant by "lag", and what is the difference between mechanical and electrical "lag"?

A. As the term implies, lag is delay: in other words, it is the time elapsing between the moment a part begins to move and the actual moment at which it moves and makes contact or carries out the function for which it is designed. Mechanical lag is due entirely to the time necessary to overcome the inertia either of a part that is stationary or of one that is moving in another direction and whose direction must be reversed. Electrical lag, on the other hand, is the time elapsing between the moment that current is switched into a piece of electrical apparatus and the moment that the apparatus actually operates. This will appear strange at first sight in view of the universal belief that anything electrical operates so swiftly that it is next to impossible to measure the time required. This is true when nothing more than the passing of a current of electricity from one point to another is concerned, but when actual work must be performed by the current, the time required is not only measurable but it may be so perceptible as to have a decided effect upon such extremely rapid functioning as that mentioned. For example, when the work to be performed by the current consists of magnetizing the core of a coil, as is necessary in automobile ignition, the time element is quite perceptible, as may be noted by referring to the oscillograph of the spark produced by an induction coil. A coil having the characteristics shown by this oscillograph would be worthless on a modern high-speed motor.

Q. How is this lag compensated for?

A. By advancing the moment that contact is made outside of the cylinder to an extent that will insure the occurrence of the spark in the cylinder at the proper moment. In other words, the current is started on its way sooner with relation to the movement of the piston. This is generally referred to as "advancing the spark".

Q. What is the proper point, in the travel of the piston on the compression stroke, for the ignition spark to occur?

A. Exactly at the upper dead center just before the piston starts downward again. The compression is then at its maximum, and firing the charge exactly at that moment results in the production of the greatest amount of power. The compression falls off very rapidly the moment the piston starts down on the power stroke, so that in a high-speed motor, the loss of even a very minute fraction of a second causes a considerable loss of energy.

Q. What will happen if it occurs too soon?

A. This will depend largely upon the type of motor and upon how much in advance of the proper time it actually occurs. In a slow-speed type running well below its normal r.p.m. rate, as where a motor with a normal speed of 1000 r.p.m. is slowed down to 600 r.p.m., owing to climbing a hill, advancing the spark to its maximum will cause hammering or pounding in the cylinders, indicating that the explosion is taking place before the piston reaches upper dead center and, consequently, that most of its energy is being expended against the rising piston, instead of helping thrust it downward as it should. When running at its normal rate, the piston speed of such a motor would be sufficient for the piston to have reached upper dead center before the burning gases had time to expand, so that their entire output of energy would be expended in producing power. In high-speed multi-cylinder motors with their diminutive cylinders and light moving parts, the piston speed is so great that even the maximum advance in the ignition timing would have no effect, even as the lowest speed of which the motor is capable.

Q. What is the result when the spark occurs too late?

A. The piston has already started on its downward stroke; the point of maximum compression has been passed, so that the full benefit of the compression is lost, and with it, a large part of the energy that would otherwise have been utilized.

Q. Has running with a late spark any other effect on the motor?

A. Yes. The gas in the combustion chamber is ignited so late that it continues to burn after the exhaust valve is opened. As the heat units it contains are not utilized as power, they are retained longer in the cylinder in the form of heat; the water jackets are compelled to absorb 50 per cent more of the heat than they should: and the whole motor is said to "run hot" or to "overheat". The late combustion of the gases does not permit of sufficient time for the normal amount of heat to escape by way of the exhaust, so that there is still burning gas in the combustion chamber when the exhaust valve closes.

Q. What other term is used in referring to a late spark?

A. Retarded ignition, or a retarded spark, meaning one that occurs later in the firing part of the cycle, i.e., later with relation to the travel of the piston itself. An *early spark*, or advanced spark, is set to occur or rather *start* to take place well before the piston has reached upper dead center on the compression stroke. The time of ignition is always based on its relation to the position of the piston, though frequently referred to in terms of degrees on the periphery of the flywheel. This is done for greater convenience in setting the ignition, as the timing of the valves is always marked on the rim of the flywheel.

Q. Why is it necessary to have the ignition occur later at one time than at another?

A. To permit of starting the motor, especially by hand. The time of ignition is advanced to compensate for the extremely rapid travel of the piston, and is designed to cause the spark to take place just as the piston reaches upper dead center on the compression stroke. In high-speed motors this advance amounts to as much as 1 inch of the stroke. It will, accordingly, be evident that if an attempt be made to start the motor with the spark advanced, ignition will take place before the piston reaches upper dead center and the motor will kick back.

Q. How much allowance is usually made for retarding the time of ignition?

A. In the ordinary type of motor in which the speed does not exceed 1500 to 1800 r.p.m., it is customary to have the latest timing allowed coincide with the upper dead center position of the piston. In some cases, the spark may be retarded to take place after the piston has traveled a short distance down on the firing stroke, or about 5 degrees to 7 degrees on the rim of the flywheel. In the majority of cases, however, "extreme retard" is at the upper dead center of the piston.

Q. How is the time of ignition, or "spark", advanced and retarded?

A. The timer, in the case of a system using vibrator coils, and the distributor, in all systems using one vibratorless coil, consists of two parts. In the case of the timer, the four contacts for a four-cylinder motor are set into the inner periphery of a circular casing and a single revolving contact, mounted on the camshaft, passes over them consecutively, sending the current through the primary winding of each of the induction coils in rotation. These coils are connected to the spark plugs of the different cylinders in the proper firing order. While this outer casing of the timer is normally stationary, it may be moved part of a revolution with relation to the position of the revolving contact. For example, if the moving member in its revolution were within 20 degrees of touching the contact corresponding to coil and cylinder No. 1, and then the casing were revolved 20 degrees in the same direction as that in which the contact is moving, it will be evident that the meeting, or contact, of the two will take place that much later, and the occurrence of the spark in the cylinder will be correspondingly delayed. If, instead of being moved away from the revolving contact, the housing is revolved against the direction of rotation of the contact, the two will come together that much sooner and the spark will be advanced, or occur earlier. This housing of the timer is connected by means of linkage with the spark lever under the steering wheel.

In the case of the distributor, the principle is exactly the same, except that the current is sent in turn by each one of the contacts, through the same induction coil, instead of having a coil for each cylinder. The method of accomplishing this result is also varied: in many magnetos the distributor consists of a revolving block carrying a single contact, while a carbon brush for each cylinder bears against it. In most of the battery-system distributors, the arrangement is very similar to a timer, but the construction and insulation are naturally carried out in very much better fashion.

Q. What is meant by an advance of 30 degrees for the ignition?

A. That the difference between the point of extreme retard, usually upper dead center of the piston, and that of maximum advance or earliest ignition for high speed, represents 30 degrees on the flywheel. Just what this corresponds to in inches on the flywheel naturally depends entirely upon the diameter of the flywheel.

Q. What is meant by the magneto setting point?

A. This is a line inscribed on the flywheel in the same manner as the usual marks for the valve timing of the motor. There is a corresponding dead line, or pointer, on the crankcase, with which these lines on the flywheel are registered to check the valve timing and the setting of the magneto. When the magneto-setting lines on the flywheel is directly opposite the dead line, or pointer, on the crankcase, the piston of cylinder No. 1 should be at upper dead center exactly on the compression stroke, i.e., on the previous down stroke, it has drawn in a fresh charge, and when the piston again reaches upper dead center it has completed compressing this charge, so that the cylinder is then ready to fire. With the piston of this cylinder at the position in question, the contact points in the breaker box of the magneto should be just opening.

Q. How is the ignition timing advanced or retarded on a magneto?

A. The magneto breaker box may be moved part of a revolution, with relation to the contact points, in exactly the same manner as previously described for a timer or distributor. This alters the relation of the cam which opens the contact points to the latter so that their separation takes place earlier or later, in accordance with the direction the breaker box is rotated.

Q. Why is it that a motor may be cranked by hand with the spark lever in the fully advanced position when starting on the magneto, and why can this not be done safely with a battery?

A. There is less actual advance in the time of ignition, as supplied by the magneto, represented by the movement of the spark lever. That is, moving the spark-advance lever to its maximum does not cause so much advance in the actual timing of the ignition when connected to the magneto for starting, as it does when connected to the battery. The speed of the magneto itself is responsible for a considerable advance in firing time when the motor is running, whereas the battery ignition has a fixed speed of operation, regardless of whether the motor is being turned over by hand or is being run full speed. Moreover, as the usual magneto setting point is upper dead center, the spark does not actually occur until slightly later, when the motor is being cranked slowly, so that, while the spark lever may be in the advanced position, the actual occurrence of the spark is not perceptibly earlier than it would be with the lever fully retarded. This makes it possible to crank the motor with the lever fully advanced on the magneto, without any danger of a backfire. This is not true to the same extent with the Ford magneto as with other types, owing to the great number of pole pieces and armature coils that it carries. With a battery, however, the occurrence of the spark is just as rapid at low speeds as at high, so that attempting to start on the battery with the lever fully advanced will invariably result in a vicious back kick; the whole force of the explosion is exerted against the rising piston. As there is a greater amount of lag in the ordinary battery system than there is with the magneto the movement of the spark lever represents more actual advance, as this is necessary to provide for the time lost in the operation of the system. This, however, applies only to old battery systems, such as the dry cells used in connection with the magneto in a dual systems and not so much so to the so-called modern battery-ignition systems which are practically as rapid as the magneto.

Q. What is meant by fixed ignition?

A. The time of ignition is not variable by means of a spark-advance lever. Only a magneto is employed (true high-tension type) for ignition and it is set to fire at the maximum advance at all times when running. Provision is usually made for retarding the time of ignition to allow for hand starting, though this is not always the case, as it is not absolutely necessary (see answer to previous question). Otherwise, the ignition is always advanced to the maximum and there is no lever on the steering wheel for varying it. This type of ignition system has been very generally employed abroad and, more particularly in France, on commercial vehicles, such as cabs, but has never found any favor here. It has not been used in this country, except on a very few makes of cars, and then only for the models of one or two seasons, the latest being about 1912.

Q. What is meant by automatically advanced ignition, and how is this accomplished?

A. Instead of relying upon a manually operated spark lever under the steering wheel, a centrifugally operated device is incorporated with the timer or distributor. As this depends upon the speed of the motor for its operation, the ignition timing is always at the point of extreme retard when the motor is stopped, so that it may be cranked by hand without taking any precautions to retard the spark. The device is practically a small centrifugal governor the weights of which expand against the action of a spring under the influence of increasing speed. These weights are connected by a lever to the timer, or distributor housing, so that, as they expand with increasing speed, they move the housing in the proper direction to advance the time of ignition. The spring automatically returns the housing to the retarded position when the motor slows down or stops. The automatic-advance device is a feature of the Atwater Kent battery system, and is also employed on one of the types of Eisemann magnetos particularly designed for use on commercial vehicles. It relieves the driver of the necessity of giving any attention to this part of the control, as the time of ignition is always proportioned exactly to the speed.

Q. When all other factors are fully up to requirements, such as proper carburetion, valves recently ground, spark-plug points not too far apart, and property fitting pistons, but the motor fails to operate with the same "snap" and power it delivered when newer, what is likely to be the cause?

A. Wear in the linkage connecting the spark-advance lever with the timer, the breaker box of the magneto, or the distributor of a modern battery system, is responsible for so much lost motion that the full movement of the spark-advance lever is no longer being transmitted to the ignition apparatus. In other words, the full advance of the spark lever on the steering wheel no longer represents the maximum advance at the timer or distributor, so that the motor is not firing as early as it should and a considerable percentage of the energy, represented by exploding a fully compressed charge, is being wasted. The effect is the same as loss of compression from any other cause.

Q. When the motor will run slowly at one moment and then speed up to the racing point without any apparent cause, then drop off again to slow speed, what is likely to be the cause?

A. Unless this is directly traceable to a partially clogged carburetor jet or a loose throttle which is jogged open and closed by the vibration, the wear mentioned above may be responsible. The spark-advance linkage has become so loose through wear that it no longer holds the timer or distributor at the point to which the lever is moved on the quadrant on the steering wheel, and the vibration of the motor moves it back and forth, advancing it at one moment and retarding it the next.

Q. When the motor does not respond at all to the movement of the spark-advance lever, what is the trouble?

A. There is an actual break between the lever on the steering wheel and the timer, or distributor, or the breaker box of the magneto. This is apt to leave the ignition timing fully advanced, making it dangerous to attempt to start the motor by cranking it on the battery.

Q. Why do some motors require a greater amount of ignition advance than others?

A. Because of their greater speed or the increased amount of lag in the ignition system which must be compensated for by the timing.

Q. What system of ignition has the greatest amount of lag, and which the least?

A. The old battery system, using a roller-contact timer and vibration coils, has the maximum amount of lag, as there is both a mechanical and an electrical lag. The high-tension magneto and the modern battery system are great improvements in this respect, and there is not much choice between them on this point. As an illustration of the effect of the motor speed on the amount of advance required, the Packard "twin-six", using a modern battery system, requires as much advance as the old Packard "four", on which a low-tension magneto was employed; while the Packard six-cylinder motor, which was fitted with a high-tension magneto, required considerably less than either of these.

Q. Why is it necessary in some ignition systems to set the spark "late", or after the piston has actually started downward on the firing stroke, when the spark lever is at the point of maximum retard, while in others it is only necessary to set it at upper dead center?

A. This is due to the difference in the amount of lag, and also to whether a battery or a magneto is employed. It is necessary to insure safety in hand cranking, as, where the lag is reduced to the minimum, the explosion would take place before the piston had reached a point where it could start downward on the firing stroke, and a back kick would result.

Q. Is the piston actually at upper dead center for ignition-setting purposes when it has ceased to rise?

A. No. There is an interval in the revolution of the crank during which the piston does not move, i.e., while the crank is moving in practically a horizontal plane, and it should complete this part of the revolution, so that the piston is just on the point of starting downward again before it reaches the proper point for ignition setting at maximum retard.

Q. To what systems does fixed ignition occur?

A. This refers to those systems in which there is no provision made for altering the time of ignition, either by a manually operated lever on the steering wheel, or by an automatic device for advancing and retarding it. In other words, the time of ignition is fixed and is always the same, regardless of the speed of the motor. It can be used only in connection with a magneto.

Q. What is the ignition-setting point for "fixed" spark?

A. This represents a mean between what would be the points of maximum retard and maximum advance in a variable system. The spark must not occur too late as the motor will develop only a percentage of its power and will overheat; nor must it be too early, as the motor is then apt to fire against the rising pistons when slowed down, as in climbing a hill or when heavily loaded. The magneto must be set, therefore, so that the spark occurs before the piston reaches upper dead center. Just how much in advance of that point the setting should be, will depend upon the motor itself. As will be noted in the table of firing orders, a much greater range of timing is allowed for on some motors than on others, so that there is no rule which will apply to all. The ignition-setting point as given by the manufacturer should be learned and the magneto set in accordance.

Q. Is fixed spark ignition in common use on American cars?

A. Although very largely used abroad, particularly in France, on commercial vehicles such as taxicabs, it has never found favor here and will be found on very few cars.

Q. Is there any other method of checking the ignition-setting point besides the marks on the flywheel or the position of the piston of one cylinder in its relation to the contacts of the interrupter?

A. In some magnetos, such as the Eisemann, there is a setting mark on the distributor of the magneto. When the piston is in the proper position for firing, it is only necessary to bring this setting mark in line with a setting screw on the stationary part of the distributor.

Q. Why do the cylinders of an automobile engine not fire consecutively?

A. Because the pistons are attached to the crankshaft in pairs in the same plane, so that when one piston of a pair is firing the other one is going down on the suction stroke.

Q. How can the firing order of a motor be determined?

A. Take out all the plugs and lay them on the cylinders so that the threaded part of the plug makes contact with the cylinder but the terminal does not. Switch battery on and turn engine over slowly, noting the order in which the sparks occur at the plugs. Watch the valve stems; after the inlet valve of cylinder No. 1 has opened and closed, it is ready to fire. The plug at which the spark then occurs is the proper plug for the first cylinder. The next plug to spark belongs to the cylinder whose inlet valve has just closed.

Q. When a motor will not start, but fires once or twice and then stops, the flywheel rocking back and forth, what is the cause?

A. Some of the spark-plug leads have been misplaced, so that after one or two explosions, the next one takes place out of sequence.

Q. How many different firing orders are possible in a motor?

A. This depends upon the number of pairs of cylinders in a motor, as the two cylinders of a pair cannot fire consecutively, so that the ignition alternates from one pair to another. For example, in a four-cylinder motor, cylinders No. 1 and No. 4 constitute one pair, in that they are attached to crankpins in the same plane, i.e., they rise and fall together; cylinders No. 2 and No. 3 constitute the other pair. Consequently, starting with cylinder No. 1, there are only two firing orders possible in a four-cylinder motor, as follows: 1-2-4-3, or 1-3-4-2. By starting with cylinder No. 2, as the first to fire, two more orders may be used, as 2-1-3-4 or 2-4-3-1. It will be apparent, of course, that the last-named combination is merely the reverse of the second one given above, i.e., 1-3-4-2. In this way, the total number possible with a motor of this type is eight firing orders.

Q. Do motors differ much in this respect, or are firing orders pretty well standardized?

A. Of the eight possible firing orders that may be used in a four-cylinder motor, only two are in common use, viz: 1-2-4-3, and 1-3-4-2. This is because the selection of cylinder No. 1 as the first to fire renders it easier to set the ignition timing, as the piston of the first cylinder is usually more accessible than the others. There are some exceptions to this, the 2-1-4-3 firing order having been used by one or two makes of cars for several years; but the two mentioned above are in such general use as to be considered practically standard.

Q. How does the firing order of a six-cylinder motor differ from that of a four?

A. It is exactly the same in principle, as the six consists of three pairs of cylinders, the pistons of which are attached in couples to the crankshaft, 120 degrees apart; in other words, a three-throw crankshaft, instead of the *two-throw* of the four-cylinder motor, in which the cranks are 180 degrees apart. *Pairs of cylinders*, in this connection, has no reference to the manner in which they are cast, but refers simply to their relation to the crankshaft, which, in turn, affects their firing order. The ignition must accordingly alternate from one pair to another in exactly the same manner as in the four, but a greater number of combinations is possible as there are three pairs of cylinders.

Q. What are some typical firing orders for six-cylinder motors?

A. As in the case of the four-cylinder motor, the use of cylinder No. 1 as the starting point has been practically standardized, hence the firing orders in general use begin with it, as: 1-5-3-6-2-4, or 1-4-2-6-3-5.

Q. How does the firing order of an eight-cylinder motor differ from that of a four?

A. In view of the fact that the usual V-type eight-cylinder motor is nothing more nor less than two four-cylinder motors working on the same crankshaft, or, as it may be better put, with a common crankshaft, the firing order is simply that of two four-cylinder motors in which the ignition alternates from one motor to the other. That is, in addition to alternating from one pair of cylinders to the other, as already described, the firing also must alternate from one group of cylinders to the other, so as to maintain the impulse balance of the motor. The explosion in a right-hand cylinder must always be followed by the firing of a left-hand cylinder.

Q. What are the usual firing orders of eight-cylinder motors?

A. The same as those for fours, i.e., 1-2-4-3 and 1-3-4-2,

alternating from one group of four cylinders to the other, as, for example, L1-R3-L2-R1-L4-R2-L3-R4, or R1-L4-R3-L2-R4-L1-R2-L3. The number of combinations is increased as the firing order may start with the first right or the first left cylinder.

Q. How does the firing order of the twelve-cylinder motor differ from that of the six?

A. It bears the same relation to the six that the eight does to the four. In other words, the twelve-cylinder motor is practically two six-cylinder motors, or two groups of six pistons each, working on a common crankshaft.

Q. Give the firing orders of two of the twelve-cylinder motors now in use?

A. The Packard twin-six firing order is R1-L6-R4-L3-R2-L5-R6-L1-R3-L4-R5-L2. That of the National twelve is R1-L6-R5-L2-R3-L4-R6-L1-R2-L5-R4-L3.

Q. What is the effect of a cylinder firing out of order?

A. The spark may either occur when there is no gas in the combustion chamber, in which case that cylinder will miss when it should fire, or it may take place in a cylinder that will be fired so as to act *against* the other cylinders instead of *with* them. In either case, irregular running will result. Where there are either eight or twelve cylinders, the fact that one is either missing or firing against the others will not stop the operation of the motor; the misfiring of a single cylinder in a twelve is scarcely perceptible, except at low speeds. But in a four or six, the misplacing of a pair of spark-plug leads will prevent the running of the motor altogether.

Q. Is it possible to have one cylinder fire out of order?

A. It will be evident that, as the cylinders in all automobile motors fire in alternate pairs, the misplacing of one spark-plug lead naturally involves the dislocation of its mate or its alternate. Consequently, one cylinder cannot fire out of order alone; two will always be affected. For example, in a four-cylinder motor, if the secondary cable to the plug of cylinder No. 1 is connected to the spark plug of cylinder No. 2, it is apparent that No. 2 also must be misplaced. Granted that there is no other fault in this respect, it will be found connected to the plug of No. 1.

Q. Give a typical example of misplaced spark-plug leads causing improper firing order and its consequences?

A. Take the instance cited above. Cable No. 1 has been connected to cylinder No. 2; cable No. 2 has been connected to cylinder No. 1. The firing order is 1-2-4-3, which means that when the piston of cylinder No. 1 is going down on the power stroke, the piston of cylinder No. 2 is drawing in a fresh charge of fuel mixture through the open inlet valve. But, as cable No. 2 is connected to cylinder No. 1, no spark takes place when the piston finishes its upward travel on the compression stroke, and no explosion results. The spark instead of occurring in cylinder No. 1, takes place in cylinder No. 2, and may ignite the incoming gas, resulting in a weak explosion or a back-fire through the inlet valve. If the placing of the inlet valve be such that the incoming charge is not fired by the spark of cylinder No. 1, but takes place in cylinder No. 2, on the suction stroke, nothing will occur in cylinder No. 2, either. When its piston comes up on the compression stroke ready for firing, the spark occurs in cylinder No. 1, and the fresh charge passes out of the exhaust valve of cylinder No. 2 without being fired. In other words, both cylinders No. 1 and No. 2 miss. Cylinders Nos. 3 and 4 being connected up in the proper order, however, they will fire as they should, but in a four-cylinder motor they are not sufficient to keep the motor turning over steadily. It will give two jerky explosions and stop. After cranking several times, cylinder No. 2 will become more or less filled with fresh gas, and a back-fire will result at every other revolution, or every time it is the turn of cylinder No. 2 to fire. Cylinder No. 1 is not likely to back-fire, since the spark is occurring in its combustion chamber on the exhaust stroke.

Q. What would be the effect if, instead of connecting to cylinder No. 1, the lead of cylinder No. 2 misconnected to No. 3?

A. As the firing order of the motor is 1-2-4-3, it will be evident (the other cables being correctly connected) that cylinder No. 1 will fire properly; No. 2 will miss; No. 4 will fire properly; and No. 3 will either miss or back-fire; so that the motor as a whole will run very jerkily; although it will continue to run on this combination, i.e., one operating cylinder in each alternate pair firing a revolution apart. In this case, when the piston of cylinder No. 2 has just finished compressing its charge and should fire it, its spark takes place in the combustion chamber of cylinder No. 3, which is then exhausting, so that it is also much more likely to miss than to back-fire. The effect will be practically the same as if no spark at all took place in either of these two cylinders, and the motor will run on one pair alone.

Q. Is it any more difficult to locate trouble due to the cables being connected in the wrong firing order, in motors having six, eight, and twelve cylinders?

A. The greater number of cylinders would naturally add to the confusion, and the fact that it is not easy to tell off-hand when one cylinder in an eight or twelve is missing, contributes to the difficulty of locating the one at fault. On the other hand, however, the spark-plug leads are so designed that it is much more difficult to connect them up in any but the right way, as they are cut to exactly the proper length and are usually numbered in addition. The contacts of the distributor are also identified in the same manner, so that the connections may be readily checked without making tests of any kind; it is only necessary to trace the cables from the distributor to the plugs. The fact that the motor will continue to run on a pair of cylinders, even though its leads should be misplaced, is misleading, as the missing, or faulty running, is likely to be ascribed to a poor spark plug or something of that nature rather than to the real cause.

Q. Can the firing order of the cylinders be disarranged in any other way than by the misplacing of the secondary cables connecting the distributor to the spark plugs?

A. No. Since the timer, in the case of a timer- and four-vibrator-coil ignition system, as on the Ford, or the distributor, as on a magneto or battery-ignition system, makes its contacts consecutively. That is, on a four-cylinder timer or distributor, the contacts occur in the order 1-2-3-4 and, unless the leads are properly connected up, the sparks will occur in the cylinders in the same order. To obtain the firing order mentioned in a previous query for a four-cylinder motor, i.e., 1-2-4-3, contact No. 1 of the distributor is connected to cylinder No. 1, contact No. 2 to cylinder No. 2, contact No. 3 to cylinder No. 4, and contact No. 4 to cylinder No. 3. The connections would naturally be the same in the case of a timer, as there is then an independent induction coil for each cylinder of the motor.

Q. Has grounding or short-circuiting of the secondary leads any effect on the firing order of the motor?

A. This may be the case where two secondary cables touch

each other and there is a short-circuit at the point of contact. For example, assume that the cables of cylinders No. 1 and No. 2 cross each other in running from the distributor to their respective spark plugs (this is poor practice in wiring, but it is nothing unusual to see it on old cars) and that there is a leak in their insulation at this point. After being in service for some time, spark-plug points burn away unevenly, so that the gap at one is less than at the other. Consequently, there will be less resistance at this plug than at the one with the wider gap; in short, the combined resistance of the leakage gap where the cables cross and that of the spark plug with the smaller opening may be less than that of the plug which has burned further apart. Then the current intended to produce a spark at this plug will leak through the insulation and produce a spark at the other plug instead, and the latter will fire out of order while the first plug will miss.

Q. Does a change in the firing order of a motor have any effect on its running? For example, assume a four-cylinder motor designed to run with the firing order of 1-2-4-3. Will such a motor run any better if the firing order be shifted to 1-3-4-2, or to 2-1-3-4? Can this be done easily?

A. So far as the actual operation of the motor is concerned, such a change would have no particular effect, as a cylinder of each alternate pair would fire consecutively. The change can be made without any difficulty, on the average motor, by providing new secondary cables, as the latter are usually cut to about the right length to reach from the distributor to the spark plugs in the firing order for which the motor is originally designed. Then connect distributor contact No. 1 to cylinder No. 2, contact No. 2 to cylinder No. 1, contact No. 3 to cylinder No. 3, and contact No. 4 to cylinder No. 4. Of course, to make the valves function to correspond to this new firing order, the camshaft must be readjusted.

Q. Why do different manufacturers adopt different firing orders, if there is no particular benefit to be derived from one as compared with another?

A. Chiefly to adapt the wiring more conveniently to the location of other essentials, such as the magneto or the distributor of a battery system, although the intake manifold design may influence the choice.

SPARK PLUGS

Q. What are the usual causes of failure of the spark plug?

A. An accumulation of carbon on the inner end of the porcelain and the shell, causing a short-circuit; broken porcelain; points burned too far apart to permit spark to pass.

Q. When there is a hissing noise at the plug, or when oil squirted on it bubbles violently with the motor running, what does it indicate?

A. Either that the porcelain of the plug is not screwed down tightly on its gasket on the shell, the porcelain is broken, or the plug itself is not tight in the cylinder.

Q. What is the cause of a discharge across the porcelain of the plug?

A. The points are too far apart, or the porcelain is broken; usually the former. This usually will be noted only on plugs having very short porcelains as where the latter are long, the distance is much greater than the maximum to which the points can be separated and any spark that would occur owing to the latter cause would take place at the safety gap.

Q. Why is it that when a good spark will occur between the points of a plug in the open air, an equally good spark cannot be obtained in the cylinder with the same distance between the points of the plug?

A. Compressing air increases its resistance to the electric current, so that a stronger current is necessary to produce the same spark under compression in the cylinder, as may be obtained in the open air with less current. This is one of the symptoms of a weak starting battery (dry cells). All the plugs will spark satisfactorily when removed from the cylinders, but it is found very difficult to start the engine by cranking. When fresh cells cannot be had, it may be overcome temporarily by adjusting the points of the plugs closer together, in this way obtaining a satisfactory spark with less current.

Q. What is the proper distance between the points of the spark plugs?

A. For good working with a magneto, this should not exceed $\frac{1}{32}$ inch. With storage-battery ignition, it may be greater, but it is good practice to employ the smaller gap with either.

Q. How rapidly do the electrodes of the spark plugs burn away, and what effect does this have?

A. This will depend upon the plugs themselves and the character of the current supply. The cheaper plugs with ordinary iron-wire electrodes will burn away very rapidly, becoming much too widely separated in a few weeks' use. With better grade plugs, using a hard alloy for the electrodes, the time will depend more or less on the magneto, or where a storage battery is employed, on the coil. Some produce a very much hotter spark than others and, consequently, burn the plugs away that much sooner; but a good plug ought not to need adjustment under two or three months of ordinary running. The fact that the plug points have burned too far apart will be evidenced by a very perceptible falling off in the power; by missing spasmodically in different cylinders at low speeds; and by an occasional discharge across the porcelain on the outside of the plug, if the latter is of the short type.

Q. What can be done when there is an escape of compression around the porcelain of the plug?

A. If the porcelain is not broken, this can be remedied by turning down the gland or packing nut which holds the porcelain in the metal shell. There is an asbestos washer under this packing nut, and tightening the latter causes it to fill completely any space between the porcelain and the shell. The nut should be given only a fraction of a turn, and the tightening should be done when the plug is cold; if it is tightened when very hot, the contraction due to the cold is liable to crack the porcelain.

Q. What is one of the commonest ways of breaking spark-plug porcelains?

A. The use of a big wrench in putting the plugs into the cylinders. Only a spark-plug wrench or a small wrench having but a 3-inch or a 4-inch handle should be employed. The leverage available with the big wrench is so great that the porcelain is crushed, frequently without the knowledge of the man using it.

Q. How can a broken porcelain be detected?

A. Usually by grasping the plug between the fingers and trying to turn the porcelain sideways or to revolve it. Any play is generally an indication that the porcelain is broken, though sometimes they will loosen up so much under the influence of vibration that they may be turned easily with the fingers. Tightening the packing nut will overcome this. Where this treatment does not locate a break, squirting a little oil on the porcelain while the motor is running will do so.

Q. How long should a good plug stay in service?

A. This is hard to average, but with occasional adjustment of the points, a good plug will frequently continüe to give satisfactory service for several thousand miles; some have been known to run for well over 10,000 miles, and still continue to operate satisfactorily. When the motor does not run properly and the plugs are suspected, it is better to try the effect of a new set and note the running of the motor carefully, before discarding the old ones, as the fault will frequently be found elsewhere. The first thing that many repair men do with a motor that is a bit off is to throw away a perfectly good set of spark plugs.

Q. What is a "series"-type plug, and what advantages has it?

A. This is a type of plug fitted with two insulated terminals, instead of one, as in the ordinary type, as it does not make a ground connection on the motor by being screwed into the cylinder. It is intended to be used in series with an ordinary plug, hence the name. The current passes through the series plug first, and then, through the second plug of the ordinary type, to the ground, thus producing two sparks instead of one. The advantages claimed for its use are a more rapid and thorough combustion of the mixture, due to the spark occuring at two widely separated places in the combustion chamber; but, in actual practice, the gain is not sufficient to warrant the extra complication, so that this type of plug is seldom used.

Q. What is the object of fitting a plug with several sparking terminals, or electrodes?

A. The current will always follow the path of least resistance and will accordingly bridge the smallest gap. The sparking will start at this gap and, when that particular electrode burns away, will shift to one of the others until that has also burned too far open, i.e., until its resistance becomes greater than that of the next one, and so on. Such a plug should stay in service longer without the necessity of adjusting the gap, but, apart from this, it has no particular advantage, as the short-circuiting of one of the gaps renders all of them useless.

Q. What threads are generally used on spark plugs, and which thread is to be preferred, from the repair man's point of view?

A. Half-inch iron pipe, metric and S.A.E. standard. Plugs with iron-pipe threads are employed only on the cheaper cars; metric-threaded plugs will be found on foreign cars only, and S.A.E. standard plugs ($\frac{7}{8}$ inch-18) on most, if not all, of the better American cars dating from about 1913 on. All work equally well so far as holding compression is concerned. The iron-pipe plug is likely to cause trouble in this way sooner than the others, as they do not depend upon the thread itself to prevent leakage. The S.A.E. standard plug is preferable, being a better mechanical product than the iron-pipe plug as the threads of the latter are not so accurate, and the plug itself is usually not so well made.

Q. When a spark plug leaks at the base, i.e., at the point where it is screwed into the cylinder, what is the cause?

A. If of the iron-pipe type which depends entirely upon its tapered thread to hold the compression, a section of the thread may have been damaged in handling. With either of the other types, the plug may not be seated snugly enough on its gasket, in which case a quarter-turn in will usually remedy the leak. If it does not, the gasket should be replaced with a new one.

Q. Is there any danger of turning an iron-pipe plug in too tightly?

A. If screwed home too tightly while the motor is hot, the plug may bind and become very difficult to remove, sometimes necessitating drilling it out and re-tapping the hole.

Q. Is the priming type of plug of any particular advantage?

A. On low-priced cars not fitted with pet cocks in the cylinders, it is an aid to starting in cold weather; though, as a matter of fact, the priming inlet of the plug is likely to become clogged with soot, so that it cannot be used for injecting gasoline into the cylinders.

Q. Why has the magnetic type of spark plug not come into general use?

A. It is an expensive type of plug to make and is more subject to derangement than the ordinary kind. It is only intended to be used with a low-tension magneto and plain-spark coil (single winding on an iron core), and this type of magneto has long since become obsolete on the automobile.

Q. Is porcelain or mica preferable, as the insulator of the plug?

A. While mica is practically unbreakable, it is liable to become saturated with oil and dirt, causing leaks that are hard to trace; so that the use of porcelain is preferable, despite its liability to breakage.

Q. Is changing from one make of plug to another likely to cause any difference in the running of a motor?

A. As the power of the motor is dependent, to a very large extent, upon the proper ignition of the charge, a change from one make of plug to another will sometimes make a very marked difference. Because of the great amount of variation in the width of the cooling jacket and the thickness of the cylinder walls, certain types of plugs are particularly adapted to certain makes of motors. In fact, the plugs have been designed especially for service on those motors. For example, some motors require an unusually long plug to permit of the sparking points extending well into the combustion chamber. In others, a plug of this type might interfere with the piston when at upper dead center. Always use the type of plug recommended by the manufacturer of the car.

Q. What is the cause of the spark plugs in a motor becoming fouled very rapidly?

A. An excess of lubricating oil is finding its way into the combustion chamber, and its burning there deposits a heavy layer of carbon on the ends of the plugs. This may be due to leaky piston rings; the use of improper oil for the motor, as where a very light oil, instead of a heavy bodied lubricant is used in a motor intended for the latter; or, in old cars, the lack of baffle plates between the crankcase and cylinders. Running the motor with an over-rich fuel mixture will also cause sooting of the plugs. The use of a heavier oil and with it a certain proportion of flake graphite, or "Gredag", will often greatly improve the compression of an old motor and effectively remedy this trouble.

Q. Is it advisable to use lubricating oil mixed with graphite in all old cars?

A. It is usually good practice, except on the Ford, as the magneto of the latter runs in oil splashed back from the crankcase, and the presence of graphite in this oil would short-circuit the magneto. In case this has been done on a Ford, it would be necessary to wash the magneto out very thoroughly with gasoline to remove all traces of the graphite, and even that might not remedy the short-circuiting.

Q. When the plug in only one cylinder of a motor continually soots up very rapidly, what is the cause?

A. That particular cylinder is being flooded with oil, and an excess amount of it is reaching the combustion chamber. A piston ring may have broken in that cylinder, the rings may have worked around until their openings are in line with one another, or the supply system may have become deranged, causing an excess of oil to reach the section of the crankcase directly under that cylinder.

Q. Is there any way of knowing to a certainty before undertaking an inspection, whether missing is due to a faulty plug or to some derangement of the carburetor?

A. If the miss continually occurs in the same cylinder, it may be put down as due to the plug or wiring of that particular cylinder, even though the plug, when tested from the outside (i.e., without removing it from the cylinder), is apparently working properly. When the missing is spasmodic, occurring first at one cylinder and then at another, it is more likely to be due to faulty carburetor adjustment or a partially clogged carburetor nozzle. A weak dry battery, however, will produce similar symptoms; failing fuel supply also will, though, in this case, the whole motor will run jerkily one moment, speed up the next, and then almost stop, only to repeat the performance.

Q. Is the usual test, made by unclipping the secondary cable from the plug and holding it near the terminal with the motor running, always conclusive as to the proper working of the plug, when a spark occurs between the two?

A. No. Since a short-circuited plug will not prevent a spark from passing between its outer terminal and the end of the cable held near it. The gap made by disconnecting the secondary cable and holding it a short distance away has simply taken the place of the gap which should exist between the plug points. This test is conclusive only as to the proper functioning of the distributor and the integrity of the wiring connecting that particular plug.

Q. When the motor misses spasmodically, and there are intermittent discharges of high-tension current at different points on the cylinders, while it is running, what is the cause?

A. Moisture. The secondary cables have become wet enough to cause leakage of the high-tension current; usually the result of the careless use of the hose in washing. The only remedy is to run the motor continuously until the heat dries things out thoroughly.

Q. When no spark plug is obtainable at any of the plugs, so that the motor cannot be started, in what order should the cause of the trouble be run down?

A. If a dry battery is used as the source of current, first see that it is not exhausted by testing with a small ammeter. Each cell should show 10 amperes or more. If one is considerably below the others, it reduces them to its level. Where a storage battery is used, switch on the lights as a test. They should burn brightly. Provided there is no failure of the current, inspect the interrupter, or contact breaker; see that its points separate when the high part of the cam strikes the arm carrying the movable point. If they do not separate, no current is induced in the secondary winding, and an adjustment of the points, to open about the space represented by an ordinary visiting card, will remedy the trouble. If the points open properly, inspect the wiring and connections between the interrupter and the battery. Should both the interrupter and its wiring be O. K., note whether the ground connection from the coil is fast. Failing all of these, inspect the high-tension distributor. This is about the order in which the trouble would be likely to occur.

Q. When the motor continues to run after the current has been turned off, what is the cause?

A. The cylinders have become so hot, due either to lack of oil or or low water in the cooling system, failure of circulation, etc., that either the spark-plug electrodes have become red hot, or particles of carbon, deposited in the combustion chamber, have become incandescent, thus firing the charge. The fuel supply must be shut off and the motor allowed to cool.

Q. Should the motor back=fire when attempting to start, even though the spark=advance lever has been pulled back as far as it will go, what is causing the trouble?

A. The linkage connecting the spark-advance lever with the breaker box of the magneto or of the interrupter (battery system) has parted at some point, so that it no longer moves the breaker box to retard the time of ignition.

REGULATION DEVICES

Interrupters and Timers

Q. How does an interrupter operate?

A. It is normally closed, short-circuiting the battery on the primary winding of the coil until just before it is necessary for the spark to occur in the cylinder; a cam then separates the contact points, and the high-tension current induced in the secondary winding of the coil jumps the gap of the plug. In the case of a magneto, the winding of the armature is short-circuited on itself to permit it to "build up", so that when the interrupter contacts are opened by the cam, the peak or highest value of the current wave generated is utilized. The opening of the circuit in either case occurs at the same time the distributor arm is passing one of its contacts.

Q. How does a timer operate?

A. Contacts, insulated from each other, their number corresponding to the number of cylinders, are located at equidistant points on the inner circumference of the timer housing, while the shaft carries a single contact which in its revolution successively touches each one of the stationary contacts. Where separate vibrator coils are used, as on the Ford, each stationary contact corresponds to one of the coils.

Q. Do interrupters and timers fail from the same causes, and what are they?

A. No. The cause of failure in one case is the reverse of that in the other. An interrupter fails when it does not open the circuit, and a timer when it does not close it. Dirt and wear are the usual causes of failure in both cases; moisture is also responsible at times. Test by having an assistant turn the engine over slowly by hand and watch the operation of the interrupter; if the cam fails to separate the contact points, true up their faces with fine sandpaper and test again. (This does not apply to Atwater Kent interrupters. See description.) Stop with the cam in the opening position and see if a sheet of ordinary paper can be slipped between the contacts. See that there is not an excess of oil in the housing, as oil on the contact point insulates them. In the case of a timer, see that the spring of the movable contact has sufficient tension to keep it pressed firmly against the stationary contacts as it revolves; note whether sufficient wear has occurred to cause poor contact even with sufficient spring pressure.

Q. How far should the contacts of an interrupter separate?

A. This differs somewhat with different systems, but in the case of the interrupters used on battery systems, it is very small, seldom exceeding a few thousandths of an inch. In the case of the Atwater Kent interrupter, this is .010 to .012 inch. (Coated catalogue paper is .005 to .007 inch thick; a thin visiting card is .010 to .015 inch thick.)

Q. How does the Atwater Kent interrupter differ from other battery interrupters?

A. The circuit is normally open and only remains closed momentarily before being opened by the dropping of the lifter into its notch on the shaft.

Q. Can this interrupter be tested in the same way as that just described?

A. No. The movement of the lifter in striking the latch to close the circuit is so rapid that it cannot be detected with the unaided eye, even though the engine be turned over very slowly by hand.

Q. What will cause this interrupter to fail?

A. Wear of the lifter to an extent where it will not engage the notches of the shaft properly, usually caused by lack of oil. Other causes of failure are the same as for other types, generally worn or burned contact points.

Q. When the contact points of an interrupter of any type burn away very rapidly, what is the cause?

A. The condenser has broken down so that it is no longer protecting the points from the full heating effect of the arc formed at the time of breaking the circuit. Use the testing-lamp outfit described in connection with lighting and starting systems. Apply one point to each of the condenser terminals; if the lamp lights, the condenser is short-circuited. The only practical remedy is replacement by the manufacturer, as even the best equipped garage is seldom fitted to take care of such work.

Q. Does discoloration always indicate burned contact points, and how often should these points require cleaning and adjustment?

A. No. According to the particular alloy used in the contacts they will assume a bright purple, an orange, or a gray tinge. The squareness of their surfaces and the contact they make when together are the best indications of whether attention is needed; if pitted or high on one side, truing up will be necessary. Unless the condenser has failed, they should not require attention oftener than once a season, or say 6000 to 8000 thousand miles' running.

Distributors

Q. What is the function of the distributor and how does it differ from that of the interrupter and timer?

A. At the same moment that the interrupter opens the primary circuit of the coil, or the timer makes it in the case of a vibrating coil, the distributor makes contact with a stationary segment representing a spark-plug terminal. The distributor accordingly is said to run synchronously with the interrupter or timer. It is practically a duplicate of the timer designed to handle a high-tension current, in that it has one revolving contact and a stationary contact for each cylinder.

Q. Does the moving member of a distributor actually make contact with the stationary contacts, as in the timer?

A. No. This is not necessary owing to the high voltage of the current. The moving member passes very close to the face of the stationary contact but does not actually touch it, thus avoiding wear. This applies, however, only to those early-type magnetos or to separate distributors employing a metal moving contact. Where carbon brushes are employed, they are pressed against a fiber disc with a metal segment countersunk flush with its face and this segment passes under each carbon brush in rotation.

Q. What are the usual causes of failure in a distributor?

A. Short-circuits, due to moisture, dirt, or carbon dust. Owing to the high voltage of the current it will leak across barely perceptible paths caused by dampness or carbon dust.

Q. What is the so-called ignition unit of the modern battery system?

A. This is a combined contact breaker and distributor similar to the contact breaker and distributor of a magneto—in other words, a magneto minus the current-generating end. It is mounted on a vertical shaft and is driven through bevel, or helical gearing, from either the camshaft or one of the auxiliary shafts of the engine (i.e., water pump or magneto-drive shaft). The contact breaker is placed directly below the distributor, the secondary cables coming out of the upper face of the latter. (See description of Westinghouse and Connecticut units.)

Q. What is the "unisparker"?

A. This is the Atwater Kent ignition unit and is similar in general design to those referred to above but it is an "open-circuit" type, while they are "closed-circuit". The term is a trade name derived from the fact that the contact breaker makes but a single spark, as compared with the vibrator coil which produces a series of sparks only one of which, however, is available for ignition. All contact breakers on magnetos and, as now used, on modern battery systems, produce but a single spark. The time of ignition is so limited on a high-speed engine, that, if this single spark fails to ignite the charge in the cylinder, subsequent sparks are of little value, as the piston is already well down on the firing stroke by the time a later spark occurs, and much of the force of the explosion is lost.

Q. How is the time of ignition advanced and retarded on the ignition unit?

A. Usually by altering the relation of the moving contact of the distributor to the stationary contacts. The distributor plate, i.e., the insulating disc carrying the stationary contacts is connected to the spark-advance lever on the steering column, and it may be moved part of a revolution backward or forward to advance or retard the time of ignition. To alter the timing, the position of the moving contact on its driving shaft may be shifted. For example, in the Delco distributor a central screw in this member is loosened, and the contact may then be moved in either direction with relation to the shaft.

Switches

Q. What is a "reversing" switch and why is it employed on ignition systems?

A. It is a double-contact switch which reverses the polarity of the current, i.e., its direction, through the contacts of the interrupter every time the switch is closed. This is done to prevent the burning away of the contact points in one direction which would cause a peak to form on the positive and a crater, or depression, on the negative. Reversing the direction of the current causes the points to become alternately negative and positive in accordance with the position of the switch.

Q. What is the nature of the trouble ordinarily to be looked for in a switch?

A. Poor contact due to wear or weakening of the spring broken or frayed connections causing a ground or short-circuit.

Q. What is an automatic switch?

A. This term is frequently applied to the battery cut-out of the lighting and starting system. On the Connecticut ignition system, it is a thermally operated switch, designed to open the circuit when the switch has inadvertently been left on after the engine has stopped.

Q. Are there any troubles peculiar to automatic type of switch?

A. None that is not equally so of any similar device such as the battery cut-out or the circuit breaker.

Coils

Q. How many different types of coils are employed in connection with ignition systems?

A. Three. The first and simplest of these is usually termed a *spark coil,* and consists of a single winding of coarse wire on a heavy iron wire core. It acts by self-induction, the circuit remaining closed long enough to permit the core to become thoroughly magnetized; the energy thus stored in the core being released when the circuit is broken again. This gives increased voltage at the spark plug and causes an "arc" rather than a spark at the latter when the terminals of the plug are separated. This type of coil is only employed in connection with low-tension or mechanically operated spark plugs, and this system is now used on stationary engines and motor boats exclusively, having long since become obsolete on the auotomobile.

The other two types are known as induction coils, and differ merely in one being fitted with a vibrator while the other does not require this attachment to operate it. The induction coil is a miniature step-up transformer. It consists of a core of iron wires on which the primary of two or three layers of No. 16 or No. 18 wire is wound almost the full length of the core, and a secondary of many thousand turns of very fine wire, such as No. 36 or 38 B. & S. gage, or even No. 40, which is almost as fine as a hair. In coils of the best construction, this fine wire is wound in pancakes, or narrow sections, several of which are necessary to complete the secondary. Their terminals are connected in series thus making practically a single winding. Heavy insulation is placed between the primary and secondary windings, and the containing case is usually filled with an insulating compound, melted into it and becoming solid when cold.

Q. Why is an induction coil termed a "step-up" transformer?

A. It literally steps up or raises the voltage of the current sent through it. The primary winding is connected to the source of current and the secondary to the spark plugs through the distributor.

Q. Is the action of the coil based on the same principle as that utilized in generators and motors?

A. The principle of induction, as explained in Part I, is the same in all three, though it is utilized in a different manner in the induction coil. Instead of a moving coil of wire cutting the lines of force of a magnet, impulses are produced either by sending a pulsating current through the primary winding or by using an alternating current.

Q. How is this pulsating current produced in the primary?

A. By placing a vibrator in series with the primary winding.

Q. Of what does the vibrator consist, and how does it act?

A. It consists of a spring-hinged armature and a pair of adjustable contact points, exactly as in an ordinary electric bell or buzzer. This armature is located directly over the end of the core of the coil and close to it. When a current passes through the primary winding, it makes the core strongly magnetic and attracts the armature. This pulls the latter away from the stationary contact point and breaks the circuit, so that the core is no longer magnetic. The spring immediately pulls the armature back and recloses the circuit, this action taking place at high speed as long as the current is on.

Q. Why is the vibrator theoretically not necessary when an alternating current is used?

A. The rise and fall of the current wave, from zero to maximum and back again in the reverse direction, produces the same effect of magnetizing and demagnetizing the core of the coil very rapidly.

Q. Is the vibrator coil as rapid in action as the induction coil used with alternating current?

A. No. Since there is a mechanical as well as an electric "lag", or delay. The inertia of the armature must be overcome before it can be pulled down by the core, and to do this on a vibrator adjusted to withstand road shocks, the core must become *saturated,* or strongly magnetic. The time necessary to overcome the inertia of the armature is the mechanical lag, while that required for the core to become saturated is the electrical lag. In combination they make the vibrator coil very much slower in action than the other type, and this is greatly accentuated by a stiff adjustment of the vibrator spring.

Q. Why is a vibrator necessary with one type of battery ignition and not with another, i.e., the so-called modern battery ignition?

A. Owing to the type of timer or interrupter, frequently erroneously termed the "commutator", employed on the two systems.

Q. What is the difference between the old-style timer and the modern interrupter?

A. In the former, a long contact is provided for each cylinder and the revolving contact member is in touch with this for quite an appreciable period of time, during all of which the vibrator of a coil of that type is in action. If the contact member of the timer were depended upon to make and break the circuit through the coil to obtain the spark in the cylinder, the stationary contacts in the housing would have to be very much shorter, and no provision for advancing and retarding the time of ignition would be available. Furthermore, the wiping contact of the ordinary style of timer is not adaptable to the extremely rapid make-and-break that is necessary for this purpose. The interrupter of the modern battery system is designed along practically the same lines as the contact breaker used in the primary circuit of a high-tension magneto or the only circuit used on a low-tension magneto. Its parts are made very small and light and with great accuracy, so that its inertia is reduced to the minimum and it will act with extreme rapidity. The gap is so small and the rapidity of action so great that the movement is often not visible to the unaided eye. It is practically a mechanical vibrator designed to give a single make-and-break at exactly the right moment as compared with the electrical vibrator, which must be started well in advance of the moment ignition is required, and which continues in action after the spark has occurred in the cylinder. Consequently, both the mechanical and the electrical lag, which make the vibrator coil comparatively slow in action, are reduced to a minimum, and the amount of current necessary is cut to a fraction of that required by the latter.

Q. How can the speed of operation of a vibrator be judged?

A. By the note it produces in action. A low note well down the scale denotes slow action; the higher the note, or buzzing, the more rapid the vibrator is acting.

Q. What effect on the ignition has the speed with which the vibrator operates?

A. A slow-moving vibrator increases the amount of lag and retards the ignition accordingly. This causes a corresponding reduction in the power of the engine, as the spark does not occur at the proper time to give the best efficiency.

Q. What is a master vibrator?

A. The vibrator type of coil on a multi-cylinder engine requires an individual coil for each cylinder, and it is often found difficult to adjust all of the vibrators so that they will act uniformly. If some are stiffer than others they will not act so rapidly, and the time of explosion in the cylinders they control will be delayed, causing uneven running of the engine. To overcome this, an extra coil with a specially made vibrator is connected in series with the timer and the other coils, so that its vibrators acts for each coil in turn, the vibrators of the other coils either being removed or screwed down hard so that they cannot act. This makes but one vibrator to adjust, instead of the four on a four-cylinder engine. As it controls all of the other coils, it is known as a master vibrator.

Q. Is the vibrator type of coil still in general use?

A. It has long since become obsolete on all cars except the Ford.

Q. As the Ford magneto produces an alternating current, why are vibrator coils necessary?

A. The Ford magneto has sixteen poles, and the armature which serves as the flywheel, carries sixteen coils, so that the number of alternations at the high speed at which the Ford motor runs, is very great. These alternations, or cycles, are so rapid that they overlap each other, as is evidenced by the steady burning of the incandescent headlights. The induction coil does not act quickly enough to be affected by the change of polarity so that a vibrator is necessary on each coil to produce a sufficiently hot spark for ignition.

Q. How can the four vibrators of the Ford coil be adjusted so as to operate uniformly?

A. The fact that they are not doing so, will be evidenced by the uneven running of the motor. Determine, by holding down the vibrators, one after the other, which cylinder or cylinders are lagging behind the others in firing. This will cut out the cylinders in turn; in fact, two may be held down at once, and the action of the remaining pair noted. When the cylinder, or cylinders, at fault have been determined, adjust one at a time by releasing the lock nut of the adjusting screw of the vibrator and turning it up or down, according to whether improvement in running is noted, or not. Usually only a small fraction of a turn one way or the other will be necessary. Turn the screw very slowly and very little at a time and, when the proper adjustment of the screw has been secured, lock in place again securely. Ordinarily, the proper adjustment may be secured simply by noting the running of the motor. When all the cylinders fire regularly and without any apparent lag, the adjustment is considered correct. To secure a finer adjustment, a small portable ammeter, reading by tenths to three or five amperes, may be used. Connect this in series with each one of the coils in turn, and note the reading at which the coil acts most rapidly. The other vibrators may then be adjusted to give the same reading. When dry batteries were relied upon for ignition, this test was employed to reduce the current consumption to the minimum but with the excess supply of current from the Ford magneto, this is not necessary, and the vibrators may be adjusted to the reading giving the most rapid action, regardless of the current consumption. This test may be employed also to check the operation of the magneto, as its current output may have fallen off to a point where it is no longer sufficient to operate the coils satisfactorily.

Q. Why is a vibrator not necessary on the alternating current generated by the ordinary type of magneto?

A. The latter has but two field poles and a single coil on a two-pole armature, so that its cycles are very much fewer in number, and there is definite drop in the current, from the maximum to an absolute zero, twice in every revolution. Assuming a speed of 1200 r.p.m., the ordinary magneto would be running 600 r.p.m., as it is driven by a half-time shaft of the motor. This is 10 revolutions per second times 2 cycles per revolution which gives an alternating current of but 20 cycles, or one which would cause an incandescent lamp to flicker very badly. The Ford magneto, on the other hand, is directly on the crankshaft. Consequently, it is turning 20 revolutions per second, and its coils produce 16 cycles per revolution, or 320 cycles per second, equivalent to 19,200 cycles per minute. For ordinary commercial lighting, only 60 cycles per second are necessary to produce a steady light. The drop to zero in the current curve of the ordinary magneto permits the core of the coil to become demagnetized, and it is then remagnetized by the subsequent rise to the maximum value in the other direction, so that no vibrator is necessary to accomplish this alternate magnetizing and demagnetizing of the core which is needed to produce the inductive effect in the coil or transformer.

Q. How many connections are there on a coil?

A. Three; one to the primary, from the battery or magneto; one from the secondary, to the distributor, in the case of a single coil, or to the spark plug in the case of a multiple coil; and one to the ground. The last named is referred to as a common-ground connection, as it grounds one side of both the primary and secondary windings of the coil.

Q. How are these connections made?

A. On the single non-vibrator coil, as used with an ordinary magneto, by means of wire cables from the magneto to the primary of the coil, from the secondary of the latter to the high-tension distributor. In the case of the Ford multiple-vibrator coils, each coil is an independent unit, having brass strap connections attached to the bottom of the coil-unit case. These straps are of spring brass and they bear against corresponding plates of brass in the bottom of the coil box fastened to the dash. Simply lifting the coil unit out of the box breaks the connection and automatically remakes it when the coil is replaced. Due to this type of connection, irregular firing of the Ford motor will frequently be found to result from the cover of the coil box not being snapped down tightly. This permits the coil units to jump around in the box owing to the jolting and vibration, and every time they are jolted up off their connections, a cylinder fails to fire, as the coil does not receive any current. Coils used in connection with modern battery systems are often grounded in the same manner as the magneto, i.e., by their attachment to a plate on the motor the ground wires of the coil being connected to this plate.

Q. When a coil fails to operate, how can it be tested for faults?

A. In the case of the single coil, used in connection with an ordinary magneto, disconnect it and test with the testing-lamp outfit described in connection with lighting and starting systems. Place one of the terminals of the testing set on the common-ground connection, then place the other in turn on the primary and the secondary leads. If the windings are intact, the lamp should light each time.

Should it fail to do so, the covers of the coil may be removed to note if a wire has broken just beneath it. This is most likely to be the case with the secondary, owing to the very fine wire used. If there is no break, either at this point or where the primary lead is connected to its winding, it will be necessary to return the coil to the manufacturer as there is an internal short-circuit, which cannot be repaired in the garage. The only difference between the method above outlined for a single coil and that of a unit-vibrator coil as used on the Ford, is to touch the test points to the brass strap connections representing the different windings.

Q. Why is but one coil used in connection with an ordinary magneto, while four are employed on the Ford?

A. Where a single coil is used, the secondary current is led to a distributor from which it is again led to the various spark plugs in the proper order of firing. No distributor is employed on the Ford, so that a vibrator coil is necessary for each cylinder. The connections from the coils to the plugs are made in the same order as they would be to a distributor.

Q. When a vibrator coil cannot be made to function properly by adjusting the contact screw, what should be done?

A. The contacts should be trued up with a very fine file, as failure to function will usually be caused by their having become badly burned away or pitted, thus making poor electrical contact. Where the above is the case and the contact points are square and true, it is only necessary to clean them by drawing a worn piece of fine sandpaper between them several times, first on one side and then on the other. See that none of the holding screws of the vibrator frame have become loosened, and that the lock nut of the movable contact holds the latter firmly in place when tightened up.

IGNITION BATTERIES

Q. What types of batteries or cells are used for ignition?

A. Dry cells and storage cells, or accumulators.

Q. What type of cell is the so-called dry cell?

A. It is a primary cell, i.e., one in which a current of electricity is generated by chemical reaction, and is technically known as an "open-circuit" battery.

Q. Of what does the dry cell consist, and how much current does it generate?

A. The elements are the zinc container and a carbon plate centrally placed in the container and insulated from it at the bottom. Around this carbon plate, which constitutes the negative element (the negative element in a primary battery carries the positive terminal and *vice versa*), is packed a depolarizing agent, usually dioxide of manganese. The active solution is sal ammoniac in water which is poured in after the cell is assembled and filled with the depolarizer, and then the cell is sealed at the top with pitch, so that it is dry in name only. No chemical action can take place without the presence of moisture.

A dry cell of the ignition type generates a current of 20 to 25 amperes (when new) at a potential of 1½ volts.

Q. Why is a depolarizing agent necessary?

A. The action of the cell generates hydrogen gas, which quickly covers the carbon plate in the form of globules, rendering it inactive. The cell is then said to be *polarized*, and the current generated drops off very rapidly. This may be illustrated by placing an ammeter across the terminals of a new cell. The ammeter reading will remain at 20 amperes for a short time and then will quickly drop until, at the end of five minutes, the instrument will show scarcely any reading at all. If the connection is broken and the cell allowed to stand for ten minutes, it will again show almost as high a reading as before; at the end of an hour or more of rest, it will give practically the same reading. The depolarizing agent has in the meantime absorbed the gas which prevented the action of the cell.

Q. Why is it termed an open-circuit cell?

A. Because it will only produce its normal output for very short periods and must be allowed to rest between each demand for current. Otherwise, it will quickly become polarized and, if tested in this condition, will apparently be dead. It cannot be used where a steady current is required but must normally stand on open circuit.

Q. How is the dry cell employed for ignition?

A. Four cells are connected in series to give current at 6 volts, and a battery of this type is ordinarily employed, either as an emergency stand-by or simply for starting purposes. Where an open-circuit type of interrupter is employed, such as the Atwater Kent, it may be used as the main source of ignition current; as this interrupter makes instantaneous contact only at the moment the spark is required in the cylinder, the battery otherwise being on open circuit.

Q. How can the life of such a battery be prolonged?

A. By connecting two or more sets of four cells each in series-multiple, i.e., each group of four is connected in series to give the required voltage, and the positive and negative terminals of each group are connected together. The amount of current then drawn from each cell is only one-half what it would be if a single set of cells were employed, or one-third what it would be where sets are in series-multiple, and so on.

Q. When a set of four dry cells will last a certain length of time, why is it that adding extra cells in series, for example, a six-cell battery, will not last longer?

A. The amount of current drawn from the cells when the circuit is closed depends upon the voltage of the entire series, and the greater the total voltage the larger the volume of current, in accordance with Ohm's law. Consequently, the six-cell battery will not last so long as the four on the same service.

Q. Why is it not good practice to connect an old set of four cells in series-multiple with a new four-cell battery, or groups of uneven numbers in the same manner, as for instance, three in one and four in another?

A. The new cells will have an output of twice that of the old ones, so that when the circuit is closed they will discharge through the latter until the amperage of all is equalized. Where uneven numbers are used in groups in series-multiple connection, the voltage of the larger will be superior to that of the smaller, and a similar action will take place on open circuit so that in a short time the maximum potential of the battery will be that of the weakest group.

Q. Why should one cell much lower than the others never be included in a dry-cell battery?

A. For the reason just given above, as well as the fact that an exhausted cell increases the resistance of the battery as a whole and decreases the current.

Q. Why will the dry-cell battery used on a dual ignition system not run the engine satisfactorily for any length of time when the magneto is not in proper working order?

A. Because the interrupter, or contact breaker, of the magneto is of the closed-circuit type, thus drawing current continuously, except when the points open to break the circuit and induce a high-tension current in the secondary of the coil. In a system of this type, the dry cells are intended only for starting purposes.

Q. Why can a modern battery system not be used with dry cells?

A. Because the interrupter is of the closed-circuit type, similar to that of the magneto, and the demand for current (usually about 3 amperes) is practically constant. It is not so much the amount of current required that affects the dry cells, as the fact that they are almost constantly on closed circuit, so that there is no opportunity for the depolarizing agent to work. This demand on the storage battery of the starting and lighting system (usually of 80- to 120-ampere-hours capacity) is negligible. At a 3-ampere discharge rate, the larger battery when fully charged and in good condition would be capable of giving practically forty hours of continuous service. Under the same conditions, new dry cells would not provide efficient ignition for more than an hour.

Q. Does the voltage, as well as the amperage, of the dry cell fall off on closed circuit?

A. The voltage is affected very slightly; a cell that is practically exhausted will show almost $1\frac{1}{2}$ volts so that a voltmeter test is no indication of the condition of the cell.

FORD IGNITION SYSTEM

While a great many of the causes of missing or breakdown, in the ignition system covered by the foregoing queries, apply to a great extent to all cars, such as loose wires, short-circuits and the like, the Ford system is distinctive. It is based on the same fundamental principles, of course, but it has many features not to be found on other cars so that there are causes of failure that could never be readily determined by experience gained on other makes of machines, though long handling of ignition apparatus would naturally be of great assistance. In working on the Ford ignition system it must be borne in mind that it is a combination of the old-time battery system with a modern generator as the source of current supply, so that the many defections due, in the earlier days to the dry battery, are now lacking.

Q. Of what does the Ford ignition system consist?

A. A multipolar alternating-current generator (magneto) built integral with the flywheel; a primary, or low-tension timer, in which a roller makes contact with the four stationary segments in the housing; four vibrator coils, one for each cylinder, plus the usual number of spark plugs and connections in the primary and secondary circuits.

Q. Is the Ford ignition system efficient and reliable, or is it advisable when much trouble is experienced with it to replace it with any of the numerous accessories and complete ignition systems that are claimed to be improvements?

A. While there are many ignition systems the parts of which are made with greater precision, and some in which the design and particularly the accessibility of the important essentials are superior, experience has proved the Ford system to be both efficient and reliable. With proper care, there should never be any necessity for replacing it with any system made by an accessory manufacturer, or for adding to it any one of the legion of devices advertised as improvements on it.

Timer

Q. What are some of the commoner causes of failure of the Ford system?

A. One of the most frequent is due to the timer and is caused by failure to lubricate it. Contrary to the usual practice, which is to prevent oil getting on the contacts of a timer as it tends to insulate the latter, the Ford timer requires plenty of oil, and should be lubricated every day. There is no fear of giving it too much oil as the excess will leak out of the housing; it will continue to operate satisfactorily even though flooded with oil, while the slightest lack of it will cause trouble.

Q. What is the nature of the trouble caused by the timer?

A. If not oiled at regular intervals, it will cause missing of various cylinders, and those that do fire will be late, as if the ignition were fully retarded, so that the motor develops very little power. An accumulation of gummed oil and dirt will produce a similar result. The timer housing should be taken off and the contacts cleaned; the roller contact also should be cleaned by squirting gasoline over it and wiping over well.

Q. Will lack of oil have any other result?

A. Besides the missing, usually most noticeable at higher speeds, failure to lubricate will result in very rapid wear of both the roller and its track (contacts in the stationary housing) so that its operation will soon become unsatisfactory, even though subsequently kept oiled.

Q. What are some other causes of faulty operation of the timer (generally referred to as a commutator)?

A. Weakening of the spring which holds the roller against its track will cause missing at low speeds, while a loosening of the spring which holds the timer housing in place is liable to cause erratic firing at all speeds. This spring is held by a single cap screw passing through the breather tube, which serves also as an oil filler for the crankcase. At its inner end it has a small boss which fits in a corresponding depression in the hub of the timer housing. This is the only thing that holds the latter in place. The loosening of this cap screw is liable to let the housing drop out of place sufficiently to prevent the roller making contact with all of the segments.

Q. Does the weather have any effect on the operation of the timer?

A. Unless precautions are taken to lubricate it properly, cold weather will make starting difficult. This is due to ordinary lubricating oil becoming congealed in the housing, thus preventing the roller from coming into good contact with the segments. An indication of this sometimes is that the motor will fire only on two or three cylinders for several minutes after being started and will thereafter fire regularly, the oil then having become liquefied again.

Q. How can the timer be removed?

A. Take out cotter pin from end of rod which attaches it to the spark-advance lever on the steering column, and detach this rod. Loosen cap screw passing through breather pipe on top of the timing gear cover. This releases the spring which holds the timer housing in place, and the latter can be easily removed. To remove roller contact, unscrew lock nut, withdraw steel brush cap and drive out the retaining pin. The brush can then be lifted from the camshaft. In replacing it, care must be taken not to alter the timing of the ignition. The upper contact of the housing represents cylinder No. 1, and the exhaust valve of that cylinder should be closed when the brush points upward. This may be determined by removing the valve mechanism cover and noting the operation of the valve in question.

Q. If parts show much wear, what should be done?

A. Replace them; as they cost so little that it is far less expense to put in new parts than to attempt to make old ones serve by truing them up.

Q. When examination reveals gummed oil and the weather is cold, what should be done?

A. Clean out the housing and the roller-contact parts with gasoline, and use a mixture of $\frac{1}{4}$ kerosene and $\frac{3}{4}$ lubricating oil in the timer, as long as the weather is cold.

Q. What other causes of trouble with the timer are more or less common?

A. Short-circuiting of the primary wires which lead to the timer, or the loosening of these wires at the terminals on the housing. The position of the timer is such that the insulation of these wires is subjected to considerable wear by reason of the movement of the housing in advancing and retarding the time of ignition. The best method of remedying this is to replace the entire set of primary wires.

Vibrator Coils

Q. Is irregular firing likely to be caused by any other part of the system than the timer?

A. The vibrators of the induction coils may get out of adjustment and cause either erratic firing, or missing of one or more cylinders altogether, in case one or more vibrators cease to operate.

Q. How can this be determined?

A. Run the motor slowly and watch the action of the vibrators; they should act regularly and with the same rapidity in each case, any stuttering or hesitation indicating either poor adjustment or points in poor condition.

Q. How can the vibrators be utilized to determine which cylinder is missing, where the cause lies in some part of the system other than the vibrators themselves?

A. Hold one vibrator down at a time with the finger; if the remaining three cylinders fire regularly, the one represented by the vibrator being held out of action is the one at fault. The cylinder can always be located quickly by holding down each vibrator in turn.

Q. When the cylinders do not all fire regularly, but there is no perceptible difference between the action of the vibrators, how can the cylinder, or cylinders, at fault be determined?

A. Hold down the vibrators in pairs, taking first Nos. 1 and 4, and then Nos. 2 and 3. This will cause the engine to run on two cylinders at a time, and any difference between the operation of the two pairs or between members of each pair will be apparent.

Q. What is the firing order of the Ford motor?

A. 1-2-4-3.

Q. How are the coil vibrators adjusted?

A. The usual method is to turn the adjusting screw up until the vibrator stops buzzing; then turn the screw down again very slowly until the points just come together and the firing of that cylinder becomes regular; then give the screw an extra quarter-turn down, and lock in place.

In adjusting K-W coils, it is important to see that the little flat-cushion spring just underneath the vibrator bridge works back and forth every time the points make and break contact. This can be determined by taking the coil unit out of the box and holding the vibrator up to the light; press down the vibrator and observe the action of the cushion spring. It is important to adjust all the units alike or the motor will not develop its full power.

Q. What is the effect of adjusting so that the contact points are too far apart; too close together?

A. If too far apart, the cylinder will not fire regularly or with its usual power. If too close together the current is likely to arc at the contact points, thus preventing the breaking of the circuit when the armature is drawn down, burning the points themselves, and sometimes putting the coil out of action entirely.

Q. Does the vibrator adjustment affect starting?

A. When the points are too close, more current is required to "make and break" the contact between them, and the motor must be turned over that much faster. For the best adjustment, the points should barely touch. If the adjustment is too light, they may not do this and a miss at that cylinder will result.

Q. If the vibrator buzzes constantly, what is the trouble?

A. There may be a short-circuit at the timer or in the wire leading from that coil to it, or the coil itself may be defective. One of the first symptoms of a defective coil is the buzzing of the vibrator, with no spark at the plug.

Q. How can a defective coil be determined?

A. To make certain that the cause is in the coil, change the location of the units in the coil box. If another unit acts the same when substituted for the one giving trouble, the fault is not in the coil but in some other part of the system. Should the coil that is shifted, however, act the same in its new location, and the one that takes its place operates properly, the coil itself causes the trouble.

Q. When there is an unusually heavy or "fat" bluish spark at the contact points, what is the cause?

A. The current may be arcing at the points, due to their being adjusted too closely; or the condenser may have broken down. To make certain that the condenser has failed, disconnect the secondary cable from the spark plug and hold the terminal about $\frac{1}{32}$ inch away from the metal end of the plug. If the condenser has failed, the spark occurring at this gap will be irregular.

Q. What will happen if the contact points are allowed to become pitted and ragged, due to the burning effect of the current?

A. They are liable to stick together and cause unnecessary difficulty in starting or occasional missing when running. They should be trued up with a very fine flat file or with an old piece of fine sandpaper. Never use emery.

Q. When the vibrator points burn badly in a very short time, what is the cause?

A. The owner of the car has probably replaced the original vibrators with cheaper substitutes having nickel or German-silver contact points. Nothing but platinum or platinum-iridium contacts will give satisfactory service, so that new parts from the makers should be installed.

Q. When the engine will suddenly lag and pound, what is the cause?

A. An intermittent short-circuit in the wiring or at the commutator. The pounding is caused by the premature explosion of the charge against the rising piston. The vibration causes the short-circuit to occur at some times, and not at others, so that the engine will run regularly for a few minutes and then pound again, until the movement once more temporarily eliminates the cause.

Q. With all the vibrators properly adjusted and the timer and wiring in good condition, what is the cause of irregular firing?

A. Provided the spark plugs are all in good condition, points not too far apart, etc., this is frequently caused by the top of the coil box coming loose. The coil units are provided with brass-strap terminals on the bottom of the wooden casing of the coil, and these terminals make contact with similar straps in the bottom of the coil box on the dash. They depend for good contact on the pressure exerted by the cover of the box, which must always be kept tightly snapped on.

Q. When missing is not traceable to any of these causes, what is likely to be the cause?

A. Something outside of the ignition system, such as a weak-valve spring, or a valve improperly seating, due to some other cause. Loss of compression at the cylinder-head gasket: run a little lubricating oil along the edge of the gasket and note whether bubbles appear. Replace with a new gasket if any leakage is apparent.

Magneto

Q. How does the Ford magneto differ from the regulation type?

A. The *magnets* are revolved instead of the *armature;* it has sixteen field poles and armature coils, and it revolves at crankshaft speed to fire a four-cylinder motor. It is not timed to the motor the same as an ordinary magneto, which is coupled to the camshaft, or other half-time shaft, and the distributor of which must rotate synchronously with the motor.

Q. Of what does it consist?

A. Two discs, one carrying sixteen magnets with their poles pointing outward, and the other sixteen coils of strap copper on oval cores, all of the coils being connected in series. The disc carrying the magnets is rotated by the flywheel to which it is attached, while the other disc is attached to the crankcase, and remains stationary. One end of the coil winding of the magneto is grounded on the supporting disc carrying the coils, while the other is led to a terminal which extends through the flywheel housing. A cable from this terminal or binding post, supplies current to the coils.

Q. Is it ever necessary to remagnetize the magnets of the field, and how can it be done?

A. Unless they have become demagnetized, due to some outside influence, it is rarely necessary to touch the magnets.

Q. How can they become demagnetized by an outside force?

A. The attachment, by mistake, of a storage battery to the magneto terminal will send a current through the coil windings in the opposite direction, and will demagnetize them. When this happens, it is not advisable to attempt to remagnetize the old magnets, as it is much cheaper and quicker to replace them. The new set is supplied mounted on a board in exactly the position they should be installed.

Q. How can the magneto be dismounted?

A. To do this, it is necessary to remove the power plant from the car. The radiator must be taken off by disconnecting its stay rod and taking out the two holding bolts at the frame, after uncoupling the hose connections. Remove the dash and loosen the steering-post bracket, fastened to the frame, permitting the dash and steering gear to be lifted off as a unit (wires having first been disconnected); take out bolts, holding front radius rod in socket underneath the crankcase; remove four bolts at the universal joint; remove pans on either side of cylinder casting; disconnect feed pipe from carburetor, and exhaust manifold from exhaust pipe, by unscrewing large brass nut; remove the bolts which hold the crankcase arms to the frame at the side; then pass a rope through the opening between the two middle cylinders, and tie it in a loose knot; through the rope pass a 2 by 4 timber or a heavy iron pipe about ten feet long; with a man at each end of this and a third at the starting crank, the whole power plant can readily be lifted out; then remove the crankcase and transmission cover, and take out the four cap screws that hold the flywheel to the crankshaft. This gives access to every part of the magneto mechanism. To take out the old magnets, simply remove the cap screw and bronze screw which holds each in place. When reassembling the magneto, great care must be taken to see that the disc or plate carrying the magnetos revolves them exactly $\frac{1}{32}$ inch of the core faces of the coils.

Q. How can it be determined whether the magnets are at fault or not?

A. Whenever there is a partial failure of the current, remove the binding post by taking out the three screws which hold it in place. Clean out any dirt or foreign matter that may have accumulated under the contact spring. If this does not materially improve the running, test by comparing with battery ignition. Connect one terminal of a six-volt storage battery to the battery terminal on the coil box (a battery of four or five fresh dry cells will serve equally well for a short test run), and ground the other terminal of the battery on the frame of the car, making certain that good electrical connection is made. Run the engine at different speeds on the magneto, and, while running, throw the switch over suddenly to the battery point; any decided acceleration in the speed will indicate that the battery is supplying a much better current for ignition, and that the magneto is at fault. If used for any length of time for testing, a dry-cell battery will not give equally accurate results, as the cells are likely to run down very quickly, and the firing will then be better on the magneto, even though the latter be weak.

Q. When the engine suddenly fails to fire altogether, what is likely to be the cause?

A. The cable leading from the binding post on the magneto has dropped off, either at the latter or at the coil; or some piece of foreign matter has come between the contact spring attached to the binding post and its contact.

Q. When the magneto gradually gets weaker and weaker in a car that has seen a great deal of service, is it certain that the magnets have weakened?

A. Not necessarily; the adjustment of the bearings may be permitting the disc carrying the magnets to revolve further away from the armature coils than intended. Any increase in this distance, even though small, will have a decided effect on the output of the magneto.

GENERAL CAUSES OF IGNITION FAILURE

Q. When the motor stops very suddenly without any apparent cause, what is likely to be the cause of the trouble?

A. A break in the current-supply circuit of the ignition system, or sudden failure of the ignition current, due to any other cause.

Q. In how many different ways may this occur?

A. A feed wire from the battery may part from its terminal, either at the battery or at the coil; the ground wire may become disconnected at either end; in a dual system, the primary cable from the magneto to the coil may become disconnected, either at the magneto or at the coil; the secondary cable from the coil to the magneto distributor may loosen and drop off, either at the coil or at the magneto; this secondary cable may become grounded between the coil and the magneto; the switch may have loosened up, through vibration, and may jar open; the magneto may have become grounded internally, so that no current is delivered to the outside circuit. The magneto cam may have loosened up on its shaft, so that it no longer revolves with the latter, and, consequently, does not open the contact points in the breaker box. The primary cable from the magneto to the coil may have become grounded on the frame or short-circuited on another cable, due to wear from chafing, thus preventing the current from reaching the coil. In a dual system, the entire system may have become grounded through the metal of the dry cells coming in contact with the metal battery box or other metal part connected to the chassis of the car.

Q. Do the foregoing constitute all of the possible causes for a sudden failure of the ignition system?

A. No brief résumé could possibly include all of the causes that may exist for a stoppage of this kind, but they include probably more than 90 per cent of all the commoner causes of such a failure, and, either as given above, or in some modified form of the same condition, will be found to represent by far the greater part of all the causes of sudden stoppage.

Q. Of the causes given above, which are the most likely, in the order of their usual occurrence?

A. The loosening of a battery connection at the terminal or at the coil, or of a magneto and coil connection at either end; grounding or short-circuiting of either the primary or secondary main connection between magneto and coil, that of the secondary being more frequent owing to the high-tension current it carries. Grounding of the dry cells in the battery box; loosening of the switch so that it jars open. With the exception of the grounding of the primary or secondary cable between the magneto and the coil, all of the above are the direct result of vibration and jarring. In old four-cylinder motors, vibration is constant, and at times very severe, so that attention should first be directed to searching for loose connections, as unless tightened up at intervals, they are very likely to shake off. Internal grounding of the magneto or loosening of the breaker-box cam, so that the interrupter does not operate; these are rarer causes of trouble, and a search for them should be deferred until after the commoner causes mentioned above have been thoroughly investigated.

Q. Where all connections have been tightened up without overcoming the trouble, how can the other possible causes of stoppage be eliminated, in tracing the real seat of the failure to run?

A. See that the dry batteries of a dual system are not touching any metal; inspect the magneto breaker box while another person slowly turns the motor over by hand, so that the operation of the interrupter may be noted. If working properly, disconnect the secondary cable from the magneto to the coil, and with the motor running, hold its terminal $\frac{1}{4}$ inch away from the coil connection. In case there is no fault here, a bright spark will result at the gap. Note whether holding the cable away from the motor has been responsible and whether, when it is dropped back on the motor again, sparking occurs at any point along its length between the cable and the metal of the motor. Disconnect the primary cable from the magneto to the coil, and, with the motor running, wipe its end on the primary terminal of the coil; sparking should result if there is no break in the cable. Take the same precaution in putting it back as with the secondary cable to see that it is not grounding at some place along its length where it touches the motor. No visible spark will be produced in this case, but the condition of the insulation of the cable itself should be the best indication of this kind. Bend the wire along its length to detect any possible breaks in the copper wire under the insulation. Disconnect the ground cable from the coil to the magneto, and, while the motor is running, hold it close to either the magneto or the coil terminal; a high-tension spark will result if the cable is all right. Note, while the motor is running, whether there is any sparking at the safety gap, on the magneto itself, if on the high-tension type, or on the coil of a dual system. Note whether the primary and secondary cables cross each other, and whether there is any sparking between the two while the motor is running. Inspect the ends of all stranded cables carefully, and see whether one or more of the fine wires have not broken through the insulation and become bent over, so as to ground the cable on some adjacent metal.

Q. Is a sudden stoppage of the motor likely to be due to any cause other than a failure of the ignition system?

A. One possible cause is the sudden and complete stoppage of the carburetor spray nozzle, but, even in this case, the failure of the motor will not be so sudden nor so complete as where the ignition current has been cut off, as the motor will continue to fire, for a few revolutions on what fuel mixture remains in the manifold. There is practically no other cause for the motor suddenly stopping.

Q. When, instead of stopping completely, the motor will fire regularly for a few minutes, hitting on all cylinders, and then begins to miss spasmodically, how can it be determined whether the fuel-supply system or the ignition is at fault?

A. The fact that, under such circumstances, the motor will fire regularly for a little while and then miss very badly, and, a few minutes later again take up its operation smoothly, is usually an indication that the ignition system is working properly, but that at intervals there is a failure of the fuel supply. One of the commonest causes of this is the exhaustion of the main supply of gasoline in the tank. On the last half-gallon or so of fuel there is no longer a regular supply to the carburetor, with the result that, with the motor running at speed, what gasoline is in the carburetor is practically exhausted in a few minutes. During this period, however, the motor will continue to run regularly. As it lowers the level in the carburetor float chamber, an insufficient supply is drawn through the nozzle and the motor misses badly, and slows down almost to the stopping point. This permits a new supply to fill the carburetor, and the motor once more runs properly. It is the extremely intermittent nature of the firing, with first one cylinder missing and then another while the carburetor is refilling itself from what little gasoline remains in the tank, that makes this appear very much as if it were due to failure of the ignition. Under conditions such as this, always inspect the fuel supply first. With an ample supply of gasoline in the tank, a partial clogging of the spray nozzle of the carburetor, due to some obstructions which is intermittently drawn up into it by the suction and again drops back, will give exactly the same symptoms of ignition trouble.

Q. Are symptoms of this nature ever due to a fault in the ignition system?

A. They will result at times from the use of a set of dry cells that is almost exhausted. The storage battery also acts in a similar manner. Both the dry cell and the storage battery recuperate very rapidly even when practically exhausted, so that they will often provide sufficient current to run the motor properly for a very short time, will then cause it to miss badly, and shortly afterward again run regularly. If, when the switch is thrown over to the magneto, the motor runs smoothly and continuously, there is no doublt that the battery is at fault, and this may be verified by testing the cells with a pocket ammeter. Should they show much less than eight amperes on test, they are the cause of the trouble and should be discarded. This may also be due to the use of a storage battery that is practically exhausted, though it would be extremely bad practice to allow a storage battery to get this low. Test with the voltmeter: if the cells show $1\frac{1}{2}$ volts each or less they are badly in need of charging, and if they will not run the motor properly, should be immediately recharged from an outside source of current.

Q. What is meant by "pre-ignition", and what causes it?

A. When the charge in the cylinder is ignited before the passing of the spark at the spark plug, it is said to be "pre-ignited", i.e., fired in advance of the proper time. As a result, the force of the explosion is partly exerted against the rising piston, as is evidenced by a heavy pounding accompanied by a decrease in the power. The cause is usually an accumulation of carbon in the form of a deposit on the piston head and results from excessive lubrication. The surplus oil finds its way into the combustion chamber and is burned. This condition is further aggravated by running with an over-rich mixture. If the motor is allowed to run very hot, these carbon deposits become incandescent, so that the fresh mixture is fired the moment it comes in contact with them. In some cases this becomes so bad that the motor cannot be stopped without shutting off the gasoline supply.

Q. How can the pounding caused by pre-ignition be distinguished from other internal noises, such as those produced by a loose crankshaft or crankpin bearing?

A. Pre-ignition takes place only after the motor has been

running long enough to become very warm and with the throttle opened to any extent the pounding is very violent, jarring the whole chassis. The noise produced is distinctive and can be identified readily once it has been experienced. Unless very loose, a bearing noise will practically disappear if the motor is allowed to idle very slowly and will always increase in proportion to the load, becoming very severe when climbing a hill.

Q. How can the condition which causes it be remedied?

A. By removing the carbon deposits. In many late-model engines this can be done most readily by removing the cylinder heads, usually a single casting. The carbon may be burned out with the oxygen-gas flame now in common use or it may be loosened by the use of kerosene in the motor. After the motor has run long enough to become hot, shut off the gasoline supply gradually, meanwhile feeding kerosene through the auxiliary-air inlet of the carburetor until the motor is running on kerosene alone; feed an excess of the latter and the carbon will be loosened and blown out through the exhaust.

Q. When a single cylinder continues to miss regularly, all the others running properly, and inspection shows every part of the ignition system to be in good condition, what is likely to be the cause?

A. Failure of its valves to operate properly. Either the inlet or the exhaust valve is not opening, or is sticking open (the result will be the same in either case). This may be caused by a weak valve spring, a bent valve stem, derangement of the valve tapper, so that it does not strike the valve-stem end, or by a piece of foreign matter, such as a piece of carbon, lodging on the seat of the valve, so that the latter cannot close. Another, though rarer cause, is a leak in the manifold, close to the inlet valve of the cylinder in question. This permits an excessive amount of air to be drawn into that particular cylinder, so that the charge is too weak to fire. In any of the above cases, the result is that fuel either does not get into the cylinder or it is exhausted before it can be fired, as with every part of the ignition system working properly, the only thing that can cause a cylinder to miss is lack of fuel.

Q. Mention one of the causes of irregular firing that is seldom suspected, except by those who have experienced it previously?

A. Excess oil finding its way into the combustion chambers in such quantities that it covers the spark-plug electrodes, thus preventing a spark from jumping the gap. The oil, particularly when fresh, is an excellent insulator and, if mixed with carbon so that it conducts the high-tension current, it does so without permitting the formation of a spark. Owing to the viscosity of heavy lubricating oil, it clings to the spark-plug points and when they are as close together as they should be ($\frac{1}{32}$ inch), it will often bridge the gap for some time, despite the vibration and succeeding compressions in the cylinder. This fault is particularly difficult to locate when not suspected, as jolting over a rough piece of road will shake the plug points free and the engine will fire regularly, again missing intermittently when on smooth going once more. When confined to one cylinder, it is usually an indication that the cylinder wall is scored, or that the lubricating-oil feed to that cylinder is deranged.

Q. What simple test can be made to determine whether the spark is occurring in each cylinder at the proper time?

A. Take out the plugs, leaving the cables attached to them, and lay the plugs on the cylinder head. Then turn the motor over slowly and note whether or not the sparks occur at the plugs in the proper sequence. Note whether there is a strong blast of air from one of the spark plug holes each time the motor is turned over; if not, pour an ounce or two of fresh oil into each cylinder. The failure to start may be due to lack of compression.

Q. When intermittent failure of the ignition is thought to be due to faults in the wiring which cannot be detected by an ordinary examination, how can the trouble be found most readily?

A. Fit a handy length of cable—one that will span practically any two points in the ignition system—with spring-clip terminals. Disconnect each wire in turn, and substitute for it this length of cable as a temporary connection; satisfactory operation with the latter indicates that the wire it replaces is at fault.

GENERATORS

Types and Requirements

Q. How many types of generators are used in starting and lighting service on the automobile?

A. Practically all are of one type, i.e., compound-wound, but this is subdivided into other types, such as differential compound-wound, cumulative compound-wound, and the like, that is, all lighting generators have a shunt and a series winding on their fields, but the relation of these windings to one another differs, depending upon the characteristics of the remainder of the system.

Q. What is a differential compound-wound generator?

A. One in which the series winding is reversed, i.e., wound in a direction opposite to that of the shunt winding so that its exciting effect on the field magnets opposes that of the shunt winding. The series winding is then termed a bucking coil because it bucks, or opposes, the exciting effect of the shunt winding on the field magnets as the speed increases. The series winding in this case is used simply for regulating the generator output.

Q. What is a cumulative compound-wound generator?

A. One in which the exciting effect of the series coil is added to that of the shunt coil, the series coil in this case having no connection with the regulation of the generator output.

Q. As one of the chief requirements of an efficiently operating system is the control of the generator output under widely varying speeds, how is a generator of the cumulative compound-wound type employed on the automobile?

A. The series winding is in practically an independent circuit in connection with the lamps of the car so that its exciting effect is not added to the field magnets except when the lights are switched on. This automatically increases the generator output in accordance with the number of lights turned on so that the lights have no effect on the battery charging rate, which remains the same whether the lights are on or off. An external regulator is employed to control the battery-charging rate.

Q. How does the generator differ from the motor?

A. Its essentials are all the same, i.e., it has a wound armature revolving in a magnetic field, commutator, brushes, etc., exactly the same as the generator.

Q. This being the case, why are the two not interchangeable?

A. To a certain extent they are, that is, when a current is sent through the generator from an outside source, it becomes motorized and will run as a motor. But the two are far from being interchangeable on the automobile, owing to the widely differing requirements for which they are designed. The generator is wound to produce a current seldom exceeding a value of 20 amperes while being driven over a wide range of speeds, and it is in constant operation. The starting motor, on the other hand, is designed to utilize an extremely heavy current, ranging up to 300 amperes or more at the moment of starting and is only used for very short periods.

Q. How are these widely varying requirements reconciled in the single=unit type, in which both the generator and the motor are combined in one machine?

A. The machine is practically two units in one, i.e., there are two totally different windings on the same magnet cores, a fine winding with shunt fields for the generator, and a very heavy simple series winding for the motor end. In some cases, as in the Delco, the different windings on the armature are brought out to independent commutators. While combined on one set of magnet cores, there is no connection whatever between the two windings in such a machine, so that when operating as a generator the motor windings are dead, and the reverse is true when being used as a starting motor.

Q. What are the characteristics of the single=unit type of machine which is simply placed in circuit with the battery by a hand=operated switch when starting and left in that relation as long as the engine is running?

A. This is a variable-potential type in which the relation that it bears to the battery and to the engine is entirely dependent upon the speed of the engine, that is, the speed at which the machine is driven. When the switch is closed, current from the battery operates the machine as a starting motor; as soon as the engine starts and attains a certain speed, the voltage of the machine overcomes that of the battery, the direction of current flow is reversed, and the battery begins to charge. Whenever the driving speed falls below a certain point, there is another reversal, and the generator once more becomes a motor until the engine speed increases.

Loss of Capacity

Q. What are the chief causes for the falling off in output of the generator?

A. In about the order of the frequency of their occurrence, these are as follows: dirty or worn commutator; worn brushes making poor contact; dirty or loose connections causing extra resistance at generator, regulator, cut-out, ground, or battery terminals; failure of cut-out to operate at proper voltage; worn or pitted contacts in regulator or cut-out; loose connections at brush holders; short-circuited coils in the armature; some of the armature-coil connections broken away from the commutator; short-circuited bars in the commutator.

Q. How can the generator output be tested?

A. The simplest method is to switch on all the lamps with the engine idle. Start the engine and speed up to equivalent of 15 miles per hour. The lights should brighten very perceptibly, the test being made indoors in the daytime with the lights directed against a dark wall, or preferably at night. A more accurate test can be made with the portable volt-ammeter, using the 30-ampere shunt. Most generators have an average current output of 10 to 12 amperes, but the normal output as given by the maker should be checked before making the test. Generators having a constant-voltage control will show a greatly increased output if the battery charge is low, running up to 20 amperes or over. On such machines, the condition of the battery should be checked either with the hydrometer or with the voltmeter before making the test. The charging current should be 10 to 12 amperes with a fully charged battery, and more in proportion when only partly charged.

Q. What other simple method is there of determining quickly whether the generator is producing its normal output or not?

A. On generators having an accessibly located field fuse (there are several makes) lift this fuse out and, with the engine running at a speed equivalent to 10 miles per hour or more, touch the fuse terminals lightly to the clips. If the machine is generating properly, there will be a bright hot spark. Should no spark appear, replace the fuse and bridge the terminals with a pair of pliers by touching the jaws to the fuse clips; if a spark appears, the fuse has blown. Before replacing with a new fuse, find the short-circuit or other cause.

Q. Granting that the fuse has not blown, that the cut-out, regulator, and wiring are all in good condition, and still the generator does not produce any current, what is likely to be the cause?

A. One of the brushes may not be touching the commutator, a brush connection may have broken, or carbon dust may have short-circuited the armature or field windings. Test for short-circuits.

Q. If the machine is generating current, and the auxiliary devices and wiring are in good condition, but the battery does not charge, what is the cause?

A. Short-circuit in the battery due to active material having been forced out of the plates, or accumulation of sediment touching plates at their lower ends. (See Battery Instructions.)

Q. Is the regulator ever responsible for a falling off in the current or for generation of excessive current?

A. Yes. Any irregularity in the operation of the regulator will affect the output of the generator.

Q. How can this be overcome?

A. This will depend upon the type of regulation employed (see Regulation). Where the method of regulation is inherent, i.e., forming part of the construction of the generator itself, such as the third-brush method, or a bucking coil, it may be remedied by cleaning and seating the brush properly or by testing the bucking-coil winding to see if its connections are tight and clean, or if it is short-circuited (see Windings). If cleaning and sanding-in the brush do not cause the generator to produce its normal output, the brush itself may be adjusted by shifting its location. Moving it backward or against the direction of rotation of the commutator will reduce the output; moving it forward or in the direction of rotation will increase the output. This refers specifically to the Delco regulation already described. To adjust properly, the portable ammeter should be put in circuit, and the effect on the reading noted as the brush is moved, clamping it back in place when the proper point is found. The brush should then be sanded-in to the commutator, as it will not have a good bearing if its original location has been disturbed.

Methods of Regulation

Q. Why is it necessary to control the output of the generator?

A. As explained in the section on electric generators, the amount of current produced depends upon the excitation of the fields, and the faster the armature revolves before the pole faces of the field magnets, the greater the amount of current that is sent through the windings of the magnets. As the speed of the automobile engine varies between such extremely wide limits, it will readily be seen that it may rise to a point where this increase in the field excitation will cause so much current to be generated that the armature windings will be literally burned up. This happened very frequently in the early attempts to produce a lighting dynamo for automobile service. Regardless of how fast the generator may be driven, it is essential that its current output does not exceed a certain safe limit.

Q. What is the usual safe limit in the majority of generators?

A. Most automobile lighting-system generators are designed to produce 10 to 15 amperes at a normal speed, i.e., sufficient to light all the lamps and still provide a slight excess for charging the battery. No matter how fast its armature revolves, it must not exceed this by more than ten to twenty-five per cent, as a rule, this being well within its factor of safety. In some instances, where a voltage system of regulation is employed, the output of the generator depends upon the condition of charge of the battery. If the battery is practically discharged, the generator will charge the battery at a rate of twenty amperes or over. As the charge proceeds, the battery voltage increases and the resistance is increased correspondingly, thus cutting down the amount of current that the generator can force into the battery.

Q. How is the current generated kept from exceeding this safe limit?

A. Mechanical methods were employed at first, a centrifugal governor being used to operate a slipping clutch. The generator was driven through this clutch, and the speed at which the armature revolved depended upon the engagement of the clutch; at low speeds both shafts would turn at the same rate. As the driving-shaft speed increased, the governor decreased the pressure on the clutch spring, and the clutch faces slipped on one another, so that the driven shaft turned proportionately slower than the driving shaft. The earliest types of governors, employed about 1903 to 1905, were not successful, but about 1908 a type was developed that worked effectively on thousands of cars. It has since been superseded by electrical methods of regulation, and practically all of those now in use are electrically operated.

Q. How many electrical methods of regulating the amount of current generated are in general use?

A. So far as their principle goes, practically all are the same.

They depend upon weakening the excitation of the fields of the generator to cut down the output. It is in the methods of accomplishing this that they differ. In the latter respect they may be divided into two general classes: those that are inherent in the design of the machine, i.e., the regulating device is actually a part of the machine itself; and those in which an external regulator is employed. Those most commonly employed are, in the first class, the bucking-coil winding and the third-brush method; in the second, an external regulator is usually combined with the battery cut-out and designed to keep either the voltage or the current at a uniform value, usually the voltage.

Q. What is a bucking=coil winding, and why is it so called?

A. We have seen that in a series-wound machine all of the current generated in the armature passes through the field windings and energizes the field magnets; in the shunt-wound machine the wires carry only a part of the current which is proportional to the resistance that the shunt winding of the fields bears to the resistance of the outside circuit. As this outside resistance (the load) increases, more current will be diverted through the path of lesser resistance, or the shunt-wound field, and the output of the machine will increase accordingly. In the compound-wound machine, the relation of the series to the shunt winding is such that it is called upon chiefly to help carry any extra load. In other words, as the demands upon the machine increase, the series winding adds its energizing effect to that of the shunt coil. A generator with a bucking-coil winding is a compound-wound machine, but the series winding is in *the opposite direction* from that of the shunt winding. Consequently, instead of adding to the field excitation caused by the latter, it *opposes or bucks* it, and the more current there is produced in the shunt field by the rise in speed, the more the series winding, or bucking coil, tends to neutralize this excess, thus keeping the amount of magnetic effect produced in the field poles practically uniform, regardless of the speed.

Q. What is the third=brush method of regulation?

A. In a conventional shunt-wound generator, the field windings are directly in shunt with the armature through the brushes; hence, a certain proportion of all the current induced in the armature windings will find its way through the field magnet windings, in proportion to their relative resistance to the outside circuit at the time. Where a third brush is employed, the main brushes are not in shunt with the fields, and they are not depended upon to supply the exciting current for the latter. The third brush instead is used for this purpose. As is well known, the output of a generator depends very largely upon the position of its brushes. In the immediate vicinity of the proper location for a brush, there is a short zone of maximum intensity. As we get away from this toward the next brush, it decreases until at a point midway between the two there is a neutral zone. The third brush is accordingly placed between the two main brushes, and its distance from the nearest main brush determines the amount of current that it diverts from the armature to the field windings. See illustration of Delco generator in section on Methods of Regulation. This method has the advantage of supplying a strong shunt field at low speeds. As the speed increases, the voltage applied to the shunt field decreases, even though that between the two main brushes may have increased.

Regulators

Q. What is a regulator, and what is its purpose?

A. It is an instrument somewhat similar to a battery cut-out, and its purpose is to regulate the output of the generator in order that the latter may not exceed safe limits at high speeds. The regulator is usually combined with the cut-out.

Q. How does the constant-voltage type of regulator operate?

A. The instrument consists of a magnet winding and a pivoted armature, normally held open by a spring and a resistance unit. The winding of the magnet has sufficient resistance to prevent the core becoming energized to a degree where it will attract the armature, unless the voltage exceeds the safe limit determined for the circuit. The voltage increases with the speed of the generator, so that when the latter is driven too fast the attraction of the magnet core for the armature becomes sufficient to overcome the pull of the spring which normally holds the contacts apart.

When the contacts come together, the field circuit of the generator is shunted through the resistance unit; this cuts down the amount of current energizing the fields, the voltage falls off, and the contacts again separate. Unless the speed of the generator is decreased, this action is rapidly repeated, so that the regulator armature vibrates at a high speed as long as the voltage is sufficiently high to energize the magnet.

Q. What is the principle on which this type of regulator operates?

A. The principle that in a circuit having considerable self-induction the amount of current which may be sent through the circuit will decrease if the current be pulsating instead of steady. Every time the contacts of the regulator open, a pulsation, or surge, of current is sent through the field windings of the generator; when they close because of the higher voltage, the current is shunted through the resistance unit, thus cutting it down. The decrease in the amount of current is in proportion to the number of pulsations per minute, i.e., the rapidity with which the vibrating contact operates. The circuit having considerable self-induction is that of the field winding of the generator, owing to its heavy iron core. (See Induction.)

Q. What is the constant=current type of regulator, and how does it differ from the constant=voltage, or potential, type?

A. It consists of an electromagnet and a spring-controlled pivoted armature, so that it is of practically the same construction as the constant-potential type, but it is connected in circuit with the armature of the generator and it is wound to operate under the influence of the current rather than the voltage. Consequently, the pivoted armature is attracted, opening the circuit when the current exceeds a certain predetermined value, usually 10 amperes. In operation, the armature vibrates the same as in the voltage regulator, but the condition of the charge of the battery has no effect on it, so that when set to limit the current to 10 amperes, it will always charge the battery at approximately that rate regardless of the condition of the battery. The only practical difference is that it is wound to actuate under the influence of changes in the current flow and is connected in the armature circuit, whereas the constant-potential regulator is influenced by variations in the voltage and is connected in the field circuit of the generator. The latter has the advantage of charging the battery at a higher rate when the charge is most needed.

Q. What other forms of regulators are employed on lighting generators?

A. The foregoing comprise practically all of the principles

employed, but the regulators differ more or less in design and operation. For example, in the Bosch-Rushmore generator, a bucking coil is employed in connection with what is termed a ballast resistor, or resistance unit. This is of iron wire, and it is based on the fact that resistance increases very rapidly with the temperature. The size of the wire is such that it allows 10 amperes to flow without undue heating, so that its resistance is practically unchanged; above this point it heats rapidly and increases in resistance so greatly that all excess current is shunted through the bucking coil. In the Splitdorf generator, the regulator is built in, projections of the pole pieces of the field being utilized in connection with special windings, instead of an independent electromagnet as in the Ward-Leonard and the Bijur. In the U.S.L. generator of the inherently regulated type, regulation is accomplished by the combination of a Gramme ring armature, a special arrangement of connections and of the field windings, and the use of only a part of the fields and armature for generating current. The regulation obtained is based on armature reaction and is similar in effect to the third-brush method. The U.S.L. external type of regulator cuts into the generator field circuit a variable resistance consisting of an adjustable carbon pile. In the Adlake regulator, which is of the constant-potential type, a solenoid operates a switch over the contacts of a variable resistance. The plunger of the solenoid is counterbalanced by a weight, which must be raised to operate the switch. It is adjustable by increasing the weight of this counterbalance.

Q. What attention does the regulation of the generator require?

A. This will depend upon the method employed in each case. Where an external regulator is employed, whether of the constant-potential or the constant-current type, the attention required is practically the same as in the case of the battery cut-out. See that the points are not sticking, and when badly burned or pitted, smooth and true up, taking off as little of the contact point as possible to effect this. When the points have become so badly pitted that this cannot be done, new parts will be necessary.

With the third-brush method, the attention required by this brush is the same as that which must be given the other brushes, i.e., sanding-in at intervals and replacement when worn too short to permit the spring to hold the brush firmly against the commutator.

Where the generator fails to produce sufficient current to keep the battery charged, all other parts of the system being in good condition and the car driven long enough in daylight to charge the battery under normal conditions, the position of the third brush may be shifted to increase the output. Care must be taken not to let it come in contact with the main brush. In the case of a bucking-coil winding, no attention is necessary, as this is an integral part of the machine itself. As the Splitdorf regulator has moving contact points, the attention necessary is the same as that required for an external regulator of this type. Special regulators, such as the U.S.L. external type, require attention covered by the maker's instructions.

Q. When the generator fails to keep the battery charged properly, a normal amount of daylight driving being given the car, is the fault most likely to be found in the regulator?

A. No. It is much more likely to be caused by a dirty commutator, worn brushes, loose connections, or some similar cause which inserts extra resistance in the charging circuit. The movement of the regulator armature is very slight, and the current handled by the contact points is small, so that it will seldom be the cause of the trouble. Other causes, such as those above enumerated, should always be sought first. (See instructions under Generator.)

Windings

Q. Are faults in the generator windings frequent?

A. They constitute one of the least frequent sources of trouble with the machine.

Q. What is likely to cause them?

A. Dousing the machine with water is likely to be one of the most frequent causes of short-circuits or grounds in the generator windings. All electrical machinery is intended to be kept dry. Except where provided with a field fuse, running the generator when disconnected from the battery or with the battery removed from the car is another cause. Excessive speed, in some instances, may generate sufficient centrifugal force to lift the armature coils out of their slots so that the insulation becomes abraded by rubbing against the pole pieces, but this is very unusual. In rare instances, a hard kink left in the wire when winding may crystallize the metal and make it break at that point, due to the vibration. Unless cleaned out at intervals, fine carbon dust from the wear of the brushes may accumulate in the interstices of the windings, and, when aggravated by moisture, this is apt to cause short-circuits.

Q. What are the usual indications of such faults?

A. With a short-circuited generator coil (armature), all other parts of the apparatus and circuits being in good condition, the charging rate will be lower than normal. The ammeter needle will vibrate violently when the engine is running at low speeds, and two or more adjacent commutator bars will burn and blacken. With an open armature coil (broken wire), the indications will be practically the same, and there will be severe sparking at the brushes, causing serious burning of the commutator bar corresponding to the open coil. A grounded armature coil will give the same general indications, and if the machine is a single-unit type, the cranking ability of the starting motor will be seriously impaired. The ammeter, however, will not vibrate as in the former cases. There will be practically no charge from the generator, and the battery will be discharged very rapidly by the starting motor.

In a single-unit machine, when the windings of the generator and the starting motor become interconnected, the indications will be practically the same as those of a grounded armature coil. If the motor windings of a single-unit machine become grounded, there will be an excessive discharge from the battery, while the motor will develop but little power.

Q. How may such faults be located?

A. With the aid of the testing-lamp outfit. Remove the brushes (when replacing them later, be sure to put each brush back in the holder from which it was taken), or the brushes may be insulated from the commutator by placing paper under them. For a grounded coil, place one test point on the commutator and the other on the frame; if grounded, the lamp will light. For interconnected motor and generator windings in a single-unit machine having two commutators, insulate the brushes as mentioned and place the test points one on each commutator. The light will burn if the two windings are connected. For a grounded-motor winding, test from the motor commutator to the frame; the light should not burn if the insulation is all right. For a break or open circuit in the field winding, touch the terminals of the latter with the test points, the commutator being insulated or the armature removed. The lamp should light. For a blown field fuse on machines so equipped, place the points on the clips; if the fuse is intact, the lamp will light.

Q. Are these tests conclusive?

A. No. They will indicate any of the faults mentioned, but they will not reveal an internal short-circuit in the windings, which cuts some of the armature or field turns out of action but does not break the circuit as a whole. Such a short-circuit reduces the output of the generator and can be determined definitely only by measuring the resistance of the windings. This requires special and expensive testing instruments, such as the Wheatstone bridge, so that where all other tests fail to reveal the cause of a falling off in the output of the generator, it should be sent to the maker for inspection.

Commutator and Brushes

Q. What does a blackened and dirty commutator indicate?

A. Sparking at the brushes or an accumulation of carbon dust due to putting lubricant on the commutator.

Q. What is the cause of sparking at the brushes?

A. Poor brush contact, due to worn brushes; brush-holder springs too loose, so that brushes are not held firmly against the commutator; excessive vibration, which may be due to a bent shaft, an unbalanced gear pinion, or improper mounting; using too much oil, or using grease in the ball bearings, which gets on the commutator and, acting as a solvent for the binder of the carbon, forms a pasty mass which prevents proper brush contact; worn or roughened commutator on which the mica needs undercutting; overload due to failure of regulator or to grounded coils in armature.

Q. What is the remedy for sparking?

A. Clean the commutator with fine sandpaper and sand-in the brushes to a true bearing on the commutator as directed in the Delco instructions. See that the brush springs have sufficient tension to keep the brushes firmly pressed against the commutator when the machine is running. If the mica protrudes above the commutator bars, it must be undercut as directed, and the commutator smoothed down again after the operation to remove any burrs.

Q. Why do some commutators need undercutting and others not?

A. Undercutting is required only on machines equipped with brushes that are softer than the mica. Copper-carbon brushes, as employed on starting motors to reduce the brush resistance, are hard enough to keep the mica worn down with the copper of the commutator itself.

Q. If, after smoothing off and undercutting the mica, the commutator still has an uneven and irregular surface, what is the remedy?

A. The armature should be removed from the machine, and the commutator trued up in the lathe, taking as light a cut as possible consistent with obtaining a true round and smooth surface.

Q. How can excessive commutator wear be prevented?

A. Inspect at regular intervals and on the first sign of sparking smooth up the surface and sand-in the brushes. Keep the commutator clean and do not permit carbon dust or oil to accumulate in the commutator and brush housing. Never replace brushes or brush springs with any but those supplied by the manufacturer for that particular model. The machine will work with any old brush and any old spring that fits, but they will prove detrimental to its operation in a comparatively short time, and its *working* under such conditions will never be satisfactory.

Q. Is discoloration of the commutator ever caused by anything else than sparking?

A. Not actual discoloration which requires cleaning, but the normal operation of the machine produces a purplish blue tinge on the bars, which is sometimes mistaken for discoloration by the inexperienced. This color, in connection with a high polish of the metal, indicates that the commutator is in the best of condition. Once the commutator takes on this high polish, it will operate for long periods without other attention than the removal of dirt by wiping with a clean rag. Sanding to remove this purple tinge is a mistake, as it only destroys the polish without having any beneficial effect.

Q. Is it necessary to lubricate the surface of the commutator?

A. No. The brushes employed are usually of what are termed a self-lubricating type and require no attention in this respect.

Q. Will any harm result from putting light grease, vaseline, or lubricating oil on the surface of the commutator?

A. As all lubricants are insulators to a greater or less extent, the efficiency of the machine will be reduced and, as the voltage is very low, but a slight falling off is necessary to represent a very substantial percentage of the maximum. The use of lubricant of any nature on the commutator also has another harmful effect in that it collects the carbon dust resulting from the wear of the brushes, causing it to lodge against them as well as between the commutator bars.

Q. Why should particular care be taken to remove all carbon dust from the commutator housing of both the generator and the motor (two-unit system) or the single unit where both functions are combined in one machine?

A. Carbon dust is an excellent conductor of electric current and, when spread over the surface of an insulator, it causes the latter to become conducting as well. Consequently, it is likely to short-circuit the commutator bars by lodging between them. It will cause leakage across fiber or other insulating bushing of brush holders when a sufficient deposit accumulates on them. It will penetrate the armature and field windings of the machine and may cause trouble by grounding or short-circuiting them. Especial care should be taken to remove all traces of carbon dust after sanding-in the brushes.

Q. How often should the commutator be inspected?

A. The commutator is the most vulnerable part of any direct-current machine, whether it be a generator or motor, and it should accordingly be inspected at more frequent intervals than any other single part of the entire system. The efficiency of both the generator and the motor depend upon it to a very great extent. Most of the failures of starting and lighting systems that are not due to poor condition of the battery may be traced directly to the commutator.

Q. What is the function of the brushes?

A. To conduct the voltage and current induced in the armature by its revolution through the lines of force created by the magnetic field, to the outer circuit, in the case of an electric generator; and to conduct the operating current to the armature windings from the battery, in the case of the starting motor.

Q. Why must the brushes bear evenly over their entire surface on the commutator?

A. Because their current-carrying capacity depends upon their size, and the latter is based upon the entire surface of the end of the brush making efficient contact. If the brush does not make uniform contact, those parts of it that do not touch the commutator will cause arcing or heavy sparking at the gap thus created, resulting in damage to both the commutator and the brush.

Q. Why are springs of different strengths used on generators and motors of different makes to hold the brushes against the commutator, though the machines are of practically the same capacity, operate at the same voltage, and are in other respects very much alike?

A. The carbon compounds of which the brushes are manufactured differ greatly in their conductivity and resistance offered to the passage of the current, and these differences call for greater or less spring pressure to hold the brush against the commutator surface in order to make efficient contact over the entire surface of the brush. Every maker has his own standard in this particular respect.

Q. Why is it not advisable to use brushes other than those supplied by the manufacturer as replacements on a machine?

A. For the reasons just given above. The manufacturer has adopted certain standards for the operation of his machines, and the brushes supplied have been made particularly to comply with those standards. No other brushes will do so well, and some will result in injury to the machine.

Q. When inspection shows that the brushes have worn down unevenly, what should be done?

A. They should be sanded-in with a strip of fine sandpaper, such as No. 00, preferably already worn if the brushes are very soft. (See instructions for doing this properly in connection with machines of different makes.) No more should be removed than is absolutely necessary to bring the end of the brush to a firm contact all over its bearing surface on the commutator; and the end of the brush, after the completion of the operation, should not show any deep scratches or pit marks. Unless the surface is smooth and true, injurious sparking will result, and the efficiency of the machine will be decreased.

Q. If, with a smooth and true surface, the brush still fails to make good contact, what is the trouble?

A. The brush has probably worn down until it is too short for the spring to exert sufficient force against it to hold it against the commutator properly, or the spring itself may be at fault. Wear of the brush beyond the point where it is any longer of service will most often be the cause.

Q. Where the brushes are true and are making good contact against the commutator, but the machine is inoperative, all other parts of the system being in good condition, what is likely to be the trouble?

A. One of the pigtails, or short flexible connections, of the brushes may have shaken out from under its spring clip. This breaks the circuit, just as a parted wire or a ruptured connection at a terminal in any other part of the system would.

Q. How often is it necessary to replace the brushes?

A. This differs so much with different makes of machines that it cannot be answered definitely, even as an average. On two-unit systems, the generator brushes will naturally require replacements much sooner than those of the starting motor, as the starting motor is only in operation for very short periods, while the generator is working constantly. On single-unit types, this naturally does not apply, as, whether the armature has one or two sets, they are always in use. Ordinarily, brushes should not require replacement under a year, and frequent instances are known of their having lasted for two years or more. It depends upon the care given the commutator and brushes quite as much as upon the mileage covered, as, if allowed to run dirty for any length of time, the brushes will wear away much faster than if kept in good condition. The best rule for the replacement of the brushes on all makes of machines is to renew them as soon as they have worn to a point where the springs no longer hold them firm against the commutator. When they have reached this condition, the vibration and jolting of the car is likely to shake them out of contact, which results in sparking.

Q. What is the "third brush", and what is its function?

A. This is an extra brush used on a generator. Its purpose is to control the amount of current supplied by the armature to the shunt-field winding as the speed increases. In other words, it regulates the output of the machine and prevents it from being burned out when the speed of the engine becomes very high.

Q. Does it differ from the other brushes in construction or in the care required?

A. It is a carbon brush of the same nature as the others used on the same machine, and the care required to keep it in good condition does not differ. However, it is mounted in an independently adjustable holder so that it may be moved backward or forward with relation to the main brushes in order to increase or decrease the output of the generator.

Q. Is it ever necessary to alter the location of the brushes of a machine?

A. Except on generators fitted with the third-brush method of regulation, on which it may be necessary to shift the main brushes slightly to avoid having the third brush come in contact with one of them when moved to change the output, it should never be necessary to shift the location of the brushes. Brush location has an important bearing on the operation of the machine, and, in designing it, the maker has fixed the location of the brushes to conform to its other characteristics. Many machines have no provision for adjusting the brushes in this respect, while some manufacturers caution the user particularly against altering their location.

Q. How much spring pressure is usually employed to hold the brushes of the generator and starting motor against the commutator?

A. This varies with different makes of machines and should be ascertained from the maker's instructions in every case in order to check up properly. In the various models of the Gray & Davis starting motors, this spring pressure ranges from $2\frac{1}{4}$ to $3\frac{1}{2}$ pounds, which is the minimum necessary. In other words, the brush must be held against the commutator with this amount of pressure in order to operate efficiently. While there will be a loss if the pressure drops below the minimum, there is no advantage in greatly exceeding it, as excess pressure simply causes greater friction loss without any compensating gain in power. Generator-brush pressures are much less than those employed on starting motors, owing to the smaller amount of current handled.

Q. How can the proper spring pressure of the brushes be checked?

A. With the aid of an ordinary spring scale of the direct-pull type, in which the pull on the hook draws the pointer down over the scale. A scale reading to five pounds is adequate for the purpose; one intended for heavy weights is not likely to be so accurate. Attach the hook of the scale to the brush and pull until the brush is just clear of the commutator. The scale will then register the pull in pounds. Where there is nothing on the brush to which to attach the hook, such as a screw, place a thin piece of wood on the brush face before passing the hook of the scale around it, to prevent injuring the contact face of the brush. In this case, the spring pressure as shown on the scale will exceed the necessary minimum, as the spring must be compressed further than it would be when in operation, in order to operate the scale. This should be allowed for when taking the reading.

Q. When is it advisable to check the spring pressure of the brushes?

A. When there is undue sparking at the commutator, while the commutator and brushes are all in proper condition, i. e., clean, and bearing uniformly over their entire surface so that the sparking is not due to any fault in either of these essentials.

Q. When the brushes and commutator are in good condition and the spring-scale test shows that the brushes are being held against the commutator with the necessary amount of pressure, what is likely to be the cause of the sparking?

A. There may be a short-circuited or open coil in the armature.

STARTING MOTOR

Q. In what way does the starting motor of a two-unit system differ from the generator?

A. It is a simple series-wound machine having but one winding of coarse wire on the fields, and all the current from the battery passes through its armature coils and field windings.

Q. Is it subject to electrical faults other than those already referred to in connection with the generator?

A. No. The care and the nature of the tests required to locate faults are the same. The commutator should be kept clean, brushes bearing firmly on commutator, and all connections kept tight. The same instructions for sanding-in brushes and keeping the commutator in good condition apply as in the case of the generator.

Q. When the starting motor fails to operate, what is likely to be the cause?

A. In the majority of instances, a low state of charge or a

wholly discharged battery will be responsible. If the battery is all right, a loose connection at the battery, switch, or motor, or a short-circuit in some part of this wiring may be the cause. Should the battery be properly charged, all wiring and connections in good condition, switch contacts clean, etc., the starting gears may be binding, owing to dirt or lack of alignment between the motor shaft and the flywheel of the engine. In this case, the motor will attempt to start when the current is first turned on, but will be held fast. Loosen the holding bolts and line up the motor, cleaning the gear teeth if necessary.

Q What is likely to be the cause of the starting motor running slowly and with very little power?

A. Exhausted battery, poor switch contacts, loose connections, partial ground or short-circuit in wiring causing leakage, improperly meshing gears, dirty commutator, brushes making poor contact owing to weak springs or worn brushes, or a ground in the motor itself. The remedies for all these faults have been given already.

Q. When the battery and all connections and wiring are in good condition, but the motor fails to crank the engine, what is likely to be the cause?

A. The engine may be too stiff. If it has been overhauled just previously, the main bearings may have been set up too tight. Test with the starting crank to see if it can be turned over easily by hand. If unusual effort is required, easing off the bearings should remedy the trouble. Should the engine not turn over as soon as the switch is closed, release immediately, as otherwise the battery will be damaged.

Q. When the engine does not start within a few seconds, why is it better to use the starting motor intermittently than to run it continuously until the engine does fire?

A. The intermittent use of the starting motor, say ten seconds at a time, with a pause of half a minute or a minute between attempts is easier on the battery. If allowed to rest for a short period, the storage battery recuperates very rapidly. Consequently, the operation of the starting motor for two minutes, divided into twelve periods of ten seconds each, will not run the battery down to anything like the extent that its continuous operation for the same length of time would. Moreover, this intermittent method of operation increases the chances of starting under adverse conditions, as, in very cold weather, every time the battery is allowed to rest, it will be able to spin the engine at its normal starting speed, whereas if the starting motor is operated continuously, the battery will become so weak that the engine will be turned over very slowly toward the end of the period in question.

Q. Why is it that a starting motor capable of turning an engine over at a speed anywhere from 75 to 150 r.p.m. will sometimes fail to start the engine, whereas hand cranking subsequently resorted to will succeed?

A. It must be borne in mind that the operation of starting an engine in cold weather involves several factors. (1) The pistons, crankpins and crankshaft (bearings) must be broken away, i.e., forcibly released from the hold that the gummed lubricating oil has on them, before they can be moved. The great difference between the power required to do this in summer and in winter is shown by the greatly increased amount of current used by the starting motor. (2) Gasoline and air must be drawn into the cylinders, to effect which in sufficient quantity to start the engine requires quite a number of revolutions. (3) The gasoline must be vaporized so that it will mix with the air, which involves more turning of the engine to create the necessary heat by compression in the combustion chambers and the friction of the moving parts. In the application of energy in any form, two factors are always involved, i.e., the unit, or quantity of power applied, and the length of time during which it is applied. The starting motor cranks the engine at a comparatively high speed for a brief period. In hand cranking, a smaller unit of power is employed, and the speed of cranking is accordingly less, but its application is continued for a much longer time. The failure of the starting motor is not due to its inferiority to hand cranking, but simply to the fact that the battery has become exhausted. Success in hand cranking where the starting motor has failed is usually due to the fact that the starting motor has done all the preliminary work, failing in the end simply because the storage battery did not have sufficient energy to finish the task. No electrical starting system can ever be any stronger than its storage battery, or source of energy.

Q. Why is it not necessary to protect the starting motor or its circuit by fuses or other protective devices as in the case of the generator?

A. A simple series-wound machine (practically ali electric starting motors are of this type) is capable of standing exceedingly heavy overloads for short periods, it being nothing unusual for these small motors to have a factor of safety of five, or even seven, for a limited time, that is, they will take five to seven times the normal amount of current for a brief period without injury. As a matter of fact, the starting motor can utilize all the current the battery is capable of supplying, provided the motor is free to move. If the engine is stuck fast or some part of the starting system has gone wrong so that the electric motor cannot turn over, then there is danger that the motor may be damaged unless the switch is opened at once. This, together with the fact that the maximum load which may be placed on the motor at different times is such a variable quantity, would make it a difficult matter to provide a fuse that would not blow unnecessarily. The only object of the fuse would be to protect the motor windings, and, as the latter can stand all the current the battery can supply, the only source of danger is the possibility of the motor being held fast so that its armature cannot revolve.

WIRING SYSTEMS

Different Plans

Q. What is the difference between the single-wire and the two-wire systems?

A. In the single-wire there is but one connection to the operating circuit by means of a wire or cable, the circuit being completed in every instance by *grounding* the other side of the circuit. For this reason the single-wire is also referred to as a grounded system. In the two-wire system, copper wires or cables are employed to complete the circuits between the generator and battery and between the battery and the starting motor, as well as to the lamps.

Q. What forms the return circuit of a single-wire system?

A. The steel frame of the chassis.

Q. How are the various circuits grounded?

A. In the case of the battery, a special ground connection is usually made by drilling the frame and fastening a clamp to it. The ground cable from the battery is attached to this clamp. The generator and starting motor are grounded internally, i.e., the end of a winding or of a brush lead that would be taken out to form the return side of a two-wire system is connected to the frame of the machine, and the latter completes the connection to the chassis through its holding bolts or other means of attachment. One side of all lamp sockets is usually grounded, so that the bulb itself completes the connection when fastened in place. Sometimes there is a special ground connection from the battery for the return side of the ignition or lighting circuits, and this ground wire is fused.

Q. What are the advantages and disadvantages of the single-wire system?

A. It greatly simplifies the wiring, as but one wire connection is necessary to the apparatus for each circuit, but this advantage renders it more susceptible to derangement through unintentional grounds or short-circuits, since the touching of any metal part of the chassis by a bare wire will cause a short-circuit. This depends to a very great extent, however, on the thoroughness with which the wiring is protected, and, with the armored cables or loom and the junction boxes used on modern installations, it is reduced to a point where both systems are practically on a par in this respect.

Q. What are the advantages and disadvantages of the two-wire system?

A. Each circuit is complete in itself thus rendering it easier to locate faults, while no one connection coming in contact with a metal part of the chassis will cause a ground. The wiring itself, however, is much more complicated, and, with the small space available on the bulb connections, it is more difficult to insulate them properly.

Q. Which system of wiring is favored?

A. The single-wire system will be found on the majority of cars, and the number of makers adopting it is steadily increasing.

Faults in Circuit

Q. What is the difference between a ground and a short-circuit?

A. So far as the effect produced is concerned, they are the same; the difference in the terms referring solely to the method of producing it. For example, if the cable of the starting motor circuit becomes abraded and the bare part touches the chassis or some connecting part of metal, this is a *ground.* But it is also a short-circuit in that the circuit to the battery is completed through a shorter path than that intended. On the other hand, if, in a two-wire system, the two cables of the same circuit become chafed close together and their bared parts touch, this is a short-circuit, but it is not a ground. For all practical purposes, however, the two terms are really interchangeable when applied to faults in the circuit.

Q. How may grounds be located in a single-wire system?

A. In any of the fused circuits, the fuse will immediately blow out. Remove the fuse cartridge and shake it; if it rattles, the fuse wire has melted and the fuse is blown. If it does not rattle, short-circuit the fuse clips with the pliers or a piece of metal; a spark will indicate the completion of the circuit and will also indicate that the fuse has blown. If, on bridging the fuse clips, the lamp lights, or other apparatus on the circuit operates, the short-circuit was only temporary. This does not mean, however, that the fault has been remedied; the vibration of the car may have shaken whatever caused it out of contact and further vibration sooner or later will renew the contact with the same result. Inspect the wiring of that particular circuit and note whether the insulation is intact throughout its length. See that no frayed ends are making contact at any of the connections and that the latter are all tight and clean. In case the lamp does not light on bridging the fuse clips, see if the bulb has blown out; if not, use the test lamp by applying one point to the terminal and the other to various points along the wiring.

Q. Does the blowing of a fuse always indicate a fault in the wiring?

A. No. A bulb, in blowing out, frequently will cause a temporary short-circuit that will blow the fuse. To determine this, apply the points of the test-lamp outfit to the bulb contacts; if the test lamp lights, the bulb is short-circuited, and a new fuse and bulb may be inserted without further inspection of the circuit. In case the test lamp does not light on this test, it does not necessarily indicate a fault in the wiring of that circuit, though inspection is recommended before putting in new fuse and bulb. The blowing out of the bulb may cause a short-circuit, which is ruptured by the current burning away the light metal parts that were in contact, such as a small piece of the filament.

Q. Can a short-circuit or ground occur without blowing a fuse?

A. Yes. No fuses are employed on starting-motor circuits owing to the very heavy current used and its great variation depending upon the conditions, such as extreme cold gumming the lubricating oil, tight bearings, binding of the pinion and gear, sprung shaft, starting motor out of alignment, or the like. On other circuits, the amount of current leaking through the fault in the circuit may not be sufficient to blow the fuse, as the capacity of the latter is such that it will carry the maximum current which the apparatus in that circuit will carry without damage—usually 5 or 10 amperes on lighting circuits and 10 amperes on generator-field circuits.

Q. How can such faults be noted?

A. The ammeter, or indicator, will show a discharge reading when the engine is idle and all lamps are switched off.

Q. What is the usual nature of such a fault?

A. The battery cut-out may have failed to open the circuit completely; a frayed end of the stranded wire at one of its connections may be making light contact which will permit a small amount of current to pass; a particle of foreign matter of high resistance may be bridging a gap either at the cut-out or some other part of the circuit; or the ignition switch may have been left on the *battery* contact so that current is flowing through the ignition coil.

Q. How may faults be located in a two-wire system?

A. With the aid of the test lamp, placing the points along the two wires of the circuit at fault from one set of terminal connections to the other, examine all connections in the circuit in question; note whether any wires have frayed ends and, if so, wind them tight together and dip in molten solder. See whether any moving part is in contact with one or both of the wires and whether the insulation of the latter has been worn off. In some two-wire systems there is a ground connection to the battery for the ignition system, in which case tests for grounds in the circuit in question must also be made. Examine the ignition switch for faults; also the switch of the circuit under test. This applies to single-wire as well as to two-wire systems.

Q. What is one of the most frequent causes of short-circuits in a two-wire system?

A. The bulbs and their sockets, owing to the very small amount of space available for the insulation. Dirt or particles of metal may be bridging the small gaps between their insulated contacts. A blown-out bulb also may be responsible, as previously mentioned.

Proper Conduction

Q. Why are different sizes of wire employed in the various circuits?

A. To permit the passage of the maximum current necessary in each circuit consistent with the minimum drop in voltage due to the resistance of the wire and its connections. The voltages employed are so low that any substantial drop due to this cause would seriously impair the efficiency of the system and particularly of the starting motor. For the latter the cables employed are not only large, but they are also made as short and direct as possible to save current as well as expense in the installation.

Q. What is the smallest wire that should be employed in automobile wiring?

A. No. 14 B. & S. gage, and this should be used only for the tail lamp, dash lamp, primary circuit of the ignition, or similar purpose. No. 10 or No. 12 is usually employed for the other lighting circuits.

Q. When, in making alterations on a car, it becomes necessary to extend a circuit, what should be done?

A. The ends of the wires should be scraped clean and bright for at least 2 inches, and a lineman's joint made with the aid of the pliers to insure having it tight. A lineman's joint is made by crossing the bared ends of the wires at their centers at right angles to each other, then wrapping or coiling each extending end tight around the opposite wire; the joint then should be soldered and well taped. A circuit should be extended only by using wire of the same size and character of insulation. None of the foregoing applies to the starting-motor circuit. It is inadvisable to lengthen this circuit if avoidable, but in the rare instances when it would be necessary, new cable of the same size or larger and with the same insulation should be cut to the proper length and the old cable discarded. All terminals should be solidly fastened to the new cable by soldering.

Q. Why is it necessary to use such heavy cable for the connection of the starting motor to the battery?

A. It is essential that the exceedingly heavy starting current be transmitted with the minimum of loss.

Q. What is considered the minimum permissible loss in the starting-system wiring?

A. One maker specifies that the starting cable must be large enough to transmit a maximum current of 400 amperes with not over one-fourth volt total loss.

Q. Why is it important to hold the voltage drop down to a maximum so small as to be negligible in almost any other application?

A. Owing to the heavy current necessary, as a drop of but $\frac{1}{4}$ volt in potential with a current of 400 amperes represents a loss of 100 watts, or close to $\frac{1}{7}$ horsepower. Of course, the current seldom reaches such a high value as this except when a motor is exceptionally stiff, as in severe cold weather or just after its bearings have been set up very tight; moreover, this loss takes place at the instant of starting only, but it is just at this time that the highest efficiency and full battery power is needed to start without spinning the engine too much.

Q. On some of the early systems whose efficiency was not of the best, how can the proper size of cable to use between the starting motor and battery be determined?

A. Test the starting motor with a high-reading ammeter (scale should read to at least 300 amperes) after having made certain by hydrometer and voltage tests that the storage battery is fully charged. (See instructions regarding this.) Carefully note ammeter reading exactly at instant of closing switch, to determine maximum current flow. Measure the length of cable between the battery and the starting motor, i.e., both sides of starting switch. Then maximum starting current times 10.7 times number of feet of cable used, divided by .25 will give the cross-section of the wire in circular mills. For example, assume that the starting motor required a maximum of 300 amperes momentarily to break away the engine, and five feet of cable are employed for the connections. Then

$$\frac{300\times10.7\times5}{25}=64{,}200 \text{ circular mills}$$

By consulting the table which gives the various size wires in circular mills and their equivalent in gage sizes, it will be noted that the closest approach to this is No. 2 cable, which is 66,373 circular mills, so that the largest size cable would have to be used. If the starting cable used on an old system which does not show particularly good efficiency is much smaller than this, it would probably be an advantage to replace it with larger cable, assuming, of course, that every other part of the system is in good condition and working properly.

Q. Why should connections be inspected frequently?

A. The vibration and jolting to which they are subjected in service is so severe that no mechanical joint can be depended upon to remain tight indefinitely.

Q. What harm does a loose or dirty connection occasion?

A. A loose connection causes the formation of an arc between its contacts whenever vibration causes the parts to separate temporarily. This wastes current and burns the metal away, leaving oxidized surfaces which are partially insulating, thus increasing the resistance at the connection. Dirt getting between the surfaces of the connector has the same effect; the resistance is increased and there is a correspondingly increased drop in the voltage of the circuit, which cuts down its efficiency.

Q. Why should all terminals be well taped when the battery, starting motor, generator, or other apparatus is temporarily disconnected for purposes of inspection or test?

A. To prevent accidental short-circuits which would be caused by these terminals coming in contact with any metal part of the chassis on a single-wire system. Such a short-circuit would ruin the battery and burn out any lamps that happened to be included in the circuit. This precaution applies with equal force to the two-wire systems, as in this case the terminals of the different wires might come together, or there might be a ground connection in the system.

PROTECTIVE AND OPERATIVE DEVICES

Q. What are the protective devices usually employed on electric systems?

A. Fuses in the separate lamp circuits, in the ground connection, and in the field circuit of the generator on some machines; battery cut-out for the charging circuit; circuit-breaker which takes the place of the fuses.

Fuses

Q. What is a fuse and what is its function?

A. A fuse consists of a piece of wire of an alloy which melts at a low temperature and which will only carry a certain amount of current without melting, the latter depending upon the diameter of the wire, i.e., cross-section and the nature of the alloy. The fuse is usually in the form of a cartridge, the wire being encased in an insulating tube having brass ends, to which the ends of the wire are soldered. These brass ends are pressed into spring clips to put the fuse in circuit. In some cases open fuse blocks are employed, the wire itself simply being clamped under the screw connectors on the porcelain block. The function of the fuse is to protect the battery and the lamps when, by reason of a ground or short-circuit in the wiring, an excessive amount of current flows.

Q. When a fuse blows out what should be done?

A. Investigate the cause before replacing it with a new one.

Q. Is it permissible to bridge the fuse gap with a piece of copper wire when no replacements are at hand?

A. Only in cases of emergency and after the short-circuit which has caused the fuse to blow has been remedied. The finest size of copper wire at hand, such as a single strand from a piece of lamp cord, should be used. If this burns out, there being no ground or short-circuit in the wiring, use two strands. Remove the wire as soon as a new fuse is obtainable.

Q. Why are fuses not employed in the starting-motor circuit?

A. In the starting circuit the current necessary is so heavy and varies so widely with the conditions that it would not be practicable to provide a protecting fuse.

Q. What does the intermittent blowing of the fuse on the same circuit indicate?

A. A short-circuit that is caused by the vibration, or jolting, of the car. The wire, lamp socket, or other part of the circuit that is at fault is shaken loose at times so that the circuit is operative, and a new fuse may be inserted without instantly blowing, as it would do were the short-circuit constant. This is often the case as the car is stopped to inspect the wiring and insert the new fuse, and standing still lets the part drop out of contact; starting up shakes it into contact once more and blows the new fuse. Loose connections, wires with abraded insulation, and bulbs loosely inserted in their sockets are apt to cause trouble of this nature.

Q. Does the blowing out of a fuse necessarily indicate a fault in the wiring or in some other part of the system?

A. No, since a bulb in burning out will frequently cause the fuse to blow out. This is due to the fact that in breaking, the end of the parted filament of the bulb may fall across the other terminal where it comes through the glass, thus causing either a short-circuit or such a reduction in the ordinary resistance as to permit a much heavier rush of current than normal, with the result that the fuse goes. To test, leave burnt-out bulb in place temporarily; short-circuit fuse clips with screw driver or pliers, just touching them momentarily; if no spark results, replace bulb with a new one and test again; if a spark occurs, remove old bulb and test again with no lamp in place; then if no spark occurs in bridging the fuse terminals, the circuit is all right, and the fuse may be replaced.

Q. When all the lighting fuses blow out at once, what does this indicate?

A. A short-circuit across the lighting-switch terminals would cause this. In some switches with exposed rear terminals, it is possible to place a screwdriver or similar piece of metal in such a position that it bridges practically all the switch terminals. If the lighting switches were all closed at the time, this would short-circuit them.

Circuit=Breaker

Q. What is a circuit=breaker, and what is its function?

A. The circuit-breaker is an electromagnet with a pivoted armature and contacts, similar in principle to the battery cut-out. All the current used in the various circuits, except that of the starting motor, passes through it, and its contacts normally remain closed. The winding of the magnet coil is such that the normal current used by the lamps or ignition does not affect it, but the passage of an excessive amount of current will energize the magnet, attract the armature, and break the circuit. The spring holding the armature away from the magnet will again close the circuit, and the circuit-breaker will vibrate until the cause has been removed. This is usually a ground or short-circuit. The function of the circuit-breaker is to protect the battery and lamps in place of the usual fuses.

Q. If the circuit-breaker operates when there are no faults in the wiring, what is likely to be the cause?

A. Its spring may have become weakened so that the vibration of the car causes it to operate on less current. The Delco circuit-breaker is designed to operate on 25 amperes or more, but, once started, a current of 3 to 5 amperes will keep it vibrating. If tests show that no faults in the wiring or connections exist, increase the spring tension with the ammeter in circuit until the reading of the latter indicates that the circuit-breaker is not operating on the current of less value than that intended. See that the contacts are clean and true.

Battery Cut-Out

Q. What is a battery cut-out?

A. It is an automatic double-acting switch which is closed by the voltage of the generator and opened by the current from the battery.

Q. Of what does it consist?

A. It is essentially a double-wound electromagnet with a pivoted armature and a pair of contacts. One winding, known as the *voltage coil,* is of fine wire and is permanently in circuit with the generator. The second winding of coarse wire is termed the *current coil* and is put in circuit by the contacts.

Q. Why is a cut-out necessary?

A. To protect the storage battery. When the generator speed falls below a certain point, it no longer produces sufficient voltage to charge the battery, and the latter then would discharge through the generator windings if not prevented. This discharge would always take place when the generator was idle, except for the cut-out.

Q. How does it operate?

A. When the generator voltage approaches the value necessary for charging, it energizes the magnet through the voltage coil and closes the contacts, cutting in the current coil, which further excites the magnet and holds the contacts firmly together. The closing of these contacts puts the battery in circuit and it begins to charge. As soon as the generator speed falls below the point necessary for charging, the battery voltage overcomes that of the generator and sends a current in the reverse direction through the current coil, causing the contacts to separate and cutting the battery out of the charging circuit.

Q. If the generator is run for any length of time at or near this critical speed, what is to prevent the cut-out from vibrating constantly instead of working positively one way or the other?

A. The resistance of the windings is so proportioned that there is a difference of 1 to 2 volts between the cutting-in and the cutting-out points.

Q. What is the result when the battery cut-out—which is variously termed a cut-out, a circuit-breaker, an automatic switch, and a reverse-current relay or an automatic relay—fails to operate?

A. If it fails to cut in, i.e., the contacts do not come together, the battery does not charge and will quickly show a falling-off in capacity, such as inability to operate the starting motor properly or to light the lamps to full brilliance. If it fails to cut out, the battery charge will be wasted through the generator windings with the same indications of lack of capacity.

Q. What is the most frequent cause of trouble?

A. Automatic cut-outs have been perfected to a point where but little trouble occurs. Freezing or sticking together of the contacts due to excessive current will most often be found to be the cause of the device failing to cut out when the generator is stopped. The points should be cleaned and trued up as described in previous instructions. Loose or dirty connections making poor contact may insert sufficient extra resistance in the circuit to prevent the device from cutting in at the proper point. Excessive vibration, particularly when the cut-out is mounted on the dash, may prevent the contacts from staying together as they should when the engine is running at or above the proper speed. See that the cut-out is solidly mounted. Temporary loss of battery capacity may be due to slow driving over rough roads at about the speed at which the cut-out is designed to put the battery in circuit.

Q. None of the above causes existing, what further tests may be made?

A. The windings may be tested as already described for the generator windings, but trouble from this source is equally rare. If the contacts are clean and true and the connections are tight, look for a loose connection elsewhere, as at the generator or battery or the ground on the frame. A loose connection vibrates when the car is moving, constantly opening and closing the circuit and causing the cut-out to do likewise, so that the battery does not charge. A wire from which the insulation has been abraded will also vibrate, owing to the movement, causing an intermittent short-circuit. With all contacts and connections in good condition, failure to cut out indicates a ground or short-circuit between the battery and cut-out; failure to cut in indicates similar trouble between the generator and the cut-out.

Q. Is a battery cut-out necessary on every electrical system?

A. No. On single-unit systems of the type of the Dyneto, in which the generator becomes *motorized* as soon as its speed and consequently its voltage drops below a certain point, the battery is always in circuit. A plain knife-blade switch, which also controls the ignition, is closed to start and left closed as long as the car is running. But the engine must not be allowed to run at a speed below which it generates sufficient voltage to charge the battery, nor must the switch be left closed when the engine is not running; otherwise, the battery will discharge through the generator windings.

Q. After having trued up points of a battery cut-out, what precautions should be taken in adjusting them?

A. To insure proper operation, they must be set to the distances given in the manufacturer's instructions. This refers not only to the gap between the contact points themselves, but also to the distance that the armature must be set from its backstop when the points are open and to the air gap between the armature and the magnet. These distances are very small in every case, and it is important that they be adjusted accurately. They differ slightly on cut-outs of different makes and also on different models of the same make. For example, in the Gray & Davis cut-out, the distance between the contact points should be .015, the air gap between the armature and its backstop not less than .010, and the armature air gap, or distance between the armature and the magnet face, .030. These dimensions refer to the flexible, or spring-arm type, while in the solid-arm type of the same make, they are .010 for the distance between the contact points and .015 for the armature air gap, it being necessary that the armature should be set parallel with the pole face of the magnet.

Q. How can these small distances be accurately determined with the facilities ordinarily found in a repair shop?

A. The manufacturers usually supply a small adjusting wrench, the different edges of which have been ground to varying thicknesses representing the proper distances for the various gaps. Lacking one of these, small pieces of strip brass or steel may be ground or filed down to the proper size and gauged with a micrometer, which should be part of the equipment of every garage. The strips should be stamped with the dimensions and name of gap for identification.

Q. How often will the point of a battery cut-out need adjustment, or truing up?

A. Service conditions vary so greatly that it is impossible to give any definite average for this, particularly as the instruments themselves also are a variable quantity, but, under ordinarily favorable conditions, they should not require attention more than once a year.

Contact Points

Q. Why is it necessary to make contact points of such an expensive metal as platinum, and why is the latter sometimes alloyed with irridium?

A. There is no other metal which withstands the oxidizing effect of the electric arc and still maintains a clean and bright conducting surface as does platinum. Irridium is added to make the platinum harder, so that it will be more durable. On cheaply made instruments in which no platinum has been used in the contacts, trouble will be experienced constantly with the contacts.

Q. Is there any substitute for platinum or any metal that approaches it in adaptability for contact points?

A. There is no substitute for platinum, and the only metal that approaches it is silver. Where contact points only separate occasionally at intervals, as in the Remy thermoelectric switch, the use of silver contacts is permissible; but in a battery cut-out, or a regulator in which the vibration of the points is more or less constant, nothing will serve so reliably as platinum.

Q. What is the cause of the platinum contacts burning into such irregular ragged forms?

A. When a current of electricity passes through a contact of this nature, the material of the positive electrode (i.e., contact point connected to the positive side of the circuit) is carried over by the current in the shape of metallic vapor, or infinitely fine particles, and deposited on the negative electrode. The positive consequently takes on the form of a sharp point, while the negative has a depression formed in it, usually referred to as a "peak and crater", which the two points resemble in miniature after long use. This peak and crater effect is much more noticeable in an old-style carbon arc lamp after it has been burning only a few hours.

Q. What can be done to prevent this?

A. The passing of the metal from one electrode to the other cannot be prevented, as it is a function of any arc or spark. It can be minimized, however, by keeping the contacts in good condition so that the sparking is reduced to a minimum.

Q. Can the formation of the pack and crater effect, which so greatly reduces the efficiency of the contacts, be avoided?

A. The use of a reversing switch in the circuit, as in the case of the magneto or the battery-type interrupter which changes the direction in which the current flows through the points every time it is turned on, will overcome this. Where there is no reversing switch in the ignition circuit or where one cannot be used, attention to the points at regular intervals will prevent this effect from reaching a stage where most of the point has to be filed away to true it up.

Q. In the use of the file, sandpaper, or emery cloth in this connection, just what is meant by truing the points up?

A. Their surfaces must be made exactly parallel to one another so that when the points come together they touch uniformly over their entire surfaces. In the hands of the unskilled user, there is a tendency to bear down sidewise with the file, thus forming rounded edges on the points. In addition to having the faces of the two points perfectly parallel, the face of each point must be at right angles to its sides. Otherwise, there is bound to be unnecessary sparking between the points, and this causes them to burn away again much sooner. It is scarcely necessary to add that as little as possible of the metal should be removed. As long as there is enough of the platinum left to make true parallel surfaces, the points need not be replaced if the means for adjustment permits utilizing them when worn far down.

Q. What is the cause of the points freezing, or sticking, together?

A. Permitting them to wear down to a point where they are in very poor condition and where the gap between parts of their surfaces causes the formation of a heavy arc, or hot flash of current, which practically welds them together. By giving them the necessary attention at regular intervals, this may be avoided.

Q. How often should the contact points need attention?

A. When new, they should run for a year or more without any attention. After they have been trued up, the succeeding interval will often depend upon the skill and care with which this has been carried out.

Switches

Q. How do switches as employed on the automobile differ in principle and operation?

A. Starting-circuit switches are either of the knife-blade or the flat-contact type, while in the majority of cases the lighting switches are of the push-button type, though knife-blade switches are used for this purpose also. In some instances, one of the brushes of the machine is made to serve as a switch, as in the Delco. Ordinarily, the switch is normally held open by a spring and is closed by foot pressure, the spring returning it to the open position as soon as released. A variation of this is the Westinghouse electromagnetically operated switch in which a solenoid takes the place of foot operation. The circuit of the solenoid is controlled by a spring push button, which is normally held out of contact. Single-unit systems, such as the Dyneto, in which the machine automatically becomes motorized when the speed drops below a certain point, are controlled by a standard single-throw single-pole knife-blade switch which is left closed as long as the machine is running.

Q. What faults may be looked for in switches?

A. Loose connections; weakening of the spring; burning of the contact faces in the knife-blade type, due to arcing caused by releasing too slowly; dirt or other insulating substance accumulating on the contact faces of the flat-contact type; failure to release through binding.

Q. Why is it important to keep the switch contact faces clean and bright?

A. Dirt or burned surfaces increase the resistance and cause a drop in the voltage at the starting motor. The energy represented by an electric current is a measure of the volume or amperes times the voltage or pressure under which it flows, and, as such low voltages are used, only a slight falling off represents a serious percentage of the total potential. With a dirty switch or one that makes poor contact, current that should be utilized in the starting motor is wasted in overcoming the resistance of the switch.

Q. Why is it inadvisable to insert an extra switch in the starting circuit, as is done in some cases by owners to insure against theft?

A. Because of the drop in voltage. The loss in switches as designed for lighting circuits is about 1 per cent, or a little over 1 volt. If the same switch is used on the low voltage of the starter system, the loss is then equivalent to about 10 per cent.

LIGHTING AND INDICATORS

Lamps

Q. How many types of bulbs are there in general use on automobiles?

A. Four: miniature and candelabra screw base, and single- and double-contact bayonet-lock base, both of the latter being of the candelabra size.

Q. Are these types equally favored?

A. No. The screw-base type, particularly in the miniature size, will be found only on old cars, and this type, generally speaking, is practically obsolete on the automobile, as the vibration tends to unscrew the lamp. Of the bayonet-lock type, the single-contact style is steadily gaining favor. Ten million bulbs for automobile lighting were produced in 1915 (S.A.E. report) and of these 67 per cent were of the single-contact type.

Q. In how many different voltages are these bulbs made?

A. Four: a 6—8-volt bulb for a 3-cell or 6-volt system; 12—16-volt bulb for 6-cell or 12-volt systems; and 18—24-volt bulbs for 9-cell systems; 3—4-volt bulbs for tail-light and dash-light use, where these lights are burned in series on a 6-volt system.

Q. Are these the only voltages in which the bulbs are made?

A. No. They are the types that are being standardized to reduce the stock of replacements that it is necessary for a garage to carry. It has been customary for the lamp manufacturer to supply bulbs made exactly for any voltage that the maker of the electric system ordered. Taking into consideration only the standard sizes now listed for use on 3-, 6-, and 9-cell systems, and the different bases regularly used, there are about twenty-four different bulbs that should be stocked by a garage. In addition, about forty other sizes are in general use, and if individual voltages had to be supplied, considering the different standard bases, a stock of over two-hundred different bulb sizes would be required.

Q. Why is the voltage of a bulb expressed as "6—8", "12—16", etc.?

A. Owing to the rise and fall of the battery voltage according to its state of charge, this variation must be provided for, or the lamps would be burned out when the battery was fully charged. Headlight bulbs for 3-cell systems are made for $6\frac{1}{2}$ volts, while the side, rear, and speedometer lights are made for $6\frac{3}{4}$ volts, owing to the lesser voltage drop in their circuits, but they will all operate satisfactorily on a potential that does not exceed 8 volts or does not drop below 6 volts.

Q. When all the lamps burn dimly, what is the cause?

A. The battery is nearly exhausted, in which case its voltage will be only 5.2 to 5.5 volts for a 3-cell system. The car should be run with as few lights as necessary to permit the generator to charge the battery quickly.

Q. What is the cause of one light failing?

A. Bulb burned out or its fuse blown; examine the fuse before replacing the bulb and if blown, examine the wiring before putting in a new bulb. Poor contact; see that the lamp is put in properly and turned to lock it in place. A double-contact bulb may have been put in single-contact socket, or *vice versa.*

Q. Why will one lamp burn much brighter than the other?

A. A replacement may have been made with a bulb of higher voltage; a 12-volt bulb will give only a dull red glow on a 3-cell system. Where the difference is not so marked as this, but still very perceptible, it may be due to the difference in the age of the lamps. As a bulb grows old in service, its filament resistance increases, so that it does not take so much current and will not burn as brightly as when new.

Q. Will the failure of a bulb cause its fuse to blow though there is no fault in its circuit?

A. This sometimes happens owing to the breaking down of the filament, causing a short-circuit when the lamp fails.

Q. Can the proper voltage bulbs needed for any system always be told simply by taking the total voltage of the battery, i.e., the number of cells times 2?

A. No. Always examine the burned out bulb and replace with one of the same kind. Many 6-cell systems use 6-volt lamps and are known as 12—6-volt systems. The battery is divided into two groups in series parallel for lighting and sometimes for charging, all the cells being in series for starting. Other arbitrary voltages are also adopted; for example, 14-volt bulbs are used on 12-cell systems, the battery being divided in the same manner, so that this would be a 24—12-volt system. The only safe way to order replacements is to give the voltage on the printed label on the old bulb and state the make of the system on which it is to be used.

Q. What type of bulb is used where the current is taken from the magneto, as on the Ford?

A. As supplied by the maker, only the headlights are wired, and they are in series, and in recent models a 9-volt bulb is used, but the above instructions for replacements will apply here also. Ordinarily, double-contact bulbs are required, unless the fixtures are insulated from one another, in which case the single-contact type can be used.

Q. Why is a bulb of a voltage lower than that of the system itself often employed on 6-, 9-, and 12-cell systems?

A. The lower the voltage, the thicker the filament can be made. A short comparatively thick filament concentrates the light and makes the bulb easier to focus; it is also much more durable than the thin filament required for higher voltages.

Q. Under what conditions will the best results be obtained from the head lamps?

A. When the bulbs are in proper focus with the lamp reflectors. The usual focal length for headlight bulbs is $\frac{13}{16}$ inch, and the focal length of the reflector is made greater than this to permit of adjustment. The center of the filament should be back of the focus of the reflector to spread the beam of light. In this position a greater number of the light rays are utilized and redirected by the reflector, producing a higher beam candlepower. If the center of the filament is forward of the focus, the lower part of the reflector will produce the most glare and throw it into the eyes of pedestrians and approaching drivers.

Q. How can the headlights be focused?

A. Place the car in position where light can be directed against a wall about 100 feet distant. Adjust the bulbs backward or forward until the spotlight on the wall is most brilliant and free from black rings and streaks. When this position is found, lock the bulb securely in place. Focus each headlight separately. See that the lamp brackets are set so that the light is being projected directly ahead.

Q. How can metal headlight reflectors be cleaned when discolored?

A. Wash by directing a gentle stream of cold water against the surfaces and allow to dry without touching them. The reflectors should never be rubbed with cloth or paper as it will scratch the highly polished surfaces. If they become very dull, it will be necessary to have them replated.

Q. What is the meaning of the identification marks usually placed on bulbs, in addition to the voltage, such as "G-6"?

A. This refers to the size and shape of the bulb. The diameter of the glass bulb is expressed in eighths of an inch and its shape by a prefixed *G* for round (globular), *T* for tubular, *S* for straight-side, etc. Thus, G-6 is a round bulb $\frac{6}{8}$ inch or $\frac{3}{4}$ inch in diameter.

Instruments

Q. What instruments ordinarily are employed in connection with electric systems on the automobile?

A. Either a double-reading ammeter, a volt-ammeter, or an indicator, the first named being employed generally. The ammeter shows whether the battery is charging or discharging or whether no current is passing; the indicator reads either *Off* or *On*; while the voltammeter gives the voltage, usually upon pressing a button to put it into operation, in addition to the readings already mentioned.

Q. On what circuits are the indicating instruments placed?

A. The charging circuit from the generator to the battery, and the lamp and ignition circuits.

Q. Why is an ammeter not used for the starting-motor circuit?

A. The current is so heavy and varies so greatly with the conditions that an ammeter designed to give an accurate reading of it would not be sensitive enough to indicate the smaller amounts of current used by the lamps, or produced by the generator for charging. Furthermore, the starting motor is intended only to be used for very short periods, while the other circuits are in constant use.

Q. Do the small ammeters employed fail very often?

A. Considering the unusually severe treatment to which they are subjected by the vibration and jolting of the car, their failure is comparatively rare, but as the conditions are so severe for a sensitive indicating instrument, too much dependence should not be placed on the ammeter reading when making tests.

Q. What are the usual causes of failure?

A. Failure to indicate—the generator, wiring, and other parts of the circuit being in good operative condition—may be caused by the pointer becoming bent, so as to bind it; the pointer may have been shaken off its base altogether by the jolting, or one of its connections may have sprung loose from the same cause.

Q. How can the ammeter reading be checked?

A. By inserting the portable testing voltammeter in circuit with it, using the 30-ampere shunt and comparing the readings. The dash ammeter must not be expected to give as accurate a reading as the finer portable instrument. Failing the latter, a spare dash ammeter may be employed in the same manner and the spare may be tested beforehand by connecting to a battery of 4 dry cells in series; if brand new, they should give a reading of 18 to 20 amperes. Do not keep the ammeter in circuit any longer than necessary to obtain the reading, as it only runs the cells down needlessly.

Q. Should an ammeter ever be used in testing the storage battery?

A. No. Because it practically would short-circuit the battery, burn out the instrument, and damage the battery itself. Nothing but a voltmeter should be employed for this purpose, as its high resistance coil permits only a small amount of current to pass. An ammeter reading from a storage battery gives no indication whatever of its condition, whereas the voltage affords a close check on the state of charge, varying from 1.75 for a completely discharged cell to 2.55 volts for a fully charged one, the readings always being taken when the battery is either charging or discharging. The voltage on discharge will not be as high as on charge, the conditions otherwise being the same.

Q. Why are indicators employed on some systems instead of ammeters?

A. As the indicator is not designed to give a quantitative reading, it need not be so sensitive as an ammeter and accordingly can be made more durable.

Q. What are the most frequent causes of failure of an indicator?

A. Usually of a mechanical nature caused by the jolting, such as the target being shaken off its bearings, broken wire, etc.

Q. When the engine is running slowly, and the ammeter or the indicator flutters constantly, going from "On" to "Off" at short intervals, in the case of the indicator, or from a small charging current to zero, in the case of the ammeter, what does this signify?

A. That the setting of the battery cut-out is very sensitive and that the engine is then running at or about the speed that the instrument should cut-in. Since the speed of an engine varies considerably when running slowly, picking up momentarily and then falling off for a longer period, there is a corresponding variation in the potential, causing the cut-out to operate intermittently. This is a condition that seldom occurs and results in no harm when it does.

Q. When the ammeter or indicator flutters in the same manner with the engine running at medium or at high speed, what does it indicate?

A. That there is a loose connection between the generator and the cut-out, or an intermittent short-circuit or ground caused by a chafed wire alternately making contact with some metal part owing to the vibration. It is much more likely to be simply a loose connection and will be found most often on the back of the cut-out itself. This should be remedied at once. If neglected, it will cause abnormal wear of the platinum points in the cut-out.

Q. When the ammeter does not indicate "Charge" though the engine is speeded up, but does register a discharge when the lights are turned on and the engine is idle, what is the nature of the trouble?

A. Either the generator is not producing current or the regulator (where an external type is employed) is not working properly. The generator brushes may not be making proper contact with the commutator, or there may be a loose, corroded, or broken connection in the generator cut-out battery circuit. Where a belt drives the generator, it may be too loose to run the machine at its proper speed.

Q. When the ammeter gives no charging indication though the lamps are off and the engine is speeded up, and gives no discharging indication though the engine is idle and lamps are switched on, what is likely to be the cause?

A. There is an open or a loose connection in the battery circuit or in the battery itself. The ammeter may be at fault. See that its indicating pointer has not become jammed nor dropped off its bearings.

Q. In case the ammeter indicates "Discharge" though the engine be idle and all lights turned off, what is the trouble?

A. There is a short-circuit or a ground somewhere in the lighting circuits or between the battery and the ammeter, as the discharge reading in such circumstances indicates a leakage of current; or the cut-out has failed to operate and still has the battery in circuit with the generator, though the engine is stopped. The ammeter pointer may be bent.

Q. When the meter indicates a charge though the engine is at rest, what is the nature of the fault?

A. The ammeter pointer has become bent or deranged so that it is stuck fast in place, showing a charge.

Q. When the ammeter charge indications are below normal, what is apt to be the cause?

A. The generator commutator or brushes may need attention, such as cleaning or sanding-in, or new brushes may be necessary. The generator speed may be too low; in case of belt drive, it may not bé getting the benefit of the full speed of the engine owing to a slipping belt. The regulator (external type) may not be functioning properly, or there may be an excessive lamp load on the generator.

Q. When the ammeter charge reading is above normal, what is likely to be the cause?

A. There may be a short-circuited cell in the battery, or a short in the charging circuit, or the regulator (external type) may not be working properly.

Q. What will cause the discharge reading of the ammeter to become abnormally high?

A. The lamp load may be excessive, as where higher candle-power bulbs are used, or more lights than originally intended are put in the circuit. There may be leakage in some part of the lighting Circuit, or the cutout contacts may be stuck together, permitting a discharge through it or through the generator.

ELECTRIC GEAR-SHIFT

Q. What is the operating principle upon which the electric gear-shifting mechanism is based?

A. That of the solenoid and its attraction for its core when a current is passed through its winding.

Q. What is the source of current supply for the electric gear-shift?

A. The storage battery of the lighting system. The operation of gear-shifting is carried out so quickly that only a nominal additional demand is made on the battery.

Q. How is the electric gear-shift controlled?

A. By a series of buttons corresponding to the various speeds and located on the steering wheel, and by a master switch.

Q. What is the object of the buttons, and what are they termed?

A. To partly close the circuit to the particular solenoid of the speed desired. They are termed "selector switches" since they permit selecting in advance the speed desired.

Q. Why is a master switch employed, and why is it so called?

A. To avoid the complication which would otherwise result from the necessity of providing two switches for each change of speed, i.e., a selector switch and an operating switch. It is termed a master switch because it controls the current supply to all of the circuits.

Q. Why is a neutral button provided in addition to the buttons for the various speeds on the selector switch?

A. To return any of the selector buttons to neutral without the necessity of going through that speed in case it is not desired to engage the speed in question after the button has been pushed. Also to open any of the selector switches that may be closed when it is desired to stop.

Q. What is the neutralizing device?

A. It is a mechanism incorporated with the shifting mechanism to open the master switch automatically after the gears have been engaged.

Q. Why is the neutralizing device necessary?

A. If it were not provided, the master switch would remain closed, causing a constant drain on the battery and rendering the mechanism inoperative after one shift had been made.

Q. How many solenoids are provided in the standard three-speed and reverse gear box?

A. One for every movement necessary.

Q. Is the current sent through a solenoid in one direction to pull the shifting bar into it and then in the opposite direction to move the bar the other way?

A. No, the current is not reversed through the same solenoid. After the left-hand solenoid, operating the first-speed gear, for example, has pulled the shifter bar to the left, a second solenoid, on the opposite end of the same bar, is energized to pull it back to the right, to shift to second or intermediate. The current is sent through a different solenoid by means of the selector switches for each shift desired.

Q. When the electric gear-shift failed to operate, where would be the most likely place to look for the cause of the trouble?

A. First see that the battery is not exhausted, then that no connections between the battery and the terminal block have parted, thus cutting off the current supply. The wiring is so simple and so strongly protected that it is very unlikely to have anything happen to it except at the connections. This is likewise true of the solenoids.

Q. In case the battery is amply charged and nothing is wrong with the connections, what procedure should be followed?

A. Use the lamp-testing set described in connection with the lighting and starting systems and test out the various circuits as shown on the wiring diagram. In using this test, it must always be borne in mind that touching the two points to the same or connecting pieces of wire or metal will always cause the lamp to light. It is useful in this way for indicating the continuity of a wire, i.e., that it has not broken under the insulation, but, until experience has been gained in its use, it will be nothing unusual to find that the points have been touched to connecting pieces of metal which have no relation to the circuit. As such metal will complete the circuit through the lamp, the latter will light, but without indicating anything of value to the trouble hunter. Always test the lamp itself before proceeding. It may have become partly unscrewed in its socket or its filament may have been broken.

BATTERY

Electrolyte

Q. Why is it necessary to refill the battery jars at regular intervals?

A. Because the heat generated in the cells evaporates the water from the electrolyte, and, if the latter is permitted to fall below the tops of the plates, they will dry out where they are exposed, and the heat of charging will then cause them to disintegrate, ruining the battery.

Q. Why should this be done at intervals of not less than two weeks?

A. Because the limited amount of electrolyte permitted by the restricted size of the cells over the plates—usually one-half inch—will be evaporated in that period by a battery that is in more or less constant use.

Q. Why should water alone and never acid or electrolyte be used to make up this loss?

A. Only the water evaporates, so that if either acid or fresh electrolyte is added, it will disturb the specific gravity of the solution in the cells and totally alter their condition.

Q. What is the reason that battery manufacturers insist that only distilled water or its nearest equivalent, rain water or melted artificial ice, be used for this purpose?

A. Because ordinary water contains impurities that are apt to harm the plates, such as iron salts, or alkaline salts that will affect both the plates and the electrolyte.

Q. What should be done to a battery that has had its efficiency impaired by being filled with impure water?

A. The cells should be taken apart, the separators discarded, the plates thoroughly washed for hours in clean running water without exposing them to the air where they would dry, the jars washed out, the plates reassembled with new separators, the jars filled with fresh electrolyte of the proper specific gravity, and the battery put on a long slow charge from an outside charging source, i.e., not on the car itself. Unless there are proper facilities for carrying this out, it will be preferable to ship the battery back to the maker so that it can be given proper treatment, particularly as it is necessary to reseal the cells.

Q. How is electrolyte prepared?

A. By adding pure sulphuric acid a very little at a time to distilled water until the proper specific gravity is reached, and then permitting the solution to cool before using. The mixture must always be made in a porcelain, hard rubber, or glass jar; never in a metal vessel. Commercial sulphuric acid or vitriol should not be employed, as it is far from pure. Never add water to acid. When the two are brought together, their chemical combination evolves a great amount of heat, and the acid will be violently spattered about.

Q. How often should distilled or rain water be added to the cells?

A. This will vary not alone with different systems but with different cars equipped with the same system, owing to the difference in conditions of operation. The only way to determine this definitely is to inspect the cells at short intervals and note how long they will operate before the electrolyte gets close enough to the tops of the plates to require additional water. This may be a week, ten days or two weeks, or even more, if the car is not run much.

Q. When a battery requires the addition of water at very short intervals to keep the level of the electrolyte one-half inch above the plates, what does this indicate?

A. It shows that the battery is being constantly overcharged, which keeps it at a high temperature, causing excessive evaporation. This will usually occur where a car is in constant use during the day but is driven very little at night. It may be remedied by adjustment of the regulation so as to reduce the output of the generator. Where this is not possible, as in the case of simple bucking-coil regulation which is entirely self-contained and permits no variation, additional resistance may be introduced in the generator-battery circuit. This may take the form of a small-resistance unit consisting of German silver or other high-resistance wire wound on a porcelain tube and mounted on the forward side of the dash. A single-pole knife switch should be placed in the circuit with the resistance so that the latter can be cut in or out of the generator circuit as circumstances may require.

Q. With conditions as in the preceding question, how can the amount of resistance to be inserted in the circuit be figured?

A. By the use of Ohm's law. In this case, it would be $R=\frac{E}{C}$, or resistance equals voltage divided by current. How much resistance to use can only be answered by the conditions of operation. Where a car is used steadily during the day and very seldom at night, it may be necessary to reduce the charge by two-thirds. In the case of a 6-volt system normally charging at 12 amperes, the generator delivers current at 7 to $7\frac{1}{2}$ volts in order to overcome the voltage of the battery when fully charged. Selecting 7 volts, we see that the resistance in circuit when the current is 12 amperes is $7 \div 12$, or .6 ohm approximately. Now when the charging current is 4 amperes, we must have $7 \div 4$, or 1.75 ohms in circuit; that is, to reduce the current from 12 to 4 amperes, a resistance of $1.75 - .6$, or 1.15 ohms must be inserted. The amount of resistance wire necessary to give this resistance or any other resistance necessary may be found in tables of wire sizes and resistances of special wire employed for this purpose. The wire is bare and must be wound on the tube so that adjacent coils do not touch. An extreme instance is cited here. It may be necessary in many cases to reduce the charging rate by a very much smaller fraction. Ordinarily the charging rate should not be altered.

Q. When the battery is constantly gassing, or "boiling", as the car owner usually puts it, what is the trouble?

A. It is being constantly overcharged. This will greatly reduce the life of the battery, and the charging rate should be reduced, as mentioned in the preceding answer. It is essential that the battery be kept fully charged; but if it is continually overcharged, this will keep the cells at an abnormal temperature which is injurious to the plates. The battery treatment will vary with the season, for the demand on it is much heavier during cold weather than in summer.

Hydrometer Tests

Q. Why should the battery be tested with the hydrometer at regular intervals of a week or so?

A. Because the specific gravity of the electrolyte is the most certain indication of the battery's condition.

Q. What should the hydrometer read when the battery is fully charged?

A. 1.280 to 1.300.

Q. What point is it dangerous to permit the specific gravity of the electrolyte to fall below, and why?

A. 1.250; because below this point, the acid begins to attack the plates and the battery plates sulphate. The lower the specific gravity, the faster sulphating takes place.

Q. What should be done when the hydrometer reading is 1.250 or lower?

A. The battery should be put on charge immediately, either by running the engine or by charging from an outside source of current until the gravity reading becomes normal.

Q. If the hydrometer reading of one cell is lower than that of the others, what should be done?

A. Inspect the cell to see if the jar is leaking; note whether electrolyte is over the plates to the depth of ½ inch and whether the electrolyte is dirty. If these causes are not apparent, the cell will have to be opened and inspected for short-circuits from an accumulation of sediment in the bottom of the jar or from buckling of the plates.

Q. Are hydrometer tests alone conclusive?

A. No. To be strictly accurate, they should be checked by voltage tests, in addition.

Q. How should these voltage readings be taken?

A. With the aid of a portable voltmeter, using the low-reading scale, i.e., 0-3 volts, and always with the battery discharging, the load not exceeding its normal low discharge rate.

Q. Why should the test not be made with the starting-motor load?

A. Because the discharge rate while the starting motor is being used is so heavy that even in a fully charged battery in good condition it will cause the voltage to drop rapidly.

Q. Why should the voltage readings not be taken while the battery is charging?

A. Because the voltage of the charging current (always in excess of six volts) will cause the voltage of a battery in good condition to rise to normal or above the moment it is placed on charge, such readings are not a good indication of the battery's condition.

Q. What should the voltage of the cells be?

A. In any battery in good condition, the voltage of each cell at the battery's normal low discharge rate (5 to 10 amperes, as in carrying the lamp load) will remain between 2.1 and 1.9 volts until it begins to approach the discharged condition. A voltage of less than 1.9 volts per cell indicates either that the battery is nearly discharged or that it is in bad condition. The same state is also indicated when the voltage drops rapidly after the load has been on a few minutes.

Joint Hydrometer-Voltmeter Test

Q. What should the hydrometer and voltmeter readings be for a fully charged battery in good condition?

A. Hydrometer 1.275 to 1.300; voltage 2 to 2.2 volts per cell.

Q. What does a hydrometer reading of 1.200 or less with a voltage of 1.9 volts or less per cell indicate?

A. This shows that excess acid has been added to the electrolyte. Under these conditions, the lights will burn dimly even though the hydrometer test alone would appear to show that the battery is more than half charged.

Q. What does a hydrometer reading in excess of 1.300 indicate?

A. It indicates that an excessive amount of acid has been added to the electrolyte, regardless of whether the voltage reading is high, low, or normal.

Q. Where a low voltage reading is found, how can it be determined whether the battery is in bad condition or merely discharged?

A. Stop the discharge by switching off the load (lamps) and put the battery on charge, cranking the engine by hand. After a few minutes of charging, note whether the voltage of each cell promptly rises to 2 volts or more. Any cells that do not are probably short-circuited or otherwise in bad condition.

Q. How can a rough test of the condition of the battery be made without the use of any instruments?

A. On systems fitted with a battery cut-out in the generator battery circuit, remove the cover of the cut-out (the generator being stationary) and momentarily close the cut-out points with the finger. The discharge shown by the ammeter the monent the points are closed should be anywhere from 10 to 20 amperes, differing, of course, with different systems. In any case, it should be equal to or greater than the maximum normal output of the generator, provided the battery is at least three-quarters charged.

Q. What effect will allowing the electrolyte to fall too low in the cells have, apart from the damage that it will cause to the plates?

A. It will tend to increase the voltage if the battery is otherwise in good condition, and this may be carried to a point where it will burn out the lamps.

Q. What is meant by "floating the battery on the line"?

A. This describes the relation of the battery to the generator and lighting circuits in systems where the current for lighting is taken directly from the generator when running, any excess over the requirements of the lamps being absorbed by the battery. The moment the generator speed falls below the point where it supplies sufficient current to supply all that is needed for the lamps, the battery automatically supplies the balance. When the generator is idle, the battery, of course, supplies the current for lighting as well as for starting.

Gassing

Q. Why should the cell tops be wiped dry from time to time and the latter as well as the terminals be washed with a weak solution of ammonia and water?

A. As the charge approaches completion, the cells gas; when overcharged they gas very freely. This gas carries with it in the form of a fine spray some of the electrolyte, and the acid of the latter will attack the terminals and corrode them. Wiping clean does not remove this acid entirely, so the ammonia solution is necessary to counteract its effect, the ammonia being strongly alkaline.

Q. Why should an unprotected light, i.e., any flame or spark, not be allowed close to a storage battery?

A. Because the gas emitted by the battery on charge is hydrogen, which is not only highly inflammable but, when mixed in certain proportions with air, forms a powerful explosive mixture.

Q. What is the cause of gassing?

A. When a battery is charged, the water of the electrolyte is decomposed by the current into gases. During the early part of the charge these gases unite with the active material of the plates, but as the charge proceeds, more gas is evolved than the plates can take care of and it bubbles up through the electrolyte. This is known as the gassing point, and the temperature of the cell also begins to rise at that point.

Q. Is gassing harmful to the battery?

A. The greatest wear on the positive plates takes place during the gassing period, and, if carried too far, they may be injured by reaching a dangerous temperature (105° F., or over) which will tend to loosen the active material.

Q. How can gassing be checked?

A. By cutting down the charge. In some systems this can be effected by the insertion of extra resistance provided for the purpose. Where this cannot be done and it is necessary to keep the car running, turn on all the lamps or start the engine once or twice to reduce the charge of the battery. As the lamps usually consume 80 to 95 per cent of the generator output, they should be sufficient to prevent a further overcharge.

Q. Can the generator be disconnected from the battery to prevent overcharge?

A. Not unless it is short-circuited, as directed in the instructions covering different systems. Otherwise, it will blow its field fuse or, where one is not provided, burn out its windings, except in cases where special provision is made to guard against this.

Sulphating

Q. Why must a battery never be allowed to stand in a fully discharged state?

A. Because the acid of the electrolyte then attacks the plates and converts the lead into white lead sulphate which is deposited on them in the form of a hard coating that is impenetrable to the electrolyte, so that the plates are no longer active. The battery then is said to be *sulphated.*

Q. Can a sulphated battery be put in good condition, and what treatment must be given it to do so?

A. If the sulphating has not gone too far, the battery may be brought back to approximately normal condition by a long heavy charge at a higher voltage than ordinary. Where the battery has become badly sulphated, it is preferable to remove it from the car and charge from an outside source of current, as it may require several days to complete the process. (Note instructions regarding the running of the generator when disconnected from the battery, as otherwise it may be damaged.) If avoidable, the car should not run with the battery removed. If the battery has not stood discharged for any length of time, the charge may be given on the car by running steadily for 8 to 10 hours with all lights off. No lamps must be turned on, as the increased voltage is liable to burn them out.

Voltage Tests

Q. What is the purpose of the voltmeter in connection with the battery?

A. It is chiefly useful for showing whether a cell is short-circuited or is otherwise in bad condition.

Q. Can the voltmeter alone be relied upon to show the condition of the cells?

A. No; like the hydrometer, its indications are not always conclusive, and it must be used in conjunction with the hydrometer to insure accuracy.

Q. What type of voltmeter should be employed for making these tests?

A. For garage use, a reliable portable instrument with several connections giving a variable range of readings should be employed. For example, on the 0-3 volt scale, only one cell should ever be tested; attempting to test any more than this is apt to burn out the 3-volt coil in the meter. The total voltage of the number of cells tested should never exceed the reading of the particular scale being used at the time, as otherwise the instrument will be ruined.

Q. Must these readings be particularly accurate?

A. Since a variation as low as .1 volt (one-tenth of a volt) makes considerable difference in what the reading indicates as to the condition of the battery, it will be apparent that the readings must not only be taken accurately, but that a cheap and inaccurate voltmeter is likely to be misleading rather than helpful.

Q. What precautions should be taken before using the voltmeter?

A. Always see that the place on the battery connector selected for the contact is bright and clean and that the contact itself is firm, otherwise the reading will be misleading since the increased resistance of a poor contact will cut down the voltage.

Q. How is the instrument connected to the battery?

A. The positive terminal of the voltmeter must be brought in contact with the positive terminal of the battery and the negative terminal of the voltmeter in contact with the negative terminal of the battery.

Q. In case the markings on the battery are indistinct, how can the polarity be determined?

A. Connect the voltmeter across any one cell. Should the pointer not give any voltage reading, butting against the stop at the left instead, the connections are wrong and should be reversed; if the instrument shows a reading for one cell, the positive terminal of the voltmeter is in contact with the positive terminal of the battery. This test can be made without any risk of short-circuiting the cell, since the voltmeter is wound to a high resistance and will pass very little current. Such is not the case with the meter, which should never be used for this purpose.

Q. When the battery is standing idle, what is the cell voltage and why is this not a good test?

A. Approximately two volts, regardless of whether the battery is fully charged or not. Voltage readings taken when the battery is on open circuit, i.e., neither charging nor discharging, are only of value when the cell is out of order.

Q. If the battery is in good condition and has sufficient charge, what should the voltmeter reading show?

A. Using the lamps for a load, the voltage reading after the load has been on for five minutes or longer should be but slightly lower (about .1 volt) than if the battery were on open circuit.

Q. When one or more cells are discharged, what will the reading show?

A. The voltage of these cells will drop rapidly when the load is first put on and sometimes even show reverse readings, as when a cell is out of order.

Q. What will the voltmeter indicate when the battery is nearly discharged?

A. The voltage of each cell will be considerably lower than if on open circuit after the load has been on for five minutes or more.

Q. How can the difference be distinguished between cells that are merely discharged and those that are in bad condition?

A. Put the battery on charge, cranking the engine by hand to start, and test again with the voltmeter; if the voltage does not rise to approximately 2 volts per cell within a short time, it is evidence that there is internal trouble which can be remedied only by dismantling the cell.

Q. What effect has the temperature on voltage readings?

A. The voltage of a cold battery rises slightly above normal on charge and falls below normal on discharge. This last is one of the chief reasons for its decreased efficiency in cold weather.

Q. What is the normal temperature of the battery and to what does this refer?

A. The normal temperature of a battery is considered at 70° F., but this refers to the temperature of the electrolyte in the battery as shown by a battery thermometer and not to the temperature of the surrounding air. If the battery has been charging at a high rate for some time, it may be normal even though the weather be close to zero at the time.

Sediment

Q. What is the cause of sediment or mud accumulating in the jars, and why must it be removed before it reaches the bottoms of the plates?

A. This sediment consists of the active material of the plates, which has been shaken out, due to the loosening caused by the charging and discharging, and aggravated by the constant vibration. It must never be allowed to reach the plates, as it is a conductor and will short-circuit them and thus ruin the battery.

Q. How long will a battery stay in service before this occurs?

A. This depends on the type of jar employed and the treatment that the battery has received. If it has been kept constantly overcharged, or if discharged to exhaustion in a very short period, as by abuse of the starting motor when the engine is not in good starting condition, or if it has been subjected to short-circuits by grounding or by dropping tools on its terminals, the plates will disintegrate much quicker than where proper treatment has been given it. With the old-style jar, only an inch or so is allowed to hold this accumulation of sediment below the plates, while in later types fully 3 inches or more are allowed in the depth of the cell for this purpose. A battery with jars of the latter type that has been cared for properly should not require washing out under two years. The procedure is the same as that given for removing the effects of impure water. The plates must never be allowed to dry.

Washing the Battery

Q. What is meant by washing the battery, and why is it necessary?

A. Washing a battery involves cutting the cells apart, washing the elements and the jars, and reassembling with new separators and new electrolyte. It is necessary to prevent the accumulation of sediment, consisting of active material shaken from the plates, to a point where it will touch them and thus cause a short-circuit.

Q. How often is it necessary to wash a battery?

A. This will depend on the type of cell in the battery and the age of the latter. If the battery has the modern-style jar with extra deep mud space, it probably will not be necessary to wash it until it has seen two to three seasons' use. With the older form of cell in which the space allowed for sediment is much less, washing doubtless will be necessary at least once a season. As the battery ages, it will be necessary to wash it oftener.

Q. What other causes besides the type of jar and the age of the battery influence the frequency with which it is necessary to wash the battery?

A. The treatment the battery has received. If it has been abused by overcharging and permitting the cells to get too hot, the active material will be forced out of the grids much sooner.

Q. How can the necessity for washing be determined?

A. The presence of one or more short-circuited cells in a battery that has not been washed for some time will indicate the necessity for it. Each cell should be tested separately with the low-reading voltmeter; a short-circuited cell will either give no voltage reading or one much below that of the others. Cut such a cell out and open it; if the short-circuit has been caused by an accumulation of sediment, the others most likely are approaching the same condition.

Q. How is a battery washed?

A. By cutting the cells apart, unsealing them, and lifting out the elements which should be immersed immediately in a wooden tub of clean pure water. The separators then are lifted out and the positive and negative groups of plates separated, but they must be marked so that the same groups may go back in the right cells. Before disposing of the old electrolyte, its specific gravity should be noted, as new electrolyte of the same density must be used. The plates should be washed in copious running water for several hours, but their surfaces must never be exposed to the air. Reassemble with new separators, fill the jars with fresh electrolyte of the same specific gravity as that discarded, and keep the elements under water until ready to place in the jars, which then should be sealed and the lead connectors burned together again.

Give a long slow charge after reassembling. The battery will not regain its normal capacity until it has been charged and discharged several times.

Connectors

Q. Why should lead connectors be employed, and why is it necessary to burn them together?

A. Any other metal will corrode quickly. Burning is necessary to make good electrical connection, except where bolted connectors are employed.

Q. When connections have become badly corroded or broken, what should be done with them?

A. They should be replaced with new lead-strap connectors supplied by the makers. If they are not obtainable and the battery must be in service meanwhile, the old ones can be cleaned by cutting away the corroded parts and burning new lead on them to bring them to normal size. If broken, burn together with lead in the same way. Heavy copper cable can be used temporarily but must be removed as soon as possible, as it will corrode quickly. Never use any other metal except lead or copper and never use light copper wire. It will either be burned up in a flash or it will cut down the amount of current from the battery, thus causing unsatisfactory operation.

Buckled Plates

Q. What is the cause of badly disintegrated or buckled plates?

A. Sudden discharge due to a short-circuit or to constant abuse of the starting motor on an insufficiently charged battery.

Q. Is there any remedy for such a condition?

A. If the plates are not badly buckled and have not lost much of their active material, the cells may be put in service again by washing and reassembling as described, but if there is any considerable loss of active material, new plates will be necessary.

Low Battery

Q. What are the indications of a low battery?

A. The starting motor fails to turn the engine over, or does so very slowly, or only a part of a revolution. The lights burn very dimly. The hydrometer shows a specific-gravity reading of 1.250 or less. Voltmeter test shows less than 5 volts for a 3-cell battery (for greater number of cells, in proportion), or 1.75 volts or less for each cell.

Q. What are the causes of a low battery?

A. The electrolyte not covering the plates, or being too weak or dirty. A short-circuit in the battery due to the accumulation of sediment reaching the bottom of the plates. An excessive lamp load, all lights being burned constantly with but little daylight running the car. Generator not charging properly.

Specific Gravity; Voltage

Q. What are the specific gravity and voltage of fully discharged and fully charged cells?

A. Total discharge: 1.140 to 1.170 on the hydrometer; and 1.70 to 1.85 volts on the voltmeter. Fully charged: 1.276 to 1.300 specific gravity; 2.35 to 2.55 volts.

Q. Are these readings always constant for the same conditions?

A. No. The charging voltage readings will vary with the temperature and the age of the cell; the higher the temperature and the older the cell the lower the voltage will be. Hydrometer readings also depend on the temperature to some extent. For every ten degrees Fahrenheit rise in temperature, the specific gravity reading will drop .003 or three points, and *vice versa*.

Q. Under what conditions should voltage tests be made?

A. Only when the battery is either charging or discharging. Readings taken when the battery is idle are of no value.

Q. Under what conditions should hydrometer tests be made?

A. The electrolyte must be half an inch over the plates and it must have been thoroughly mixed by being subjected to a charge. Hydrometer readings taken just after adding water to the cells are not dependable.

Q. When should acid be added to the electrolyte?

A. As the acid in a battery cannot evaporate, the electrolyte should need no addition of acid during the entire life of the battery under normal conditions. Therefore, if no acid has leaked or splashed out and the specific gravity is low, the acid must be in the plates in the form of sulphate and the proper specific gravity must be restored by giving the battery an overcharge at a low charging rate.

Q. What does a specific gravity in some cells lower than in others indicate?

A. Abnormal conditions, such as a leaky jar, loss of acid through slopping, impurities in the electrolyte, or a short-circuit.

Q. How can it be remedied?

A. Correct the abnormal conditions, and then overcharge the cells at a low rate for a long period, or until the specific gravity has reached a maximum and shows no further increase for 8 or 10 hours. If, at the end of such an overcharge, the specific gravity is still below 1.270, add some specially prepared electrolyte of 1.300 specific gravity. Electrolyte should not be added to the cells under any other conditions.

Q. Is an overcharge beneficial to a battery?

A. The cells will be kept in better condition if a periodical overcharge is given, say once a month. This overcharge should be at a low rate and should be continued until the specific gravity in each cell has reached its maximum and comparative readings show that all are alike. To carry this out properly will require at least 4 hours longer than ordinarily would be necessary for a full charge. If the plates have become sulphated due to insufficient charging, it may be necessary to continue the overcharge for 10 to 15 hours longer. Should the specific gravity exceed 1.300 at the end of the charge, draw off a small amount of electrolyte with the syringe from each cell and replace with distilled water. If below 1.270, proceed as mentioned above for addition of acid.

Charging from Outside Source

Q. What is meant by charging from an outside source?

A. A source of direct current other than the generator on the car.

Q. Why is it necessary to charge the battery from an outside source?

A. When the battery has become sulphated, has been standing idle for any length of time, or has been run down from any other cause so that it is out of condition, a long charge at a uniform rate is necessary, and it would seldom be convenient to run the car for 8 or 10 hours steadily simply to charge the battery; frequently, a longer charging period than this is necessary.

Q. How is charging from an outside source effected?

A. This will depend upon the equipment at hand and the nature of the supply, i.e., whether alternating or direct current. If the current is alternating, a means of converting it to direct current is necessary, such as a motor-generator, a mercury-arc rectifier, chemical or vibrating type of rectifier. These are mentioned about in the order of the investment involved. In addition, a charging panel is needed to complete the equipment, this panel being fitted with switches, voltmeter, and ammeter, and a variable resistance for regulating the charge. Where direct-current service is obtainable at 110 or 220 volts, the rectifier is unnecessary.

Q. How can a battery be charged from direct-current service mains without a special charging panel?

A. By inserting a double-pole single-throw switch and 10- or 15-ampere fuses on taps from the mains and ordinary incandescent lamps in series with the battery to reduce the voltage, Fig. 409.

Q. How many lamps will be needed?

A. This will depend upon their character and size, as well as upon the amount of charging current necessary. For a 10-ampere charge for a 6-volt storage battery, seven 110-volt 100-watt (32 c.p.) carbon-filament lamps, or their equivalent, will be needed; i.e., fourteen 110-volt 50-watt (16 c.p.) carbon-filament lamps; eighteen 110-volt 40-watt tungsten lamps, or twenty-eight 110-volt 25-watt tungsten lamps. For a 12-volt or 24-volt battery the number of lamps will have to be decreased in proportion in order not to cut the voltage of the supply current below that of the battery. For 220-volt d.c. supply mains, if a three-wire system is employed, the taps should be taken from the center wire and one outside wire; this will give 110 volts. If the service is 220-volt two-wire, more lamps will be needed to reduce the voltage, which should exceed that of the battery by only $1\frac{1}{2}$ to 2 volts except where a high voltage charge to overcome sulphating is being given, in which case it may be slightly higher.

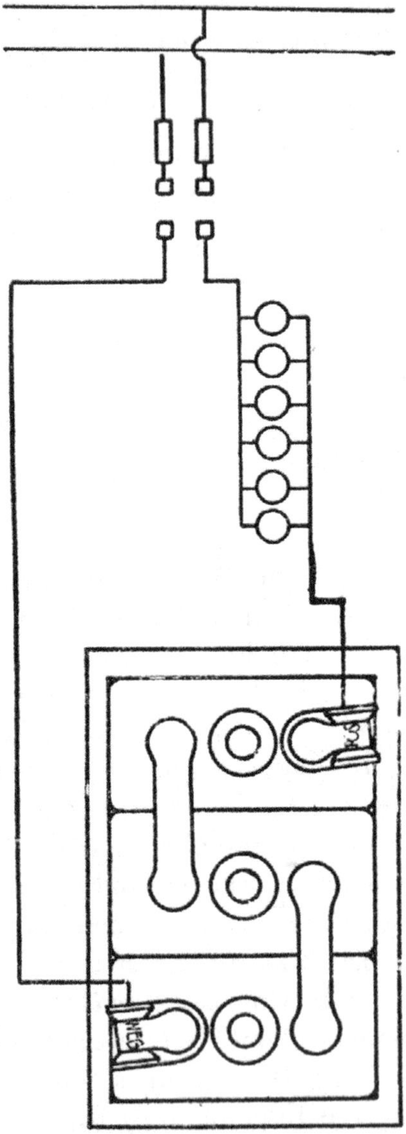
Fig. 409. Diagram of Connections for Charging Six-Volt Storage Battery from Lighting Circuit

Q. Where no outside source of current is available, or where no rectifier is at hand to convert alternating current, how can the battery be given the long charge necessary?

A. Run the engine. Supply it with plenty of oil and provide hose connections from the water supply to the filler cap on the radiator and a drain from the lower petcock. Open the latter and turn on just sufficient water to keep the engine reasonably cool; increase if necessary as it runs hotter.

Q. What precaution must be taken always before putting the battery on charge from an outside source?

A. The polarity of the circuit must be tested in order to

make certain that the battery will be charged in the proper direction.

Q. How can this be done?

A. If a suitable voltmeter is at hand, i.e., one of the proper voltage for the 110-volt current, connect it to the mains. If the needle does not move over the scale but shows a tendency to butt against the stop pin at the left, reverse the connections. The needle will then give a proper reading and the positive connection to the meter must be used for the positive side of the battery. Should no voltmeter of the right voltage be available, connect two short wires with bared ends to the fused end of the switch. Dip the bared ends of the wire in a glass of water, being careful to keep them at least an inch part. When the switch is closed, fine bubbles will be given off by the wire connected to the negative side. The battery terminals are stamped *Pos.* and *Neg.*, and the connections should be made accordingly.

Intermittent and Winter Use

Q. What should be done with an idle battery?

A. If it is to be idle for any length of time, as where the car is to be stored, it should be given a long overcharge as described above before being put out of service. Fill the cells right to the top with distilled water to allow for evaporation and absorption of acid by the plates. Give the battery a freshening charge at a low rate once a month. Discharge the battery and re-charge before putting it into service again. If it has stood out of service for a long period, the battery will be found at a low efficiency point and will not reach its maximum capacity again until it has had several charges and discharges.

Q. Does cold weather have any effect on the storage battery?

A. It causes a falling off in its efficiency. If not kept charged, the electrolyte will freeze under the following conditions: battery fully discharged, sp. gr. 1.120,20° Fahrenheit; battery three-quarters discharged, sp. gr. 1.160, temperature zero; half discharged, sp. gr. 1.210, 20 degrees below zero; one quarter discharged, sp. gr. 1.260, 60 degrees below zero. When storing away for the winter, the battery must either be kept charged or put where the temperature does not go lower than 20 degrees above zero.

Edison Battery

Q. Is it ever necessary to wash out an Edison battery?

A. No. The cells are permanently sealed, as the active material cannot escape from its containers.

Q. Do all of the foregoing instructions apply to the Edison as well as to the lead-plate battery?

A. No. The Edison requires very little attention, practically the only care necessary being to keep the cars replenished with distilled water at intervals.

Charging rates for Edison cells are given in the article on Electric Automobiles. S.A.E.-standard instructions for lead-plate cells are also given in the same article.

GLOSSARY

THE following glossary of automobile terms is not intended in any sense as a dictionary and only words used in the articles themselves have been defined. The definitions have been made as simple as possible, but if other terms unfamiliar to the reader are used, these should be looked up in order to obtain the complete definition.

A

A. A. A.: Abbreviation for American Automobile Association.

Abrasive: Any hard substance used for grinding or wearing away other substances.

Absorber, Shock: See "Shock Absorber".

Accelerate: To increase the speed.

Acceleration: The rate of change of velocity of a moving body. In automobiles, the ability of the car to increase in speed. Pickup.

Accelerator: Device for rapid control of the speed for quick opening and closing of the throttle. Usually in the form of a pedal, spring returned, the minimum throttle opening being controlled by the setting of the hand throttle.

Accessory: A subordinate machine that accompanies or aids a more important machine; as, a horn is an accessory of an automobile.

Accumulator: A secondary battery or storage battery. It usually consists of chemically prepared lead plates combined with an acid solution. Upon being charged with an electric current from a primary source, a chemical change takes place which enables the plates in their turn to give a current of electricity when used as a source of power, the plates at the same time returning to their original chemical state.

Acetone: A liquid obtained as a by-product in the distillation of wood alcohol, and used in connection with reservoirs for storing acetylene for automobile lights, as it dissolves many times its own volume of acetylene gas.

Acetylated Alcohol: Alcohol which has been denatured by the addition o. acetylene, which also increases its fuel value. See "Alcohol, Denatured".

Acetylene: A gaseous hydrocarbide used as an illuminant; is usually generated for that purpose by the action of water on calcium carbide.

Acetylene Generator. A closed vessel in which acetylene gas may be produced by the action of water on calcium carbide and which supplies the gas under uniform pressure.

Acetylene Lamp: A lamp which burns acetylene gas.

Acetylite: Calcium carbide which has been treated with glucose. It is used to obtain a more uniform and slower production of acetylene gas than can be obtained with the untreated calcium carbide.

Acid: In connection with automobiles the term usually means the liquid or electrolyte used in the storage battery. See "Electrolyte".

Acid Cure. Method of rapid vulcanization of rubber without heat. Used in tire repairs. The agent is sulphur chloride.

Acidimeter. An instrument for determining the purity of an acid.

Active Material: Composition in grids that forms plates of a storage battery. It is this material in which the chemical changes occur in charging and discharging.

Adapter: Device by which one type of lamp burner may be used instead of the one for which the lamp was designed. Usually a fitting by which a gas or oil lamp may be converted into an electric lamp.

Adhesion: That property of surfaces in contact by virtue of which one of them tends to stick to the other. It is used as synonymous with friction. The adhesion of wheels acts to prevent slipping.

Adjustment: The slackening or tightening up of parts to compensate for wear, reduce friction, or secure better contact.

Admission: In a steam engine, the letting in of the steam to the cylinder; in gas engine, the letting in of mixture of gas and air to the cylinder.

Advanced Ignition: Usually called *advancing the spark*. Setting the spark of an internal-combustion motor so that it will ignite the charge at an earlier part of the stroke.

Advance Sparking: A method by which the time of occurrence of the ignition spark may be regulated, by completing the electric circuit at the earlier period.

Advancing the Spark: See "Advanced Ignition".

Aerodynamics: The science of atmospheric laws, i.e., the effects produced by air in motion.

After-Burning: Continued burning of the charge in an internal-combustion engine after the explosion.

After-Firing: An explosion in the muffler or exhaust passages.

A-h: Abbreviation for *ampere hour.*

Air Bottle: A portable container holding compressed air or carbon dioxide for tire inflation.

Air-Bound: See "Air Lock"

GLOSSARY

Air Compressor: A machine for supplying air under pressure for inflating tires, starting the motor, etc.

Air Cooled: Cooled by air direct. Usually referring to the cylinder of an engine, whose heat caused by the combustion within it is carried away by air convection and radiation.

Air Cooling: A system of dispersing by air convection the heat generated in the cylinder of an internal-combustion motor.

Air Intake: An opening in a carbureter to admit air.

Air Leak: Entrance of air into the mixture between carbureter and cylinder.

Air Lock: Stoppage of circulation in the water or gasoline system caused by a bubble of air lodging in the top of a bend in the pipe.

Air Pump: A pump operated by the engine or by hand to supply air pressure to the oil tank or gasoline tank; sometimes called *pressure pump.*

Air-Pump Governor: A device to regulate the speed of the air pump so as to give a uniform air pressure.

Air Resistance: The resistance encountered by a surface in motion. This resistance increases as the square of the speed, which makes it necessary to employ four times as much power in order to double a given speed.

Air Tube: See "Pneumatic Tire".

Airless Tire: Name of special make of nonpuncturable resilient tire.

A. L. A. M.: Abbreviation for Association of Licensed Automobile Manufacturers, now out of existence.

A. L. A. M Horsepower Rating: The horsepower rating of an automobile found by the standard horsepower formula approved by the Association of Licensed Automobile Manufacturers. Since the dismemberment of this organization, the formula is usually called the S.A.E. rating. This formula is h.p. =bore of cylinder (in inches) squared × No. of cylinders ÷ 2.5, at a piston speed of 1000 feet per minute.

Alarm, Low-Water: See "Low-Water Alarm".

Alcohol: A colorless, volatile, inflammable liquid which may be used as fuel for internal-combustion engines.

Alcohol, Denatured: Alcohol rendered unfit for drinking purposes by the addition of wood alcohol, acetylene, and other substances.

Alignment: The state of being exactly in line. Applied to crankshafts and transmission shafts and to the parallel conditions of the front and rear wheels on either side.

Alternating Current: Electric current which alternates in direction periodically.

Ammeter: An instrument to measure the values of current in an electric circuit directly in amperes. Also called *ampere meter.*

Amperage: The number of amperes, or current strength, in an electric circuit.

Ampere: The practical unit of rate of flow of electric current, measuring the current intensity.

Ampere Hour: A term used to denote the capacity of a storage battery or closed-circuit primary battery. A battery that will deliver three amperes for six hours is said to have an eighteen-ampere-hour capacity.

Ampere Meter: See "Ammeter".

Angle-Iron Underframe: An underframe constructed of steel bars whose cross section is a right angle.

Anneal: To make a metal soft by heating and cooling. To draw the temper of a metal.

Annular Gear: A toothed wheel upon which the teeth are formed on the inner circumference.

Annular Valve: A circular valve having a hole in the center.

Annunciator: An installation of electric signals or a speaking tube to allow the passengers in an enclosed car to communicate with the driver.

Anti-Freezing Solution: A solution to be used in the cooling system to prevent freezing in cold weather; any harmless solution whose freezing point is somewhat below that of water may be used.

Anti-Friction Metal: Various alloys of tin and lead used to line bearings, such as Babbitt metal, white metal, etc.

Anti-Skid Device: Any device which may be applied to the wheels of a motorcar to prevent their skidding, such as tire coverings with metal rivets in them, chains, etc.

Apron: Extensions of the fenders to prevent splashing by mud or road dirt.

Armature: In dynamo-electric machines, the portion of a generator in which the current is developed, or in a motor, the portion in which the current produces rotation. In most generators in automobile work, the armature is the rotating portion. In magnetic or electromagnetic machines the armature is the movable portion which is attached to the magnetic poles.

Armature Core: The iron portion of the armature which carries the windings and serves as part of the path for the magnetic flux.

Armature Shaft: The shaft upon and with which the armature rotates.

Armature Winding: Electrical conductors, usually copper, in an armature, and in which the current is generated, in case of a generator, or in which they produce rotation in a motor.

Artillery Wheel: A wheel having heavy wood spokes.

Aspirating Nozzle: An atomizing nozzle to make the liquid passing through it pass from it in the form of a spray.

Assembled Car: A car whose chief parts, such as engine, gearset axles, body, etc., are manufactured by different parts makers, only the final process of putting them together being carried out in the car-making plant.

Atmospheric Line: A line drawn on an indicator diagram at a point corresponding with the pressure of the atmosphere.

Atmospheric Valve: See "Suction Valve".

Atomizer: A device by which a liquid fuel, such as gasoline, is reduced to small particles or to a spray; usually incorporated in the carbureter.

Auto: (1) Popular abbreviation for automobile. (2) A Greek prefix meaning self.

GLOSSARY

Auto-Bus: An enclosed motor-driven public conveyance, seating six or more people; usually has a regular route of travel.

Autocar: A motorcar or automobile; a trade name for a particular make of automobile.

Auto-Cycle: See "Motorcycle".

Autodrome: A track especially prepared for automobile driving, particularly for races.

Autogenous Welding: See "Welding, Autogenous".

Auto-Igniter: A small magneto generator or dynamo for igniting gasoline engines, the armature of which is connected with the flywheel by gears or by friction wheels, so that electric current is supplied as long as the engine revolves.

Autoist: One who uses an automobile.

Automatic Carbureter: A vaporizer or carbureter for gasoline engines whose action is entirely automatic.

Automatic Cut-Out: See "Cut-Out, Automatic".

Automatic Spark Advance: Automatic variation of the instant of spark occurrence in the cylinder. Mechanical advancing and retarding of the spark to correspond with and controlled by variations in crankshaft speed.

Auto-Meter: Trade name for special make of combined speedometer and odometer.

Automobile: A motor-driven vehicle having four or more wheels. Some three-wheeled vehicles are properly automobiles, but are usually called *tricars*.

Automobilist: The driver or user of an automobile.

Auto Truck: A motor-driven vehicle for transporting heavy loads; a heavy commercial car.

Auxiliary Air Valve: Valve controlling the admission of air through the auxiliary air intake of a carbureter.

Auxiliary Air Intake: Opening through which additional air is admitted to the carbureter at high speeds.

Auxiliary Exhaust: Ports cut through cylinder walls to permit exhaust gases to be released from the cylinder when uncovered by the piston. These are sometimes used as an additional scavenging means for the regular exhaust valves.

Auxiliary Fuel Tank: See "Fuel Tank, Auxiliary".

Auxiliary Spark Gap: See "Spark Gap, Outside".

Axle: The spindle with which a wheel revolves or upon which it revolves.

Axle, Cambered: An axle whose ends are slanted downwards to camber the wheels.

Axle, Channel: An axle which is U-shaped in cross section.

Axle, Dead: Solid, fixed, stationary axle. An axle upon which the wheels revolve but which itself does not revolve.

Axle, Dropped: An axle in which the central portion is on a lower level than the ends.

Axle, Floating: A full-floating axle. A live axle in which the shafts support none of the car weight, but serve only to turn the wheels.

Axle, I-Beam: An axle whose cross section is in the shape of the letter I.

Axle, Live: An axle in which are comprised the driving shafts that carry the power of the motor to the driving wheels.

Axle, Semi-Floating: A live axle in which the driving shafts carry all of the car weight as well as transmitting the driving torque.

Axle, Three-Quarters Floating: A live axle in which the shafts carry a part of the weight of the car, while the housing carries the balance of the weight. It is intermediated by a floating axle and the semi-floating axle.

Axle, Trussed: An axle in which downward bending is prevented by a truss.

Axle, Tubular: An axle formed of steel tubing. Usually applied to the front axles, but sometimes used in referring to tubular shafts of rear axles.

Axle Casing: That part of a live axle that encloses the driving shafts and differential and driving gears. Axle housing.

Axle Housing: See "Axle Casing".

Axle Shaft: The member transmitting the driving torque from the differential to the rear wheels.

B

Babbitt: A soft metal alloy used for lining the bearings of shafts.

Back-Firing: An explosion of the mixture in the intake manifold or carbureter caused by the communication of the flame of explosion in the cylinders. Usually due to too weak a mixture. Popping.

Back Kick: The reversal of direction of the starting, caused by back-firing.

Backlash: The play between a screw and nut or between the teeth of a pair of gear wheels.

Back Pressure: Pressure of the exhaust gases due to improper design or operation of the exhaust system.

Baffle Plate: A plate used to prevent too free movement of a liquid in the container. In a gas engine cylinder, a plate covering the lower end of the cylinder to prevent too much oil being splashed into it. The plate has a slot through which the connecting rod may work.

Balance Gear: See "Differential Gear".

Balancing of Gasoline Engines: Insuring the equilibrium of moving parts to reduce the vibration and shocks.

Ball-and-Socket Joint: A joint in which a ball is placed within a socket recessed to fit it, permitting free motion in any direction within limits.

Ball Bearing: A bearing in which the rotating shaft or axle is carried upon a number of small steel balls which are free to turn in annular paths, called *races*.

Balladeur Train: A French name for a sliding change-speed gear.

Barking: The sound made by the explosions caused by after-firing.

Base Bearing: See "Main Bearing".

Base Explosion: See "Crankcase Explosion".

Battery: A combination of primary or secondary cells, as dry cells or storage cells.

Battery, Dry: See "Dry Battery".

Battery, Storage: See "Accumulator".

Battery Acid: The electrolyte in a storage battery.

GLOSSARY

Battery-Charging Plug: Power terminals to which the leads of a storage battery may be connected for charging the battery.

Battery Gage: (1) Voltmeter or ammeter or voltammeter for testing the specific gravity of the electrolyte in a secondary battery.

Battery Syringe: A syringe used to draw out a part of the electrolyte or solution from a storage battery cell to test its density and specific gravity.

Baumè: A scale indicating the specific gravity or density of liquids and having degrees as units. Gasoline of a specific gravity of .735 has a gravity of 61 degrees Baumé.

Bearing: A support of a shaft upon which it may rotate.

Bearing, Annular Ball: A ball bearing consisting of two concentric rings, between which are steel balls.

Bearing, Ball: A bearing in which the rotating shaft and the stationary portion of the bearings are separated from sliding contact by steel balls. A steel collar fitted to the shaft rolls upon the balls, which in turn roll upon steel collar attached to the stationary portion of the bearing.

Bearing, Cup and Cone: A ball bearing in which the balls roll in a race, which is formed between a cone-shaped fixed collar and a cup-shaped shaft collar.

Bearing, Main: The bearing in which rotates the crankshaft of an engine.

Bearing, Plain: A bearing in which the rotating shaft is in sliding contact with the bearing supporting it.

Bearing, Radial: A bearing designed to resist loads from a direction at right angles to the axis of the shaft.

Bearing, Roller: A bearing in which the journal rests upon, and is surrounded by, hardened steel rollers which revolve in a channel or race surrounding the shaft.

Bearing, Thrust: A bearing designed to resist loads or pressures parallel with the axis of the shaft.

Bearing Cap: That portion of a plain bearing detachable from the stationary portion, and which holds the bearing bushing and shaft.

Bearing Surface: The projected area of a bearing in a perpendicular plane to the direction of pressure.

Beau de Rochas Cycle: The four-stroke cycle used in most internal-combustion engines. This cycle was proposed by M. Beau de Rochas and put into practical form by Dr. Otto. See "Four-Cycle".

Belt and Clutch Dressing: A composition to be applied to belts and clutches to prevent them from slipping.

Belt Drive: A method of transmitting power from the engine to the countershaft or jack shaft by means of belts.

Benzine: A petroleum product having a specific gravity between that of kerosene and gasoline. Its specific gravity is between 60 degrees and 65 degrees Baumé.

Benzol: A product of the distillation of coal tar. Coal tar benzine. Used as a rubber solvent and in Europe as a motor fuel.

Berline Body: A limousine automobile body having more than two seats in the back part.

Bevel-Gear: Gears the faces of whose teeth are not parallel with the shaft, but are on a beveled edge of the gear wheel.

Bevel-Gear Drive: Method of driving one shaft from another at an angle to the first. The chief method of transmitting the drive from the propeller shaft to the rear axle shafts.

B. H. P.: An abbreviation for brake horsepower.

Bicycle: A two-wheeled vehicle propelled by the pedaling of the rider.

Binding Posts: See "Terminals".

Bleeder: A by-pass in the sight-feed of a mechanical oiling system by which the oil delivered through that feed is allowed to pass out instead of going to the bearings.

Blister: A defect in tires caused by the separation of the tread from the fabric.

Block Chain: A chain used in automobiles, bicycles, etc., of which each alternate link is a steel block.

Blow-Back: The backward rushing of the fuel gas through the inlet valve into the carbureter.

Blower Cooled: A gas engine cooled by positive circulation of air maintained by a blower.

Blow-Off: A blow-out caused by the edge of the bead of tire becoming free from the rim and allowing the tube to protrude through the space thus formed.

Blow-Out: The rupture of both the inner tube and outer casing of a pneumatic tire.

Blow-Out Patch: See "Patch, Tire Repair".

Body: (1) The superstructure of an automobile; the part that resembles and represents the body of a horse-drawn vehicle. (2) In oils, the degree of viscosity. The tendency of drops of oils to hang together.

Body Hangers: Attachments to or extensions of the frame for holding the body of the vehicle. They should be properly called frame hangers.

Boiler: A vessel in which water is evaporated into steam for the generation of power.

Boiler, Fire-Tube: A tubular steam boiler in which the end plates are connected by a number of open ended thin tubes, the spaces around which are filled with water, the hot gases passing through the tubes.

Boiler, Flash: A steam boiler in which steam is generated practically instantaneously. There is practically no water or steam stored in the boiler. A flash generator.

Boiler, Water-Tube: A steam boiler in which the water is carried in metal tubes, around which the hot gases circulate.

Boiler Alarm: See "Low-Water Alarm".

Boiler Covering: A non-conducting substance used as a covering for boilers to prevent loss of heat by radiation.

Boiler-Feed Pump: An automatic and self-regulating pump for supplying a boiler with feed water.

Boiler-Feed Regulator: A device to make the feed-water supply of the boiler automatic.

Bonnet: (1) The hood or metallic cover over the front end of an automobile. See "Hood". (2) The cover over a pump-valve box, or a slide-valve casing. (3) A cover to enclose and guide the tail end of a steam-engine-valve spindle or the cover of a piston-valve casing. (4) The pan underneath the engine in an automobile.

Boot: A covering to protect joints from dirt and water or to prevent the leakage of grease. (2) Space provided for baggage at the rear of a car.

Bore: The inside diameter of the cylinder.

Boss: An enlarged portion of a part to give a point for attachment of another part.

Bottom: The meshing of gears without clearance.

Bow Separator: A part to prevent chafing of the bows of a top when folded.

Boyle's Law of Gases: A law defining the volume and pressure of gases at constantly maintained temperatures. It states that the volume of a gas varies inversely as the pressure so long as the temperature remains the same; or, the pressure of a gas is proportional to its density.

Brake: An apparatus for the absorption of power by friction, and by clamping some portion of the driving mechanism to retard or stop the forward motion of the car.

Brake, Air-Cooled: A brake whose parts are ridged to present a large surface for transferring to the air the frictional heat generated in them.

Brake, Band: A brake which contracts upon the outside of a drum attached to some part of the driving mechanism.

Brake, Constricting Band: A form of brake applied by tightening a band around a pulley or drum.

Brake, Differential: A brake acting upon the differential gear.

Brake, Double-Acting: A brake which will hold when the drum is rotating in either direction.

Brake, Drum, and Band: See "Brake, Band".

Brake, Emergency: A brake intended to be used in case the service brake does not act to a sufficient extent.

Brake, Expanding-Band: A drum brake in which the braking force is exerted by a band forced outward against the inner rim of a pulley.

Brake, External-Contracting: A brake consisting of a drum affixed to a rotating part, the outer surface of which is encircled by a contracting band.

Brake, Foot: A brake designed to be operated by the driver's foot. A pedal brake. Usually the service brake.

Brake, Front-Wheel: A brake designed to operate on the front wheels of the car.

Brake, Gearset: A brake designed to act on the transmission shaft and attached to the gearbox.

Brake, Hand: A brake designed to be operated by means of a hand lever. Usually the emergency brake.

Brake, Hub: A brake consisting of a drum secured to one of the wheels. This is the usual type.

Brake, Internal: A brake in which an expanding mechanism is contained within a rotating drum, the expansion bringing pressure to bear on the drum.

Brake, Internal-Expanding: A brake consisting of a drum, against the inside of which may be expanded a band or a shoe.

Brake, Motor: A brake in an electric vehicle which acts upon the armature shaft of the motor.

Brake, Service: A brake designed to be used in ordinary driving. It is usually operated by the driver's foot.

Brake, Shoe: A brake in which a metal shoe is clamped against a revolving wheel.

Brake, Transmission: A brake designed to act upon the transmission shaft.

Brake, Water-Cooled: A brake through which water may be circulated to carry off the frictional heat.

Brake Equalizer: A mechanism applied to a system of brakes operated in pairs to assure that each brake shall be applied with equal force.

Brake Horsepower: The horsepower supplied by an engine as shown by the application of a brake or absorption dynamometer.

Brake Housing: A casing enclosing the brake mechanism.

Brake Lever: The lever by which the brake is applied to the wheel.

Brake Lining: The wearing surface of a brake; usually arranged to be easily replaced when worn.

Brake Pedal: Pedal by which the brake is applied.

Brake Pull Rod: A rod transmitting the tension from the lever or pedal to the movable portion of the brake proper.

Brake Ratchet: A device by which the brake lever or brake pedal can be set in position and retained there; usually consists of a notched quadrant with which a movable tongue on the lever head or pedal engages.

Brake Rod: The rod connecting the brake lever with the brake.

Brake Test: A test of a motor by means of a dynamometer to determine its power output at different speeds.

Braking Surface: The surface of contact between the rotating and stationary parts of a brake.

Braze: To join by brazing.

Brazing: The process of permanently joining metal parts by intense heat.

Breaker Strip: A strip of canvas placed between the tread and body of an outer tire casing to increase the wearing qualities.

Breather: An opening in the crankcase of a gas engine to permit pressure therein to remain equal during the movement of the pistons.

British Thermal Unit. The ordinary unit of heat. It is that quantity of heat required to raise the temperature of one pound of pure water one degree Fahrenheit at the temperature of greatest density of water.

Brougham Body: A closed-in automobile body having windows at the side doors, and in front, but with no extension of the roof over the front seat.

Brush Holder: In electrical machinery, an arrangement to hold one end of a connection flexible in contact with a moving part of the circuit.

B. T. U.: Abbreviation for *British Thermal Unit.*

Buckboard: A four-wheeled vehicle in which the body and springs are replaced by an elastic board or frame

Buckling: Irregularities in the shape of the plates of storage cells following a too rapid discharge.

Bumper: (1) A contrivance at the front of the car to minimize shock of collision; it consists of plungers working in tubes and gaining elasticity from springs. (2) A bar placed across the end of a car, usually the front end, to take the shock of collision and thus prevent damage to the car itself. A rubber or leather pad interposed between the axle and frame of a car.

Burner, "Torch" Igniter: A movable auxiliary vaporizer for starting the fire in steam automobile burners.

Bushing: A bearing lining. Usually made of anti-friction metal and capable of adjustment or renewal.

Bus-Pipe: A manifold pipe.

Butterfly Valve: A valve inserted in a pipe, usually circular and of nearly the same diameter as the pipe, designed to turn upon a spindle through its diameter and thus shut off or permit flow through the pipe. Usually employed for throttle valves and carbureter air valves.

Buzzer: (1) A name sometimes applied to the vibrator or trembler of a jump-spark ignition coil. (2) A device used in place of a horn, and consisting of a diaphragm which is made to vibrate rapidly by an electromagnet.

By-Pass: A small valve to provide a secondary passage for fluids passing through a system of piping.

C

C: Abbreviation for a centigrade degree of temperature.

Calcium Carbide: A compound of calcium and carbon used for the generation of acetylene by the application of water.

Calcium Chloride: A salt which dissolved in water is used as an anti-freezing solution.

Cam: A revolving disk, irregular in shape, fixed on a revolving shaft so as to impart to a rod or lever in contact with it an intermittent or variable motion.

Cam, Exhaust: A cam designed to operate the exhaust of an engine.

Cam, Ignition: A cam designed to operate the ignition mechanism. In magnetos it operates the make-and-break device.

Cam, Inlet: A cam designed to operate the inlet valve of an engine.

Camber: (1) The greatest depth of curvature of a surface. (2) The amount of bend in an axle designed to incline the wheels.

Camber of Spring: The maximum distance between the upper and lower parts of a spring under a given load.

Cambered Frame: A narrowing of the front of a motor car to permit of easier turning.

Cam Gear: The gear driving the camshaft of a gas engine. In a four-cycle engine this is the same as the two-speed gear.

Camshaft: A shaft by which the valve cams are rotated; also known as the *secondary shaft.*

Camshaft, Overhead: The camshaft carried along or above the cylinder heads, to operate overhead valves.

Camshaft Gears: The gears or train of gears by which the camshaft is driven from the crankshaft. Half-time gears, timing gears, distribution gears.

Canopy: An automobile top that can not be folded up.

Capacity of a Condenser: The quality of electricity or electrostatic charge. Of a storage battery, the amount of electricity which may be obtained by the discharge of a fully charged battery. Usually expressed in ampere hours.

Cape Hood: An automobile top which is capable of either being folded up or extended.

Car: A wheeled vehicle.

Carbide: See "Calcium Carbide".

Carbide Feed: A type of acetylene generator in which the calcium carbide is fed into the water.

Carbon Bridge: Formation of soot between points of spark plug.

Carbon Deposit: A deposit upon the interior of the combustion chamber of a gasoline engine composed of carbonaceous particles from the lubricating oil, too rich fuel mixture, or road dust.

Carbon Remover: A tool or solution for removing carbon deposits from the cylinder, piston, or spark plug of a gasoline engine.

Carbonization: The deposit of carbon.

Carbureter: An appliance for mixing an inflammable vapor with air. It allows air to be passed through or over a liquid fuel and to carry off a portion of its vapor mixed with the air, forming an explosive mixture.

Carbureter, Automatic: A carbureter so designed that either the air supply alone or both the air and gasoline supplies are regulated automatically.

Carbureter, Constant-Level: A carbureter the level of the gasoline in which is maintained automatically at a constant height. A float-feed carbureter.

Carbureter, Exhaust-Jacketed: A carbureter whose mixing chamber is heated by the circulation of exhaust gas.

Carbureter, Multiple-Jet: A carbureter having more that one spray nozzle or jet.

Carbureter, Water-Jacketed: A carbureter whose mixing chamber is heated by the circulation of water from the cooling system.

Carbureter Float: A buoyant part of the carbureter designed to float in the gasoline and connected to a valve controlling the flow from the fuel tank, designed to maintain automatically a constant level of the gasoline in the flow chamber.

Carbureter Float Chamber: A reservoir containing the float and in which a constant level of fuel is maintained.

Carbureter Jet: The opening through which liquid fuel is ejected in a spray from the standpipe of a carbureter nozzle.

Carbureter Needle Valve: A valve controlling the flow of fuel from the flow chamber to the standpipe.

Carbureter Nozzle: See "Carbureter Jet".

Carbureter Standpipe: A vertical pipe carrying the nozzle.

Carburetion: The process of mixing hydrocarbon particles with the air. The action in a carbureter.

Cardan Joint: A universal joint or Hooke's coupling.

GLOSSARY

Cardan Shaft: A shaft provided with a Cardan joint at each end.

Casing: The shoe or outer covering of a double-tube automobile tire.

Catalytic Ignition: See "Ignition,Catalytic".

Cell: One of the units of a voltaic battery.

Cell, Dry: See "Dry Cell".

Cell, Storage: See "Accumulator".

Cellular Radiator: A radiator in which the openings between the tubes are in the form of small cells. The same as a *honeycomb radiator.*

Cellular Tire: A cushion tire which is divided into compartments or cells.

Center of Gravity: That point in a body, which, if the body were suspended freely in equilibrium, would be the point of application of the resultant forces of gravity acting upon the body.

Center Control: The location of the gear-shift and emergency brake levers of a car in the center of a line parallel to the front of the front seat.

Centigrade Scale: The thermometer scale invented by Celsius. Used universally in scientific work.

Century. In automobiling, a hundred-mile run.

C. G. S. System: Abbreviation for centimeter-gram-second system of measurement; the standard system in scientific work.

Chain, Drive: A heavy chain by which the power from the motor may be transmitted to the rear wheels of an automobile.

Chain, Roller: A sprocket chain, the cross bars of whose links are rollers.

Chain, Silent: See "Silent Chain".

Chain, Tire: A small chain fastened about the tire to increase traction and prevent skidding.

Chain Wheel: A sprocket wheel for the transmission chains of a motor-driven vehicle.

Change-Speed Gear: See "Gear, Change-Speed".

Change-Speed Lever: See "Lever, Change-Speed".

Charge: The fuel mixture introduced into the cylinder of a gas engine. The act of storing up electric energy in an accumulator.

Charging: The passing of a current of electricity through a storage cell.

Charles' Law of Gases: See "Gases, Gay Lussac's Law of".

Chassis. The mechanical features of a motor car assembled, but without body, fenders, or other superstructure not essential to the operation of the car.

Chauffeur: In America this term means the paid driver or operator of a motor car. The literal translation from the French means stoker or fireman of a boiler.

Check. Steering: See "Steering Check".

Check Valve: An automatic or non-return valve used to control the admission of feed water in the boiler, etc.

Choke: The missing of explosions or poor explosions due to too rich mixture.

Circuit, Primary: See "Primary Circuit".

Circuit, Secondary: See "Secondary Circuit".

Circuit Breaker: A device installed in an electric circuit and intended to open the circuit automatically under predetermined conditions of current flow.

Circulating Pump: A pump which keeps a liquid flowing through a series of pipes which provides a return circuit. In a motor car, water and oil circulation is maintained by circulating pump.

Circulation Pump: A mechanically operated pump by which the circulation of water in the cooling system is maintained.

Circulating System: The method or series of pipes through which a continuous flow of water or oil is maintained and in which the liquid is sent through the system over and over.

Clash Gear: A sliding change-speed gear.

Clearance: (1) The distance between the road surface and the lowest part of the under-body of an automobile. (2) The space between the piston of an engine when at the extremity of its stroke, and the head of the cylinder.

Clearance, Valve: See "Valve Clearance".

Clearance Space: The space left between the end of the cylinder and the piston plus the volume of the ports between the valves and the cylinder.

Clevis: The fork on the end of a rod.

Clevis Pin: The pin passing through the ends of a clevis and through the rod to which the clevis is joined.

Clincher Rim: A wheel rim having a turned-in edge on each side, forming channels. Into this the edge or flange of the tire fits, the air pressure within locking the tire and rim together.

Clincher Tire: A pneumatic tire design to fit on a clincher rim.

Clutch: A device for engaging or disconnecting two pieces of shafting so that they revolve together or run free as desired.

Clutch Cone: A clutch whose engaging surfaces consist of the outer surface of the frustrum of one cone and the inner surface of the frustrum of another.

Clutch, Contracting-Band: A clutch consisting of a drum and band, the latter contracting upon the former.

Clutch, Dry-Plate: A clutch whose friction surfaces are metal plates, not lubricated.

Clutch, Expanding-Band: A clutch consisting of a drum and band, the latter expanding within the former.

Clutch, Jaw: A clutch whose members lock end to end by projections or jaws in one entering corresponding depressions in the other.

Clutch, Multiple-Disk: A clutch whose friction surfaces are metal plates or disks, alternate disks being attached to one member and the rest to the other member of the drive.

Clutch Brake: A device designed to stop automatically the rotation of the driven member of a clutch after disengagement from the driving member.

Clutch Lining: The wearing surface of a clutch. This may be easily removed and replaced when worn.

Clutch Pedal: The pedal by which the clutch may be disengaged, engagement being obtained automatically by means of a spring.

GLOSSARY

Clutch Spring: A spring arranged to either hold a clutch out of gear or throw it into gear.

Coasting: The movement of the car without constant applications of the motive power, as in running downhill with the aid of gravity or on the level, through the momentum obtained by previous power applications.

Cock, Priming: A small cock, usually operated by a lever, for admitting gasoline to the carbureter to start its action.

Coil, Induction: See "Spark Coil".

Coil, Non-Vibrator: A coil so designed that it will supply a sufficient spark for the ignition with one make and break of the primary circuit.

Coil, Primary: See "Primary Coil".

Coil, Secondary: See "Secondary Spark Coil".

Coil, Spark: See "Spark Coil".

Coil, Vibrator: A spark coil with which is incorporated an electromagnetic vibrator to make and break the primary circuit.

Coil Vaporizer: An auxiliary vaporizer to assist in starting a steam boiler. It is a coil of tubing into which liquid gasoline is admitted and burned to start the generation of gas in the main burner.

Cold Test: The temperature in degrees Fahrenheit at which a lubricant passes from the fluid to the solid state.

Combustion Chamber: That part of an explosive motor in which the gases are compressed and then fired, usually by an electric spark.

Combustion Space: See "Clearance" and "Clearance Space".

Commercial Car: A motor-driven vehicle for commercial use, such as transporting passengers or freight.

Commutator: In the ignition system of an explosive motor, the commutator is a device to automatically complete the circuit of each of a number of cylinders in succession.

Commutator of Dynamo or Motor: That part of a dynamo which is designed to cause the alternating current produced in the armature to flow in one direction in the external circuit; in a motor, to change the direct current in the external circuit into alternating current.

Compensating Carbureter: An automatic attachment to a carbureter controlling either air or fuel admission, or both, so that the proportion of one to the other is always maintained under any vibration of power required.

Compensating Gear: See "Differential Gear".

Compensating Joint: See "Universal Joint".

Compound Engine: A multiple-expansion steam engine in which the steam is expanded in two stages, first in the high-pressure cylinder and then in the low-pressure cylinder.

Compression: (1) That part of the cycle of a gas engine in which the charge is compressed before ignition; in a steam engine it is the phase of the cycle in which the pressure is increased, due to compression of the exhaust steam behind the piston. (2) The greatest pressure exerted on the gas in the compression chamber.

Compression Chamber: The clearance volume above the piston in a gas engine; also called "Compression Space".

Compression Cock: See "Compression-Relief Cock".

Compression Line: The line on an indicator diagram corresponding to the phase of the cycle in which the gas is compressed.

Compression-Relief Cock: A small cock by which the compression chamber of an internal-combustion motor may be opened to the air and thus allow the compression in the cylinder to be relieved to facilitate turning by hand, or *cranking*.

Compression Space: See "Compression Chamber".

Compression Tester: A small pressure gage by which the degree of compression of the mixture in a gas-engine cylinder may be tested.

Compressor, Air: See "Air Compressor".

Condenser: (1) In a steam motor, an apparatus in which the exhaust steam is converted back into water. (2) A device for increasing the electric capacity of a circuit. Used in an ignition circuit to increase the strength of the spark.

Cone Bearing: A shaft bearing in which the shaft is turned to a taper and the journal turned to a conical or taper form.

Cone Clutch: A friction clutch in which there are two cones, one fitting within the other.

Connecting Rods: The part of an engine connecting the piston to the crank, and by means of which a reciprocating motion of the piston is converted into the rotary motion of the crank.

Constricting Band Brake: See "Brake, Constricting Band".

Constricting Clutch: A friction clutch in which a band is tightened around a drum to engage it.

Contact Breaker: A device on some forms of gasoline motors having an induction coil of the single jump-spark type, to open and close the electric circuit of the battery and coil at the proper time for the passage of the arc or spark at the points of the spark plug.

Contact Maker: See "Contact Breaker".

Continental Drive: Double-chain drive.

Control: The levers, pedals, etc., in general with the speed and direction of a car is regulated by the driver. In speaking of right, left, or center control, the gearshift and emergency brake levers only are meant.

Control, Spark: Method of controlling the power of an engine by varying the point in the stroke at which ignition takes place.

Control, Throttle: Method of governing the power of the engine by altering the area of the passage leading to the admission valve so that the amount of the fuel introduced into the cylinder is varied.

Controller, Electric: Apparatus for securing various combinations of storage cells and of motors so as to vary the speed of the car at will.

Converter: A device for changing alternating current into direct current for charging storage batteries, etc. Converters may be any of three kinds: rotary, electrolytic, or mercury-vapor. The mercury-vapor converter is most widely used.

Convertible Body: An automobile body which may be used in two or more ways, usually as an open or closed carriage, or in which several seats may be concealed, and raised to increase the seating capacity.

Cooling Fan: Fan used in automobiles to increase the current of air circulating around the cylinders, or through the radiator.

Cooling System: The parts of a gas engine or motor car by which the heat is generated in the cylinder by the combustion of the fuel mixture. See "Water Cooling" and "Air Cooling".

Cork Inserts: Pieces of cork inserted in friction surfaces of clutches or brakes to give softer action.

Cotter Pin: A split metal pin designed to pass through holes in a bolt and nut to hold the former in place.

Coulomb: The unit of measure of electrical quantity. Sometimes called "Ampere Second". It is equivalent to the product of the current in amperes by the number of seconds current has been flowing.

Counterbalance: Weights attached to a moving part to balance that part.

Countershaft: An intermediate or secondary shaft in the power-transmission system.

Coupé: An enclosed body seating one or two passengers and the driver, all within.

Coupling, Flexible: See "Universal Joint".

Cowl: That portion of the body of the car which forms a hood over the instrument board or dash.

Cowl Tank: A fuel tank carried under the cowl and immediately in front of the dash.

Crank: A lever designed to convert reciprocating motion into rotating motion or *vice versa;* usually in the form of a lever formed at an angle with the shaft, and connected with piston by means of connecting rod.

Crank, Starting: A handle made to fit the projecting end of the crankshaft of a gas engine, so that the engine may be started revolving by hand.

Crankcase: The casing surrounding the crank end of the engine.

Crankcase Explosion: Explosion of unburned gases in the crankcase.

Crank Chamber: The enclosed space of small engines in which the crank works.

Cranking: The act of rotating the motor by means of a handle in order to start it. Turning the flywheel over a few times causes the engine to take up its cycle, and after an explosion it continues to operate.

Crankpin: The pin by which the connecting rod is attached to the crank.

Crankshaft: The main shaft of an engine.

Crankshaft, Offset: A crankshaft whose center line is not in the same plane as the axis of its cylinders.

Creeping of Pneumatic Tires: The tendency of pneumatic tires to push forward from the ground, and thus around the rim, in the effort to relieve and distribute the pressure.

Cross Member: A structural member of the frame uniting the side members.

Crypto Gear: See "Planetary Gear".

Crystallization. The rearrangement of the molecules of metal into a crystalline form under continued shocks. This is often the cause of the breaking of the axles and springs of a motor car.

Cup, Priming: A small cup-shaped device provided with a cock, by which a small quantity of gasoline can be introduced into the cylinder of a gasoline engine.

Current: The rate of flow of electricity; the quantity of electricity which passes per second through a conductor or circuit.

Current Breaker: See "Contact Breaker".

Current Indicator: A device to indicate the direction of current flow in a circuit; a polarity indicator.

Current Rectifier: A device for converting alternating current into direct current. See "Converter".

Cushion Tire: See "Tire, Cushion".

Cut-Off, Gas Engine: That point in the cycle of an internal-combustion engine at which the admission of the mixture is discontinued by the closing of the admission valve.

Cut-Off, Steam Engine: That point in the cycle of a steam engine, or that point on an indicator diagram, at which the admission of steam is discontinued by the closing of the admission valve.

Cut-Out, Automatic: A device in a battery charging circuit designed to disconnect the battery from the circuit when the current is not of the proper voltage.

Cut-Out, Muffler: A device by which the engine is made to exhaust into the air instead of into the muffler.

Cut-Out Pedal: Pedal by means of which the engine is made to exhaust into the air instead of into the muffler.

Cycle: A complete series of operations beginning with the drawing in of the working gas, and ending after the discharge of the spent gas.

Cycle, Beau de Rochas: See "Beau de Rochas Cycle".

Cylinder: A part of a reciprocating engine consisting of a cylindrical chamber in which a gas is allowed to expand and move a piston connected to a crank.

Cylinder Bore: See "Bore".

Cylinder Cock: A small cock used to allow the condensed water to be drained away from the cylinder of a steam engine, usually called a *drain cock.*

Cylinder Head: That portion of a cylinder which closes one end.

Cylinder Jacket: See "Jacket, Water".

Cylinder Oil: Lubricant particularly adapted to the lubrication of cylinder walls and pistons of engines.

D

Dash: The upright partition of a car in front of the front seat and just behind the bonnet.

Dash Adjustment: Connections by which a motor auxiliary may be adjusted by a handle on the dash. Usually applied to carbureter adjustments.

Dash Coil: An induction coil for jump-spark ignition, having an element for each cylinder, with dash connections to the commutator on the engine or camshaft.

Dash Gage: A steam, water, oil, or electric gage placed upon the dash of the car.

GLOSSARY

Day Type of Engine: The two-cycle internal-combustion engine with an air-tight crankcase.

Dead Axle: See "Axle, Dead".

Dead Center: The position of the crank and connecting rod in which they are in the same straight line. There are two positions, and in these positions no rotation of the crankshaft is caused by pressure on the piston.

Decarbonizer: See "Carbon Remover".

Deflate: Reduction of pressure of air in a pneumatic tire.

Deflector: In a two-cycle engine, the curved plate on the piston head designed to cause the incoming charge to force out the exhaust gases and thus assist in scavenging.

Deflocculated Graphite: Graphite so finely divided that it remains in suspension in a liquid.

Demountable Rim: A rim upon which a spare tire may be mounted and carried, and so arranged that it may be easily and quickly taken off or put on the wheel.

Denatured Alcohol: See "Alcohol, Denatured".

Densimeter: See "Hydrometer".

Depolarizer: Material surrounding the negative element of a primary cell to absorb the gas which would otherwise cause polarizing.

Detachable Body: A body which may be detached from and placed upon the chassis.

Detachable Rim: See "Demountable Rim".

Diagram Indicator: See "Indicator Card".

Diagram, Jeantaud: A diagrammatic representation of the running gear of an automobile, showing it turning corners of various radii for the purpose of determining the front-axle and steering connections.

Diesel Gas Engine: Four-cycle internal-combustion engine in which the explosion of the charge is accomplished entirely by the temperature produced by the high compression of the mixture.

Differential, Bevel-Gear: A balance gear in which the equalizing action is obtained by means of bevel gears.

Differential, Spur-Gear: A differential gear in which the equalizing action is obtained by spur gears.

Differential Brake: See "Brake, Differential".

Differential Case: See "Differential Housing".

Differential Gear: A mechanism to permit driving the wheels and yet allow them to turn a corner without slipping. An arrangement such that the driving wheels may turn independently of each other on a divided axle, both wheels being under the control of the driving mechanism. Sometimes called *balance, compensating,* or *equalizing gear.*

Differential Housing: The case that encloses the differential gear.

Differential Lock: A device which prevents the operation of the differential gear, so that the wheels turn as if they were on a solid shaft.

Dimmer: An arrangement for lowering the intensity of, or reducing the glare from headlights.

Direct Current: A current which does not change its direction of flow, as the current from a battery or a direct-current generator. Distinguished from an alternating current, which reverses its direction many times a minute.

Direct Drive: Transmission of power from engine to the final driving mechanism at crankshaft speed.

Discharge: In a storage battery, the passage of a current of electricity stored therein. In the ignition circuit, the flow of high-tension current at the spark gap.

Disk Clutch: A clutch in which the power is transmitted by a number of thin plates pressed face to face.

Distance Rod: See "Radius Rod".

Distribution Shaft: See "Camshaft".

Distributor: That part of the ignition system which directs the high-tension current, to the respective spark plugs in the proper firing order.

Double Ignition: A method of ignition which comprises two separate systems, either of which may be used independently of the other, or both together as desired. Usually distinguished by two current sources and two sets of plugs.

Drag: That action of a clutch or brake which does not completely release.

Drag Link: That rod in a steering gear which forms the connection between the mechanism mounted on the frame and the axle stub, and transmits the movements of steering from steering post to wheels.

Drive Shaft: The shaft transmitting the motion from the change gears to the driving axle; the torsion rod.

Driving Axle: The axle of a motor car through which the power is transmitted to the wheels.

Driving Wheel: The wheel to which or by which the motion is transmitted.

Dry Battery: A battery of one or more dry cells.

Dry Cell: A primary voltaic cell in which a moist material is used in place of the ordinary fluid electrolyte.

Dual Ignition: An ignition system comprising two sources of current and one set of spark plugs.

Dust Cap: A metal cap to be screwed over a tire valve to protect the latter from dust and water.

Dynamo: The name frequently applied to a dynamo-electric machine used as a generator. Strictly, the term *dynamo* should be applied to both motor and generator.

Dynamometer: The form of equalizing gear attached to a source of power or a piece of machinery to ascertain the power necessary to operate the machinery at a given rate of speed and under a given load.

E

Earth: See "Ground".

Economizer, Gas: An appliance to be attached to a float-feed carbureter to improve the mixture by automatically governing the amount of air in the float chamber.

Eccentric: A disk mounted off-center on a shaft to convert rotary into reciprocating motion.

Economy, Fuel: The fuel economy of a motor is the relation between the heat units in the fuel used in the motor and the work or energy given out by the motor.

Efficiency: The proportion of power obtained from a mechanism as compared with that put into it.

Efficiency of a Motor: The efficiency of a gasoline motor is the relation between the heat units consumed by the motor and the work of energy in foot-pounds given out by it. Electrical efficiency of a motor is the relation between the electrical energy put into the motor and the mechanical energy given out by it.

Ejector: An apparatus by which a jet of steam propels a stream of water in almost the same way as an injector, except that the ejector delivers it into a vessel having but little pressure in it.

Electric Generator: A dynamo-electric machine in which mechanical energy is transformed into electrical energy; usually called *dynamo.*

Electric Horn: An automobile horn electrically operated.

Electric Motor: A dynamo-electric machine in which electrical energy is transformed into mechanical energy.

Electric Vehicle: An automobile propelled by an electric motor, for which current is supplied by a storage battery carried in the vehicle.

Electrolyte: A compound which can be decomposed by electric current. In referring to storage batteries, the term electrolyte means the solution of sulphuric acid in water in which the positive and negative plates are immersed.

Electromagnet: A temporary magnet which obtains its magnetic properties by the action of an electric current around it and which is a magnet only as long as such current is flowing.

Electromotive Force: A tendency to cause a current of electricity to flow; usually synonymous with *potential, difference of potential, voltage,* etc.

Element: The dissimilar substances in a battery between which an electromotive force is set up, as the plates of a storage battery.

Emergency Brake: A brake to be applied when a quick stop is necessary; usually operated by a pedal or lever.

En Bloc: That method of casting the cylinders of a gasoline engine in which all the cylinders are made as a single casting. Block casting; monoblock casting.

End Play: Motion of a shaft along its axis.

Engine, Alcohol: An internal-combustion engine in which a mixture of alcohol and air is used as fuel.

Engine, Gasoline: An internal-combustion motor in which a mixture of gasoline and air is used as fuel.

Engine, Kerosene: An internal-combustion engine in which a mixture of kerosene and air is used as fue

Engine, Steam: An engine in which the energy in steam is used to do work by moving the piston in a cylinder.

Engine Primer: A small pump to force fuel into the carbureter.

Engine Starter: An apparatus by which a gasoline engine may be started in its cycle of operations without use of the starting crank. It belongs usually to one of four classes: (1) Mechanical or spring actuated, such as a coil spring wound up by the running of the engine or a strap around the flywheel; (2) fluid pressure, such as compressed air or exhaust gases induced into the cylinder to drive the piston through one cycle; (3) the electric system, in which a small motor is used to turn the engine over; (4) combinations of these.

Epicyclic Gear: See "Planetary Gear".

Equalizing Gear: See "Differential Gear".

Exhaust: The gases emitted from a cylinder after they have expanded and given up their energy to the piston; the emission of the exhaust gases.

Exhaust, Auxiliary: See "Auxiliary Exhaust".

Exhaust Horn: An automobile horn in which the sound is produced by the exhaust gases.

Exhaust Lap: The extension of the inside edges of a slide valve to give earlier closing of the exhaust. Also called *inside lap.*

Exhaust Manifold: A large pipe into which the exhaust passages from all the cylinders open.

Exhaust Port: The opening through which the exhaust gases are permitted to escape from the cylinder.

Exhaust Steam: Steam which has given up its energy in the cylinder and is allowed to escape.

Exhaust Stroke: The stroke of an internal-combustion motor during which the burned gases are expelled from the cylinder.

Exhaust Valve: A valve in the cylinder of an engine through which the exhaust gases are expelled.

Expanding Clutch: A clutch in which a split pulley is expanded to press on the inner circumference of a ring which surrounds it, and thus transmits motion to the ring.

Expansion, Gas Engine: That part of the cycle of a gas engine immediately after ignition, in which the gas expands and drives the piston forward.

Expansion, Steam Engine: That portion of the stroke of the steam engine in which the steam is cut off by the valves and continues to perform work on the piston, increasing in volume and decreasing in pressure.

Explosive Motor: See "Internal-Combustion Motor".

F

Fan, Cooling: A mechanically operated fan for producing a current of air for cooling the radiator or cylinder of a gas engine.

Fan, Radiator: A mechanically operated rotary fan used to induce the flow of air through the radiator to facilitate the cooling of the water.

Fan Belt: The belt which drives the cooling fan.

Fan Pulley: A pulley permanently attached to the fan and over which the fan belt runs to drive it.

Fat Spark: A short, thick, ignition spark.

Feed Pump: A pump by which water is delivered from the tank to the boiler of a steam car.

Feed Regulator: A device to maintain a uniform water level in a steam boiler by controlling the speed of the feed pump.

Feed-Water Heater: An apparatus for heating the boiler-feed water, either by means of a jet of steam or steam-heated coils.

Fender: A mud guard or shield over the wheels of a car.

Field, Magnetic: Space in the neighborhood of the poles of a magnet in which the magnetism exerts influence. Field also refers to the coils which produce the magnetism in an electromagnet.

Fierce Clutch: A clutch which cannot be engaged easily. A grabbing clutch.

Filler Board: Woodwork shaped to fill the space between the lower edge of the windshield and the dash.

Fin: Projections cast on the cylinders of a gas engine to assist in cooling.

Final Drive: That part of a car by which the driving effort is transmitted from the parts of the transmission carried on the frame to the transmission parts on the rear axle. The propeller shaft in a shaft-drive car.

Fire Test: A test of a lubricant to determine the temperature at which it will burn.

Firing: (1) Ignition of the charge in a gas engine. (2) The act of furnishing fuel under the boiler of a steam engine.

First Speed: That combination of transmission gears which gives the lowest gear ration forward. Slow speed; low speed.

Flash Boiler: A boiler arranged to generate highly superheated steam almost instantaneously, by allowing water to come in contact with very hot metal surfaces.

Flash Generator: See "Flash Boiler".

Flash Point: The temperature at which an oil will give off a vapor that will ignite when a flame comes in contact with it.

Flash Test: A test to determine the flash point of oils.

Flexibility: In an engine the ability to do useful work through a range of speeds.

Flexible Coupling: See "Universal Joint".

Flexible Shaft: A pliant shaft which will transmit considerable power when revolving.

Flexible Tubing: A tube for the conduction of liquids or gases, which may be bent at a small radius without leaking.

Float Carbureter: A carbureter for gasoline engines in which a float of cork or hollow metal controls the height of the liquid in the atomizing nozzle. Sometimes called *float-feed carbureter*.

Float Valve: An automatic valve by which the admission of a liquid into a tank is controlled through a lever attached to a hollow sphere which floats on the surface of the liquid and opens or closes the valve according as it is high or low.

Floating Axle: See "Axle, Floating".

Floating the Battery on the Line: Charging the battery while it is giving out current.

Flooding: Excessive escape of fuel in a carbureter from the spraying nozzle.

Flushing Pin: In a float-feed carbureter, a pin arranged to depress the float in priming. Also called *primer* and *tickler*.

Flywheel: A wheel upon the shaft of an engine which, by virtue of its moving mass, stores up the energy of the gas transmitted to the flywheel during the impulse stroke and delivers it during the rest of the cycle, thus producing a fairly constant torque.

Flywheel Marking: Marks on the race of a flywheel to indicate the time of valve opening and closing and thus assist in valve setting.

Foaming: See "Priming".

Fore Carriage: A self-propelled vehicle in which the motor is carried on the forward trucks, and propelling and steering is done with the forward trucks.

Fore-Door Body: An automobile body having doors in the forward compartment.

Four-Cycle or Four-Stroke Cycle: The cycle of operations in gas engines occupying two complete revolutions or four strokes.

Four-Wheel Drive: Transmission of driving effort to all four wheels.

Fourth Speed: That combination of transmission gears which gives the fourth from the lowest gear ratio forward. Usually the highest speed.

Frame: The main structural part of a chassis. It is carried upon the axles by the springs and carries the different elements of the car.

Frame Hangers: See "Body Hangers".

Free Wheel: A wheel so arranged that it can rotate more rapidly than the mechanism which drives it.

Friction: The resistance existing between two bodies in contact which tends to prevent their motion on each other.

Friction Clutch: A device for coupling and disengaging two pieces of shafting while in motion, by the friction of cones or plates on one another.

Friction Disk: The thin plate used in a disk or friction clutch. See "Disk Clutch".

Friction Drive: A method of transmitting power or motion by frictional contact.

Fuel: A combustible substance by whose combustion power is produced. Gasoline and kerosene are the chief automobile fuels.

Fuel Economy. See "Economy, Fuel".

Fuel Feed, Gravity: See "Gravity Fuel Feed".

Fuel Feed, Pressure: See "Lubrication, Force-Feed."

Fuel Feed, Vacuum. See "Vacuum Fuel Feed".

Fuel-Feed Regulator: A device in the fuel system of steam motor by which the rate of flow of fuel to the burner is automatically regulated.

Fuel Level: The height of the top of the fuel in the float chamber of a carbureter.

Fuel-Level Indicator: An instrument either permanently connected to the fuel tank or which may be inserted thereon to indicate the quantity of fuel in the tank.

Fuel Tank, Auxiliary: A tank designed to hold a supply of fuel in addition to that carried in the main shaft.

Fuse: A length of wire in an electric circuit designed to melt and open the circuit when excess current flows through it and thus prevent damage to other portions of the circuit.

Fusible Plug: A hollow plug filled with an alloy which melts at a point slightly above the temperature of the steam in a boiler, as when the water runs low, thus putting out the fire and preventing the burning out of the boiler.

GLOSSARY

G

Gage: (1) Strictly speaking, a measure of, or instrument for determining dimensions or capacity. Practically, the term refers to an instrument for indicating the pressure or level of liquids, etc. (2) The distance between the forward or rear wheels measured at the points of contact of the tires on the road. Tread; track.

Gage Cock: A small cock by which a pipe leading to a gage may be opened or closed.

Gage Lamp: Lamp, usually electric, placed above or near the gages to enable them to be read at night.

Gage, Oil: See "Oil Gage".

Gage, Tire: See "Tire-Pressure Gage".

Gap: In automobiles, the spark gap.

Garage: A building for storing and caring for automobiles.

Garage, Portable: A garage which may be moved from one place to another either as a whole or in sections.

Gas: Matter in a fluid form which is elastic and has a tendency to expand indefinitely with reduction in pressure.

Gas Economizer: See "Economizer".

Gas Engine: An internal-combustion motor in which a mixture of gas and air is used as fuel. The term is also applied to the gasoline engine.

Gas Engine, Otto: A four-stroke cycle engine developed by Otto and using the hot-tube method of ignition.

Gas Generator: An apparatus in which a gas is generated for any use.

Gas Lamp: See "Acetylene Lamp".

Gases, Boyle's Law of: See "Boyle's Law of Gases".

Gases, Gay Lussac's Law of: Called *Charles's Law* and the *Second Law of Gases*. Law defining the physical properties of gases at constantly maintained pressure. It states that at constant pressure the volume of gas varies with the temperature, the increase being in proportion to the change of temperature and volume of the gas.

Gasket: A thin sheet of packing material or metal used in making joints, piping, etc.

Gasoline: A highly volatile fluid petroleum distillate; a mixture of fluid hydrocarbons.

Gasoline-Electric Transmission: A system of propulsion in which a gasoline engine drives an electric generator, and the power is transmitted electrically to motors which drive the wheels.

Gasoline Engine: An internal-combustion motor in which a mixture of gasoline and air is used as a fuel.

Gasoline Primer: The valve on the carbureter of a gasoline engine by which the action of the engine can be started.

Gasoline-Tank Gage: A fuel-lever indicator for gasoline.

Gasoline Tester: A hydrometer graduated to indicate the specific gravity of gasoline, usually in degrees Baumé.

Gate: A plate which guides the gearshift lever in making speed changes.

Gather: Convergence of the forward portions of the front wheels. Toeing in.

Gay Lussac's Law of Gases: See "Gases, Gay Lussac's Law of".

Gear, Balance: See "Differential Gear".

Gear, Bevel: See "Bevel Gear".

Gear, Change-Speed: An arrangement of gear wheels which transmits the power of the motor to the differential gear at variable speeds independently of the motor speed.

Gear, Differential: See "Differential".

Gear, Fiber: A gear cut from a vulcanized fiber blank.

Gear, Helical: A gear whose teeth are not parallel to the axis of the cylinders.

Gear, Internal: A gear whose teeth project inward toward the center from the circumference of gear wheel.

Gear, Planetary: See "Planetary Gears".

Gear, Progressive: See "Progressive Change-Speed Gears".

Gear, Rawhide: A gear cut from a blank made up of compressed rawhide.

Gear, Selective: See "Selective Change-Speed Gears".

Gear, Timing: See "Timing Gears".

Gear, Worm: A helical gear designed for transmitting motion at angles, usually at right angles and with a comparatively great speed reduction.

Gearbox: The case covering the change-speed gears.

Gear Shifting: Varying the speed ration between motor and rear wheels by operating the change-speed gears.

Gear-Shift Lever: A lever by which the change-speed gears are shifted.

Geared-Up Speed: A speed obtained by an arrangement of gears in the gearset such that the propeller shaft rotates more rapidly than the crankshaft.

Gearset: See "Gear, Change-Speed".

Generator, Acetylene: See "Acetylene Generator".

Generator, Electric: See "Electric Generator".

Generator, Steam: A steam boiler.

Generator Tubing: Tubing by which acetylene is conducted from the generator to the lamp.

Gimbal Joint: A form of universal joint.

Gong: A loud, clear sounding bell, usually operated either electrically or by foot power

Governor: A device for automatically regulating the speed of an engine.

Governor, Dynamo: A method of automatic control of the generator (usually an ignition generator, in automobile work) by which its speed is maintained approximately constant.

Governor, Hydraulic: A governor applied to engines cooled by a pump circulation of water in such a way that the throttle opening is controlled by the pressure of the water.

Governor, Spark: A method of automatically controlling the speed of the engine by varying the time of ignition. See "Governor".

Grabbing Clutch: See "Fierce Clutch".

Gradometer: An instrument for indicating the degree of the gradient or the per cent of the grade. It consists of a level with a graduated scale.

Graphite: One of the forms in which carbon occurs in matter. Also known as *black lead*

and *plumbago*. Used as a lubricant in powdered or flake form in the cylinders of explosive engines.

Gravity-Feed Oiling System: See "Lubrication, Gravity".

Gravity Fuel Feed: Supply of fuel to the carbureter from the tank by force of gravity.

Grease and Oil Gun: A syringe by means of which grease or oil may be introduced into the bearings of the machinery.

Grease Cup: A device designed to feed grease to a bearing by the compression of a hand screw.

Grid: A lead plate formed in the shape of a gridiron to sustain and act as a conductor of electricity for the active material in a storage battery.

Grinding Valves: See "Valve Grinding".

Gripping Clutch: See "Fierce Clutch".

Ground: An electric connection with the earth, or to the framework of a machine.

H

Half-Motion Shaft: See "Half-Time Shaft".

Half-Time Gear: See "Timing Gears".

Half-Time Shaft: The cam shaft of a four-cycle gas engine. It revolves at one-half the speed of the crankshaft.

Hammer Break: A make-and-break ignition system in which the spark is produced when the moving terminal strikes the stationary terminal like a hammer.

Header: A pipe from which two or more pipes branch. Manifold.

Heater, Automobile: A device for warming the interior of an automobile, usually electric, or by means of exhaust gases or jacket water.

High Gear: That combination of change-speed gears which gives the highest speed.

High-Tension Current: A current of high voltage, as the current induced in the secondary circuit of a spark coil.

High-Tension Ignition: Ignition by means of high-tension current.

High-Tension Magneto: A magneto which delivers high-tension current.

Honeycomb Radiator: A radiator consisting of many very thin tubes, giving it a cellular appearance.

Hood: (1) That part of the automobile body which covers the frame in front of the dash. The engine is usually under the hood. (2) The removable covering for the motor.

Hooke's Coupler: See "Universal Joint".

Horizontal Motor: A motor the center line of whose cylinder lies in a horizontal plane.

Horn, Automobile: A whistle or horn for giving warning of the approach of the automobile.

Horsepower: The rate of work or energy expended in a given time by a motor. One horsepower is the rate or energy expended in raising a weight of 350 pounds one foot in one second, or raising 33,000 pounds one foot in one minute.

Horsepower, Brake: The power delivered at the flywheel of an internal combustion engine as ascertained by a brake test.

Horsepower, Rated: The calculated power which may be expected to be delivered by a motor. In America the term usually refers to the horsepower as calculated by the S.A.E. formula.

Hot-Air Intake: The pipe or opening conveying heated air to the carburetor.

Hot-Head Ignition: The method of igniting the charge in a gas-engine cylinder by maintaining the head of the combustion chamber at a high temperature from the internal heat of combustion, as in the Diesel engine.

Hot-Tube Ignition: An ignition device formerly used for gas engines in which a closed metal tube is heated red-hot by a Bunsen flame. When the compressed gases in the cylinder are allowed to come in contact with this, ignition takes place.

Housing: A metallic covering for moving parts.

H.P.: (1) Abbreviation for *horsepower.* (2) Abbreviation for *high pressure.*

Hub Cap: A metal cap placed over the outer end of a wheel hub.

Hydrocarbons: Chemical combinations of carbon and hydrogen in varied proportions, usually distillates of petroleum, such as gasoline, kerosene, etc.

Hydrometer: An instrument by which the specific gravity or density of liquids may be ascertained.

Hydrometer Scale, Baumé's: An arbitrary measure of specific gravity.

I

I-Beam: Sometimes called *I-Section*. A structural piece having a cross section resembling the letter I. I-Beam front axle.

Igniter: An insulated contact plug without sparking points, used in make-and-break ignition with low-tension magneto.

Igniter, High-Speed: An igniter having a short spark coil for high-speed engines.

Igniter, Jump-Spark: A system of ignition in which is used a current of high pressure, which will jump across a gap in the high-pressure circuit, causing a spark at the gap.

Igniter, Lead of: Amount by which the ignition is advanced. See "Advanced Ignition".

Igniter, Primary: The apparatus in a primary circuit for making and breaking the circuit.

Igniter Spring: A spring to quickly break the circuit of a primary igniter.

Ignition, Advancing: See "Advanced Ignition".

Ignition, Battery: A system which gets its supply of current from a storage battery or dry cells. This system usually consists of a battery, a step-up coil, and a distributor for sending the current to the different spark plugs.

Ignition, Catalytic: Method of ignition for explosive motors based on the property of some metals, particularly spongy platinum, of becoming incandescent when in contact with coal gas or carbonized air.

Ignition, Double: See "Double Ignition".

Ignition, Dual: See "Dual Ignition".

Ignition, Fixed: Ignition in which the spark occurs at a given point in the cycle and cannot be changed from that point at the will of the operator except by retiming the ignition system. Fixed spark.

Ignition, Generator: Ignition current which is furnished by a combination lighting generator and magneto. The generator is fitted with an interrupter and distributor. Sometimes refers to system in which a generator charges a battery and the latter furnishes the ignition current in connection with a coil and distributor.

Ignition, High-Tension: Sometimes called jump-spark. Ignition which is effected by means of a high-tension or high-voltage current which is necessary to jump a gap in the spark plug.

Ignition, Hot-Head: See "Hot-Head Ignition".

Ignition, Jump-Spark: See "Ignition, High-Tension".

Ignition, Low-Tension: See "Ignition, Make-and-Break".

Ignition, Make-and-Break: A system in which the spark is produced by the breaking or interruption of a circuit, the break occurring in the combustion space of the cylinder. The current used is of low-voltage, hence the synonym, low-tension ignition.

Ignition, Magneto: Ignition produced by an electric generator, called a magneto, which is operated by the gas engine for which it furnishes current. Dynamo ignition. Generator ignition.

Ignition, Master Vibrator: A system which uses as many non-vibrator coils as there are cylinders, and one additional coil, called the master vibrator, for interrupting the primary circuit for all coils. The master vibrator also is used with vibrator coils in which the vibrators are short-circuited.

Ignition, Premature: Ignition occurring so far before the top dead center mark that the explosion occurs before the piston has reached upper dead center.

Ignition, Primary: An ignition system in which a low-tension current flows through a primary coil, the circuit being mechanically opened, allowing a high-tension spark to jump across the gap. See "Primary Coil".

Ignition, Retarding. Setting the spark of an internal-combustion motor so that the ignition will occur at a later part of the stroke.

Ignition, Self: Explosion of the combustible charge by heat other than that produced by the spark. Incandescent carbon will cause this. Motor overheating because of lack of water is another cause.

Ignition, Single: A system using but one source of current.

Ignition, Synchronized: Ignition by means of which the timing in each cylinder of a multicylinder engine is the same. In synchronized ignition the spark occurs at the same point in the cycle in each cylinder. This type of ignition is obtained with a magneto and is lacking in a multi-coil system using vibrator coils.

Ignition, Timing of: The adjustment of the ignition system so that ignition will take place at the desired part of the cycle.

Ignition, Two-Independent: See "Ignition, Double".

Ignition, Two-Point: A system comprising two ignition sources, or a double-distributor magneto, and two sets of spark plugs, both of which spark at the same time.

Ignition Distributor: See "Distributor."

Ignition Switch: A control or switch for turning the ignition current on and off voluntarily

I. H. P.: Abbreviation for *indicated horsepower.*

Indicated Horsepower: (1) The horsepower developed by the fuel on the pistons, in contradistinction to brake horsepower. See "Horsepower, Brake". (2) The horsepower of an engine as ascertained from an indicator diagram.

Indicator: An instrument by which the working gas in an engine records its working pressure.

Indicator Card: A figure drawn by means of an indicator by the working gas in an engine. Also called *indicator diagram.*

Induction Stroke: The downstroke of a piston which causes a charge of mixture to be drawn into the cylinder.

Inflammation: The act or period of combustion of the mixture in the cylinder.

Inflate: To increase the pressure within a tire by forcing air into it.

Inflator, Mechanical Tire: A small power-driven air-pump for inflating the tire; either driven by gearing, chain, or belt from the engine shaft, or by friction from the flywheel.

Inherent Regulation: Expression applied to electric generators which use no outside means of regulating the output, the regulation being affected by various windings of the armature and fields.

Initial Air Inlet: See "Primary Air Inlet".

Initial Pressure: Pressure in a cylinder after the charge has been drawn in but not compressed.

Injector: A boiler-feeding device in which the momentum of a steam jet, directed by a series of conical nozzles, carries a stream of water into the boiler, the steam condensing within and heating the water which it forces along.

Inlet, Valve: The valve which controls the inlet port and so allows or prevents mixture from passing to the cylinder.

Inlet Port: Passage or entrance in the cylinder wall through which the fuel mixture is taken. Sometimes called intake port.

Inlet Manifold: Sometimes called intake manifold or header. A branched pipe connected to the mixing chamber at one end and at the branch ends to the cylinders so as to communicate with the inlet ports.

Inlet Manifold, Integral: A manifold or header cast integral with the cylinder.

Inner-Tire Shoe: A piece of leather or rubber placed within the tire to protect the inner tube.

Inner Tube: A soft air-tight tube of nearly pure rubber, which fits within a felloe upon the casing.

Inside Lap: See "Exhaust Lap".

Intake Manifold: The large pipe which supplies the smaller intake pipes from each cylinder of a gas engine.

Intake Pipe: Sometimes made synonymous with inlet manifold. Correctly, the pipe from the carbureter to the inlet manifold.

Intake Stroke: See "Induction Stroke".

Intensifier: See "Outside Spark Gap".

Intermediate Gear: A gear in a change-speed set between high and low. In a three-speed set it would be second speed. In a four, either second or third.

GLOSSARY

Intermediate Shaft: See "Shaft, Intermediate".

Internal-Combustion Motor: Any prime mover in which the energy is obtained by the combustion of the fuel within the cylinder.

Internal Gear: See "Gear, Internal".

Interrupter: See "Vibrator".

J

Jack: A mechanism by which a small force exerted over a comparatively large distance is enabled to raise a heavy body. Used for raising the automobile axle to remove the weight from the wheels.

Jacket, Water: A portion of the cylinder casting through which water flows to cool the cylinder.

Jacket Water: The cooling water circulating in a water-cooling system.

Jackshaft: Shaft used in double-chain drive vehicles. Shaft placed transversely in the frame and driving from its ends chains which turn the rear wheels mounted on a dead axle.

Jeantaud Diagram: See "Diagram, Jeantaud".

Joint Knuckle: See "Swivel Joint."

Joule's Law of Gases: See "Gases, Joule's Law of".

Jump Spark: A spark produced by a secondary jump-spark coil.

Jump Spark, Circuit Maker: A mechanically operated switch by which the circuit in a jump-spark ignition system is opened and closed.

Jump-Spark Coil: An electrical transformer and interrupter, consisting of a primary winding of a few turns of coarse wire surrounding an iron core, and a secondary winding consisting of a great number of turns of very fine wire. The condenser is usually combined with this. Also known as *secondary spark coil.*

Jump-Spark Igniter: See "Igniter, Jump-Spark".

Jump-Spark Plug: See "Spark Plug".

Junction Box: A portion of an electric-lighting system to which all wires are carried for the making of proper connections.

Junk Ring: A packing ring used in sleeve-valve motors. It has the same functions as a piston ring. See "Piston Ring".

K

Kerosene: A petroleum product having a specific gravity between 58° and 40° Baumé. It is used as a fuel in internal-combustion engines and can often be used in gasoline engines by starting the engine on gasoline, then switching to kerosene.

Kerosene Burner: A burner especially adapted to use kerosene as a fuel.

Kerosene Engine: An engine using kerosene as fuel.

Key: A semicircular or oblong piece of metal used to hold a member firmly on a revolving shaft so as to prevent the member from rotating.

Key, Baldwin: A key with an oblong section.

Key, Woodruff: A key with a semicircular section.

Keyway: Slot in a rotating member used to hold the key

Kick Switch: Ignition switch mounted so that the driver can operate it with the foot.

Kilowatt: An electrical unit equal to 1000 watts.

Knuckle Joint: See "Swivel Joint".

L

Labor: The jerky operation of an engine. The engine is said to labor when it cannot pull its load without misfiring or jerking.

Lag, Combustion: The time between the instant of the spark occurrence and the explosion.

Lag, Ignition: The time between the instant of spark occurrence and the time at which the spark mechanism producing it begins to act.

Lamp, Trouble: Sometimes called inspection lamp. A small electric bulb carried in a suitable housing, and attached to a long piece of lamp cord. Used for inspecting parts of the car.

Lamp Bulb: The incandescent bulb used in a lamp.

Lamp Bracket: A support for a lamp.

Lamp Lighter: An apparatus for lighting gas lamps by electricity. The lamps are usually so arranged that by pushing the button the gas is turned on and the spark made at the same time.

Landaulet: A type of car which may be used as an open or closed car. The rear portion of the body may be folded down like a top.

Landaulet Body: An automobile body resembling a limousine body, but having a cover fitted to the back, which may be let down, leaving the back open. The top generally extends over the driver.

Lap: To make parts fit perfectly by operating them with an abrasive, such as ground glass, between the rubbing surfaces. To finish.

Lap of Steam Valves: In the slide valve of a steam engine, the amount by which the admission edges overlap the steam port when the valve is central with the cylinder case.

Layshaft: A countershaft or secondary shaft of a gearset operated by the main or shifter shaft.

Lead, or Lead Wire: Any wire carrying electricity.

Lead: In a steam engine the amount by which the steam port is opened when the piston is at the start of its stroke.

Lead Battery: See "Accumulator".

Lead of Igniter: See "Igniter, Lead of".

Lead of Valve: In an engine the amount by which the admission port is opened when the piston is at the beginning of the stroke; according as this is greater or less, the admission of working fluid is varied through several fractions of the stroke.

Lean Mixture: Fuel after leaving the carbureter, which contains too much air in proportion to the gasoline. Sometimes called thin mixture, rare mixture, or weak mixture.

Lever, Brake: See "Brake Lever."

Lever, Change-Speed: Lever by which the different combinations of change gears are made so as to vary the speed of the driving wheels in relation to the speed of the engine; also called gearshift lever.

Lever, Spark: Lever by which the speed and power of the engine are controlled by adjusting the time of ignition.

Lever, Steering: See "Steering Lever".

Lever, Throttle: A lever by which the speed and power of the engine are controlled by adjusting the amount of mixture admitted to the cylinder.

Lever Lock: An arrangement for locking the gearshift lever in free position so that with the engine running the driving axle will not be driven.

Lift: The distance through which a poppet valve is moved in opening from fully-closed to fully-open position.

Lifting Jack: See "Jack".

Lighting Outfit, Electric: An outfit for electrically lighting an automobile. This usually consists of a dynamo, storage battery, and lamps and switchboard, with the necessary wiring and cut-outs.

Limousine Body: An enclosed automobile body having the front and sides with side doors. The top extends over the seat of the driver.

Liner: One or more pieces of metal placed between two parts so they may be adjusted by varying the thickness of the liner. Sometimes called a shim. Also refers to a tool used for lining up parts.

Liner, Laminated: A liner or shim made in a number of parts, the thickness being varied by removing or adding parts.

Lines of Force: See "Field, Magnetic".

Link Motion: In a steam engine, the name for the arrangement of eccentric rods, links, hangers, and rocking shafts by which the relative motion and position of the slide valves are changed at will, providing for varying rates of expansion of the steam and thus varying the speed for either forward or backward motion.

Live Axle: See "Axle, Live".

Lock, Auto Safety: A device arranged so that it is impossible to start the motor car except by the proper combination or key.

Lock Nut: A nut placed on a bolt immediately behind the main nut to keep the main nut from turning.

Lock Switch: A switch in the ignition circuit so arranged that it can not be thrown on except by the use of a key.

Lock Valve: A valve capable of being secured with lock and key.

Long-Stroke: A gas engine whose stroke is considerably greater than its bore.

Lost Motion: Sometimes called play or backlash. Looseness of space between two moving parts.

Louver: A slit or opening in the side of a hood or bonnet of a motor car. Used to allow air from the draft to escape. A ventilator.

Low Gear: The lowest speed gear. First speed in a change-speed set.

Low-Speed Adjustment: A carbureter adjustment which regulates the mixture when the motor is operating slowly, with little throttle opening.

Low-Speed Band: The brake or friction band which controls the low speed of a planetary change-speed set.

Low-Tension Current: A current of low voltage or pressure, such as is generated by dry cells, storage battery, or low-tension magneto.

Low-Tension Ignition: See "Ignition, Make-and-Break".

Low-Tension Magneto: A magneto which initially generates a current of low voltage.

Low-Tension Winding: The winding of a transformer or induction coil through which the primary or low-tension current flows.

Low Test: Gasoline which has a high density, thus giving a low reading on the Baumé scale. Low-grade gasoline.

Low-Water Alarm: An automatic arrangement by which notice is given that the water in the boiler is becoming too low for safety.

Lubricant: An oil or grease used to diminish friction in the working parts of machinery.

Lubrication: To supply to moving parts and their bearings grease, oil, or other lubricant for the purpose of lessening friction.

Lubrication, Circulating: A system in which the same oil is used over and over.

Lubrication, Constant-Level: A system in which the level in the crankcase is kept to a predetermined level by means of a pump.

Lubrication, Force-Feed: Method of lubricating the moving parts of an engine by forcing the oil to the points of application by means of a pump.

Lubrication, Gravity: Method of supplying oil to moving parts of an engine by having a reservoir at a certain height above the highest point to be lubricated and allowing the oil to flow to the points of application by gravity.

Lubrication, Non-Circulating: A system in which the same oil is used but once.

Lubrication, Pressure-Feed: See "Lubrication, Force-Feed".

Lubrication, Sight-Feed: System of lubrication in which the oil pipe to different points of application is led through a glass tube in plain sight; usually at a point on the dashboard.

Lubrication, Splash: Method of lubricating an engine by feeding oil to the crankcase and allowing the lower edge of the connecting rod to splash into it.

Lubricator: A device containing and supplying oil or grease in regular amounts to the working parts of the machine.

Lubricator, Force-Feed: A pump-like device which automatically forces oil to the moving parts.

M

Magnet: A piece of iron or steel which has the characteristic properties of being able to attract other pieces of iron and steel.

Magnet, Horseshoe: A magnet shaped like the letter U.

Magnet, Permanent: A magnet which when once charged retains its magnetism.

Magnetic Field: See "Field, Magnetic".

Magnetic Spark Plug: A spark plug used in a make-and-break system of ignition in which contact is obtained by means of a magnet.

Magneto: See "Ignition, Magneto".

GLOSSARY

Magneto: See "Magneto-Electric Generator".

Magneto, Double-Distributor: A magneto with two distributors feeding two sets of spark plugs, two in each cylinder and both sparking at once. See "Ignition, Two-Point."

Magneto, High-Tension: A magneto has two armature windings and requires no outside coil for the generation of high-tension current.

Magneto, Induction: A type of magneto in which the armature and fields are stationary and a rotator or spool-shaped piece of metal is used to break the lines of force.

Magneto, Low-Tension: See "Low-Tension Magneto".

Magneto, Rotating Armature: A magneto in which the armature winding revolves.

Magneto Bracket: A shelf or portion of the crankcase web used to support the magneto.

Magneto Coupling: A flexible joint which connects the magneto with a revolving motor shaft.

Magneto Distributor: See "Distributor".

Magneto-Electric Generator: A machine in which there are no field magnet coils, the magnetic field of the machine being due to the action of permanent steel magnets. Usually contracted to *magneto.*

Main Bearing: A bearing, used for supporting the crankshaft.

Manifold: A main pipe or chamber into which or from which a number of smaller pipes lead to other chambers. See "Intake Manifold", "Exhaust Manifold", and "Inlet Manifold".

Manometer: A device for indicating either the velocity or the pressure of the water in the cooling system of a gasoline motor.

Master Vibrator: A single vibrator which interrupts the current to each of a set of several spark coils in order.

Mean Effective Pressure: The average pressure exerted upon a piston throughout its stroke.

M. E. P.: Abbreviation for *mean effective pressure.*

Mercury Arc Rectifier: A mercury vapor converter. See "Mercury Vapor Converter".

Mercury Vapor Converter: An apparatus for converting alternating current into direct current by means of a bubble of mercury in a vacuum. The vapor of mercury possesses the property of allowing the flow of current in one direction only. Its principal use is for charging storage batteries.

Mesh: Two gears whose teeth are so positioned that one gear will drive the other are said to be in mesh.

Misfire: Failure of the mixture to ignite in the cylinder; usually due to poor ignition or poor mixtures.

Miss: The failure of a gas engine to explode in one or more cylinders. Sometimes called misfiring.

Mixing Chamber: A pipe or chamber placed between the carbureter and inlet manifold. Sometimes integral with the carbureter or manifold.

Mixing Tube: A tubular carbureter for a gas or gasoline engine.

Mixing Valve: A device through which air and gas are admitted to form an explosive mixture. The carbureter of a gasoline engine combines the mixing valve and vaporizer.

Mixture: The fuel of a gas engine, consisting of sprayed gasoline mixed with air.

Monobloc: Cast *en bloc* or in one piece. Refers usually to cylinders, which are cast two or more at once.

Motocycle: A trade name for a special make of motorcycle.

Motor, Electric: See "Electric Motor".

Motor, Gasoline: See "Gasoline Motor".

Motor, High-Speed: A gas engine whose rotative speed is very high and whose power output goes up with the speed to an unusual degree.

Motor, Horizontal: A gas engine whose cylinder axis lies in a horizontal plane.

Motor, I-head: A gas engine which has cylinders, a section of which resembles the letter I. This type has the valves in the head.

Motor, L-Head: A gas engine in which a section of cylinders resembles the letter L. The valves in this type are all on one side.

Motor, Long-Stroke: See "Long-Stroke Motor".

Motor, Non-Poppet: A gas engine whose valves are not of the poppet type. In this class is the Knight sleeve valve, the rotary valve, and the piston valve.

Motor, Overhead Valve: A motor with cylinders whose valves are in the head.

Motor, Piston Valve: A gas engine using valves which are in the form of pistons.

Motor, Poppet: A gas engine using poppet-type valves. See "Poppet Valve".

Motor, Revolving Cylinder: A motor whose cylinders revolve as a unit.

Motor, Rotary Valve: One in which the valves consist of slots cut out along cylindrical rods which rotate in the cylinder casting.

Motor, Sliding Sleeve: The Knight type motor in which thin sleeves slide up and down in the cylinder, the sleeves having ports which register with the inlet and exhaust manifolds.

Motor, T-Head: A gas engine with the valves on opposite sides of the cylinders, a section of which resembles the letter T.

Motor, V-Type: A motor whose cylinders are set on the crankcase so as to form an angle of 45 to 90 degrees between them.

Motor, Vertical: A motor with the cylinder axis in a vertical plane.

Motorcycle: A bicycle propelled by a gasoline engine.

Mud Guard: Metal or leather strips placed over the wheels to catch the flying mud and to prevent the clothing from coming in contact with the wheels when entering and leaving the car.

Muffler Cut-Out: See "Cut-Out, Muffler".

Muffler Cut-Out Pedal: See "Cut-Out Pedal".

Muffler Exhaust: A vessel containing partitions, usually perforated with small holes and designed to reduce the noise occasioned by the exhaust gases of an engine, by forcing the gases to expand gradually.

Muffler Explosion: Explosion of unburned gases in exhaust passages of the muffler, usually due to poor ignition or poor mixture.

Multiple Circuit: A compound circuit in which a number of separate sources or electrically operated devices, or both, have all their positive poles connected to a single positive conductor and all their negative poles to a single negative conductor.

N

N.A.A.M.: Abbreviation for National Association of Automobile Manufacturers.

Naphtha: A product of the distillation of petroleum used to some extent for marine engines.

Needle Valve: A valve in a carbureter used for regulating the amount of gasoline to flow in with the mixture.

Negative Plate: Plate of a storage battery to which current returns from the outside circuit.

Negative Pole: That pole of an electric source through which the current is assumed to enter or flow back into the source after having passed through the circuit external to the source.

Neutral Position: The position of the change-speed lever which so places the gears that the motor may run idle, the car remaining still.

Non-Deflatable Tire: See "Tire, Non-Puncturable".

Non-Freezing Solution: A solution placed into the radiator of a motor car to prevent the water therein from freezing. Alcohol and glycerine are the usual anti-freezing agents. See "Anti-Freezing Solution".

Non-Puncturable Tire: See "Tire, Non-Puncturable".

Non-Skid Device: See "Anti-Skid Device".

O

Odometer: (1) The mileage-recording mechanism of a speedometer. (2) An instrument to be attached to an automobile wheel to automatically indicate the distance traveled.

Odometer, Hub: A speed-recording device which is placed on the hub cap of a wheel.

Offset: Off center, as a crankshaft in which a line vertically through the crankpins does not coincide with a line vertically through the center of the cylinder.

Ohm: (1) Unit of electrical resistance. (2) Amount of electrical resistance. Such resistance as would limit the flow of electricity under an electromotive force of one volt to a current of one ampere.

Ohm's Law: The law which gives the relation between voltage, resistance, and current flow in any circuit. Expressed algebraically, $C=\frac{I}{R}$ where C is the current flowing in amperes, I the voltage and R the ohmic resistance.

Oil Burner: A burner equipped with an atomizer for breaking up liquid fuel into a spray.

Oil Engine: An internal-combustion motor using kerosene or other oil as fuel.

Oil Gage: (1) A gage to indicate the flow of oil in the lubricating system. (2) Used to show the level of oil in a compartment in the base of a gas engine.

Oil Gun: A cylinder with a long point and a spring plunger for squirting oil or grease into inaccessible parts of a machine.

Oil Pump: A small force pump providing a constant positive supply of oil under pressure; usually considered to be more reliable than a lubricator.

Oiler: An automobile device for oiling machinery.

Opposed Motor: A gasoline engine whose cylinders are arranged in pairs on opposite sides of the crankshaft, both connecting rods of each pair being connected to the same crank, so that the shock of the explosion in one will be balanced by the cushioning effect of the compression in the other. In general these motors are two-cylinder, horizontal.

Otto Cycle: See "Four-Stroke Cycle".

Outside Spark Gap: See "Spark Gap, Outside".

Overcharged: The state of the storage battery when it has been charged at too high a rate or for too great a length of time.

Overhead Camshaft: A camshaft which is placed above the cylinder of a gas engine.

Overhead Valves: See "Motor, Overhead Valve".

Overheating: The act of allowing the motor to reach an excessively high temperature due to the heat of combustion being not carried away rapidly enough by the cooling devices, or to insufficient lubrication. Overheating of a bearing is due to insufficient lubrication.

P

Packing: The material introduced between the parts of couplings, joints, or valves, to prevent the leakage of gas or liquids to or from them.

Panel, Charging: A small switchboard for charging a storage battery.

Parallel Circuit: See "Multiple Circuit".

Patch, Tire-Repair: Rubber strips for making repairs in punctured or ruptured tires.

Petcock: A control cock which when open allows gas or liquid to escape from the chamber to which it is attached.

Petrol: Word used in England for gasoline.

Picric Acid: Acid which may be added to gasoline to increase the motor efficiency. Gasoline will absorb about five per cent of its weight of picric acid.

Pin, Taper: A conically shaped pin.

Pinch: A cut in an inner tube caused by the tube being caught or pinched between the outer casing and the rim.

Pinion: (1) The smaller of any pair of gears. (2) A small gear made to run with a larger gear.

Piston: The hollow, cylindrical portion attached to the connecting rod of a motor. The reciprocating part which takes the strain caused by the explosion.

Piston Air Valve: A secondary air valve in the piston of earlier types of gas engines to compensate the imperfect operation of surface carbureters used with those engines and to secure the injection of a sufficient quantity of air to insure the combustion of the charge.

Piston Head: The top of the piston.

GLOSSARY

Piston Pin: A pin which holds the connecting rod to the piston.

Piston Ring: (1) A metal ring inserted in a groove cut into a piston assisting in making the latter tight in the cylinder. There are usually three rings on each piston. (2) Rings about the circumference of a piston, whose diameter is slightly greater than that of the piston. These are to insure closer fit and prevent wearing of the piston, as the wear is taken up by the rings which may be easily removed.

Piston Rod: Usually called connecting rod. The rod which connects the piston with the crankshaft.

Piston Skirt: The portion of a piston below the piston pin.

Piston Speed: The rate at which the piston travels in its cylinder.

Piston Stroke: The complete distance a piston travels in its cylinder.

Pitted: Condition of a working surface which has become covered with carbon particles which have been imbedded in the metal.

Planetary Gear: An arrangement of spur and annular gears in which the smaller gears revolve around the main shaft as planets revolve around the sun.

Planetary Transmission: A transmission system in which the speed changes are obtained by a set of planetary gears.

Plate: Part of a storage battery which holds active material. See "Negative Plate".

Pneumatic Tire: A tire fitted to the wheels of automobiles, consisting usually of two tubes, the outer of India rubber, canvas, and other resilient wear-resisting material, and the inner composed of nearly pure rubber which is inflated with compressed air to maintain the outer tube in its proper form under load.

Polarizing: Formation of gas at the negative element of a cell so as to prevent the action of the battery. This formation of gas is caused by the violent reaction taking place in a circuit of low resistance.

Pole Piece: A piece of iron attached to the pole of a magneto used in an electric generator.

Poppet Valve: A disk or drop valve usually seating itself through gravitation or by means of springs, and frequently opening by suction or cams.

Port: An opening for the passage of the working fluid in an engine.

Portable Garage: See "Garage, Portable".

Positive Connection: A connection by which positive motion is transmitted by means of a crank, bolt, or key, or other method by which slipping is eliminated.

Positive Motion: Motion transmitted by cranks or other methods in which slipping is eliminated.

Positive Plate: Plate in a storage battery, from which the current flows to the outside circuit.

Positive Pole: The source from which electricity is assumed to flow; the opposite of negative pole. In a magnet the positive pole is the end of the magnet from which the magnetic flux is assumed to emanate.

Pounding in Engine: Pounding noise at each revolution, usually caused by either carbon deposit, loose or tight piston, loose bearing or other part, or pre-ignition.

Power Stroke: The piston stroke in a gas engine in which the exploded gases are expanding, thus pushing the piston downward.

Power Tire Pump: A pump which is operated by a gas engine and is used to inflate the tires of a motor car.

Power Unit: The engine with fuel, cooling, lubrication, and ignition systems, without the transmission or running gears. Sometimes the gearset and driving shaft are included by the term.

Pre-Ignition: See "Premature Ignition".

Premature Ignition: Ignition of fuel before the proper point in the cycle.

Pressure-Feed: See "Lubrication, Force-Feed".

Pressure Gage: A gage for indicating the pressure of a fluid confined in a chamber, such as steam in a boiler, etc.

Pressure Lubricator: A lubricating device in which the oil is forced to the bearings by means of a pump or other device for maintaining pressure.

Pressure Regulator: A device for maintaining the pressure of the steam in the principal pipe at a constant point irrespective of the fluctuations of pressure in the boiler.

Primary Air Inlet: The main or fixed air intake of a carbureter.

Primary Circuit: The circuit which carries low-tension current.

Primary Coil: A self-induction coil consisting of several turns of wire about an iron core.

Primary Spark Coil: An induction coil which has only a single winding composed of a few layers of insulated copper wire wound on a bundle of soft iron wires, known as the *core*, also as a *wipe*, or *touch, spark coil*.

Primer: A pin in a float-feed valve so arranged that it may depress the float in priming a gasoline engine. Also called *tickler* and *flushing pin*.

Priming: (1) The carrying of water over with the steam from the boiler to the engine, due to dirty water, irregular evaporation, or forced steaming. (2) Injecting a small amount of gasoline into the cylinder of a gasoline engine to assist in starting.

Priming Cock: A control cock screwed into the cylinder and which when open communicates with the combustion chamber allowing gasoline to be poured into the cylinder.

Progressive Change-Speed Gears: Change-speed gears so arranged that higher speeds are obtained by passing through all the intermediate steps and *vice versa*.

Prony Brake: A dynamometer to indicate the horsepower of an engine. A band encircles the flywheel of the engine and is secured to a lever, at the other end of which is a scale to measure the pull.

Propeller Shaft: The shaft which turns the rear axle of a motor car. The drive shaft.

Pump, Centrifugal: A pump with a hollow hub and curved blades which by centrifugal force throw water or oil into the system requiring it.

Pump, Circulation: See "Circulation Pump".

Pump, Fuel-Feed: A mechanically operated pump for insuring positive feed of fuel to the burner of a steam engine or carbureter of a gas engine.

Pump, Oil: See "Oil Pump".

Pump, Plunger: Sometimes called piston pump. One containing a piston which forces a liquid to a system.

Pump, Power Tire: See "Tire Pump".

Pump, Steam Boiler-Feed: See "Boiler-Feed Pump".

Pump, Water Circulating: See "Circulation Pump".

Pump Gear: A pump composed of two gears in mesh placed in a housing. When the gears revolve they carry oil or water, as the case may be, on their teeth, which deliver it to an outlet.

Puncture: The perforation of an inflated rubber automobile tire by some sharp substance on the roadbed.

Puncture-Closing Compound: A viscous compound placed within the inner tire tube to close the hole caused by a puncture.

Push Rod: A rod which operates the valves of a poppet-valve motor. A rod which imparts a pushing motion.

R

Race: (1) The parts upon which the balls of a ball bearing roll. (2) When referring to a gas engine, to operate at high speed without a load.

Racing Body: A low, light automobile body, having two seats with backs as low as possible; designed for large fuel capacity and very high speed.

Radiator: A device consisting of a large number of small tubes, through which the heated water from the jacket of the engine passes to be cooled, the heat being carried away from the metal of the radiator by air.

Radiator, Cellular: See "Honeycomb Radiator".

Radiator, Tubular: A radiator consisting of many tubes, through which water passes to be cooled.

Radiator Protector: See "Bumper".

Radius Rod: A bar in the frame of an automobile to assist in maintaining the proper distance between centers. Also called *distance rod.*

Rawhide Gear: Tooth gears, built up of compressed rawhide, used for high-speed drive. Sometimes a metal gear is merely faced with rawhide for the purpose of reducing noise.

Reach Rod: See "Radius Rod".

Reciprocating Parts: The parts such as pistons and connecting rods which have a reciprocating motion.

Rectifier, Alternating-Current: See "Current Rectifier".

Relief Cock: See "Compression-Relief Cock".

Removable Rim: See "Demountable Rim".

Resiliency: That property of a material by virtue of which it springs back or recoils on removal of pressure, as a spring.

Resistance, Electrical: (1) A part of an electric circuit for the purpose of opposing the flow of the current in the circuit. (2) The electrical resistance of a conductor is that quality of a conductor by virtue of which the conductor opposes the passage of electricity through its mass. Its unit is the *ohm.*

Retard: With reference to the ignition system, causing the spark to occur while the piston is retarding or moving downward on the working stroke.

Retarding Ignition: See "Ignition, Retarding".

Retarding the Spark: See "Ignition, Retarding".

Retread: To replace the tread of a pneumatic tire with a new one.

Reverse Cam: On a gasoline engine a cam so arranged that by reversing its motion or shifting it along its shaft it will operate the valves and cause the engine to reverse.

Reverse Gear: In a steam engine, a device by which the valves may be set to effect motion of the car in either direction. In a gasoline automobile, the reversing gear is usually incorporated with the change-speed gears.

Reverse Lever: A lever by which the direction of movement of the driving wheels may be reversed without reversing the engine. This is usually combined with the change-speed levers.

Rheostat: A device for regulating the flow of current in a closed electrical circuit by introducing a series of graduated resistances into the circuit.

Rim: The portion of a wheel to which a solid or pneumatic tire is fitted. A circular, channel-shaped portion attached to the wheel felloe.

Rim, Demountable: A rim which may be removed from the wheel easily in order that another with an inflated tire may take its place.

Rim, Quick-Detachable: A rim made of two or more parts so that the tire may be detached and attached quickly.

Rim, Removable: See "Demountable Rim".

Road Map: A map of a section or locality showing the best roads for motor-car travel, and usually the best stopping places and repair stations.

Roadster: A small motor car designed to be fairly speedy; usually has carrying capacity for an extra large quantity of fuel and supplies; generally seats two persons, with provision for one or two more, by the attachment of a rumble seat in the rear.

Rocker Arm: A pivoted lever used to operate overhead valves in a T-head motor.

Rod, Radius: See "Radius Rod".

Rod, Steering: See "Steering Rod".

Roller Bearings: See "Bearing, Roller".

Roller Chain: A chain whose links are provided with small rollers to decrease the friction and the noise.

Rotary Valve: A type of valve somewhat similar to the Corliss engine valve used on automobile motors.

Rumble: A small single seat to provide for an extra passenger on a two-seated vehicle. Usually detachable.

Runabout: A small two-seated vehicle, usually of a lower power and lower speed, as well as lower operating radius, than a roadster.

Running Board: A horizontal step placed below the frame and used to assist passengers in leaving and entering a motor car.

Running Gear: The frame, springs, motor, wheels, speed-change gears, axles, and machinery of an automobile, without the body; used synonymously with *chassis*.

S

Safety Plug: See "Fusible Plug".

Safety Valve: A valve seated on the top of a steam boiler, and loaded so that when the pressure of the steam exceeds a certain point the valve is lifted from the seat and allows the steam to escape.

Saturated Steam: The quality of the steam when no more steam can be made in the closed vessel without raising the temperature or lowering the pressure.

Scavenging: The action of clearing the cylinder of an internal-combustion motor of the burned-out gases.

Score: To burn, or abrade a moving part with another moving part.

Screw: An inclined plane wrapped around a cylinder; a cylinder having a helical groove cut in its surface.

Searchlight: A headlight designed to throw a very bright light on the road. Electricity or acetylene is usually used as an illuminant, and the lamp has a parabolic reflector and may be turned to throw the light in any direction.

Secondary Battery: See "Accumulator".

Secondary Current: A current in which the electromotive force is generated by induction from a primary circuit in which a variable current is flowing. The high-tension current of a jump-spark ignition system.

Secondary Circuit: The circuit which carries high-tension current.

Secondary Spark Coil: An induction coil having a double winding upon its core. The inner winding is composed of a few layers of insulated wire of large size, and the outer winding consists of a great many layers of very small insulated copper wire. Also known as a *jump-spark coil*.

Seize: Refers to moving parts which adhere because of operation without a film of oil between the working surfaces.

Selective Change-Speed Gears: Change-speed gears so arranged that any desired speed combination can be obtained without going through the intermediate steps.

Self-Firing: Ignition of the mixture in a gas engine due to the walls of the cylinder or particles attached to them becoming overheated and incandescent.

Self-Starter: See "Engine Starter".

Separator, Steam: A device attached to steam pipes to separate entrained water from live steam before it enters the engine, or to separate the oily particles from exhaust steam on its way to the condenser.

Series Circuit: A compound circuit in which the separate sources or the separate electrical receiving devices, or both, are so placed that the current supplied by each, or passed through each, passes successively through the other circuits from the first to the last.

Set Screw: A small screw with a pointed end used for locking a part in a fixed position to prevent it from turning.

Setting Valves: See "Valve Setting".

Shaft, Intermediate: The shaft placed between the first and third motion gearing and acting as a carrier of motion between the two.

Shaft Drive: System of power transmission by means of a shaft.

Shim: See "Liner".

Shock Absorber: A device attached to the springs or hangers of motor cars to decrease the jars due to rough roads, instead of allowing them to be transmitted to the frame of the carriage.

Short Circuit: A shunt or by-path of comparatively small resistance around a portion of an electric circuit, by which enough current passes through the new path to virtually cut out the part of the circuit around which it is passed, and prevent it from receiving any appreciable current.

Sight Feed: An indicator covered with glass which shows that oil is flowing in a system. A telltale sight. A check on the oiling system.

Side-Bar Steering: See "Steering, Side-Bar".

Side-Slipping: See "Skidding".

Silencer: See "Muffler, Exhaust".

Silent Chain: A form of driving chain in which the links are comprised of sections which so move over the sprocket that practically all noise is eliminated. Silent chains are used specially for driving timing gears, gearsets, etc.

Skidding: The tendency of the rear wheels to slide sideways to the direction of travel, owing to the slight adhesion between tires and the surface of the roadbed, also called *side-slipping*.

Skip: See "Miss".

Sleeve Valve: A form of valve consisting of cylindrical shells moving up and down in the cylinders of such a motor as the Silent Knight.

Sliding Gears: A change-speed set in which various gears are placed into mesh by the sliding on a shaft of one or more gears.

Sliding Sleeve: See "Motor, Sleeve-Valve".

Slip Cover: A fabric covering for the top when down or for the upholstery of a motor vehicle.

Smoke in Exhaust: Smoky appearance in the exhaust due to too much oil, too rich mixture, low grade of fuel, or faulty ignition.

Solid Tire: See "Tire, Solid".

Sooting of Spark Plug: Fouling of the spark plug with soot, due to poor mixture, impure fuel, or improper lubrication.

Spare Wheel: An extra wheel complete with inflated tire, carried on the car for quick replacement of wheel with damaged tire.

Spark, Advancing: See "Advanced Ignition".

Spark Coil: A coil or coils of wire for producing a spark at the spark plug. It may be either a secondary or primary spark coil.

Spark Gap: A break in the circuit of a jump-spark ignition system for producing a spark within the cylinder to ignite the charge. The spark gap is at the end of a small plug called the *spark plug*.

Spark Gap, Extra: See "Spark Gap, Outside".

Spark Gap, Outside: A device to overcome the short circuiting in the spark gap due to fouling and carbon deposits between the points of the high-tension spark plug. It is a form of condenser, or capacity in which the air acts as the dielectric between two surfaces at the terminals of a gap in a high-tension circuit.

Spark Intensifier: See "Spark Gap, Outside".

Spark Lever: See "Timing Lever".

Spark Plug: The terminals of the secondary circuit of a jump-spark ignition system mounted to leave a spark gap between the terminals projecting inside the cylinder for the purpose of igniting the fuel in the cylinder by means of a spark crossing the gap between them.

Spark Plug, Pocketing: Mounting the spark plug in a recess of the cylinder head to reduce the sooting of the sparking points.

Spark Plug, Sooting of: See "Sooting of Spark Plug".

Spark Regulator: A mechanism by which the time of ignition of the charge is varied by a small handle on or near the steering wheel.

Spark, Retarding: See "Ignition, Retarding".

Spark Timer: See "Timer, Ignition".

Speaking Tube: See "Annunciator".

Specific Gravity: The weight of a given substance relative to that of an equal bulk of some other substance which is taken as a standard of comparison. Air or hydrogen is the standard for gases, and water is the standard for liquids and solids.

Specific Heat: The capacity of a substance for removing heat as compared with that of another which is taken as a standard. The standard is generally water.

Speed-Change Gear: A device whereby the speed ratio of the engine and driving wheels of the car is varied.

Speed Indicator: An instrument for showing the velocity of the car.

Speedometer: A device used on motor cars for recording the miles traveled and for indicating the speed at all times.

Speedometer Gears: Gears used to drive a shaft which operates the speedometer.

Speedometer Shaft: A flexible shaft which operates a speedometer.

Spiral Gear: A gear with helically-cut teeth.

Splash Lubrication: See "Lubrication, Splash".

Spline: A key.

Spontaneous Ignition: See "Self-Firing".

Sprag: A device to be let down (usually at the rear of the car) to prevent its slipping back when climbing a hill.

Spray Nozzle: That portion of a carbureter which sprays the gasoline.

Spring: An elastic body, as a steel rod, plate, or coil, used to receive and impart power, regulate motion, or diminish concussion.

Spring, Cantilever: A type of spring which appears like a semi-elliptic reversed; and which is flexibly attached in the center, rigidly at one end, and by a shackle at the other.

Spring, Elliptic: A spring, elliptic in shape, and consisting of two half-elliptic members attached together.

Spring Semi-Elliptic: A spring made up of a number of leaves, the whole resembling a portion of an ellipse.

Spring, Supplementary: See "Shock Absorber".

Spring, Underslung: A spring which is fastened under the axle instead of over it.

Spring Hangers: See "Body Hangers".

Spring Shackle: A link attached to one end of a spring which allows for flattening of the spring.

Sprocket: A wheel with teeth around the circumference, so shaped that the teeth will fit into the links of a chain which drives or is driven by the sprocket.

Starboard: The right-hand side of a ship or vessel.

Starter, Engine: See "Engine Starter".

Starting, Gas Engine: The operation necessary to make the engine automatically continue its cycle of events. It usually consists of opening the throttle, retarding the spark, closing the ignition circuit, and cranking the engine.

Starting Crank: A crank by which the engine may be given several revolutions by hand in order to start it.

Starting Device: See "Engine Starter".

Starting on Spark: In engines having four or more cylinders with well-fitting pistons, it is often possible to start the motor after it has stood idle for some time by simply closing the ignition circuit, provided that the previous stopping of the engine was done by opening the ignition circuit before the throttle was closed, leaving an unexploded charge under compression in one of the cylinders.

Steam: The vapor of water; the hot invisible vapor given off by water at its boiling point.

Steam Boiler: See "Boiler".

Steam Condenser: See "Condenser".

Steam, Cycle of: A series of operations of steam forming a closed circuit, a fresh series beginning where another ends; that is, steam is generated in the boilers, passes through the pipes of the engine, doing work successively in its various cylinders, escaping at exhaust pressure to the condenser, where it is converted into water and returned to the boiler, to go through the same operations once more.

Steam Engine: A motor depending for its operation on the latent energy in steam.

Steam Gage: See "Pressure Gage".

Steam Port: See "Admission".

Steering, Side-Bar: Method of guiding the car by means of an upright bar at the side of the seat.

Steering Angle for Front Wheels: Maximum angle of front wheels to the axle when making a turn; should be about 35°.

Steering Check: A device for locking the steering gear so that the direction will not be changed unless desired.

Steering Column: See "Steering Post".

Steering Gear: The mechanism by which motion is communicated to the front axle of the vehicle, by which the wheels may be turned to guide the car as desired.

Steering Knuckle: A knuckle connecting the steering rods with the front axle of the motor.

Steering Lever: A lever or handle by which the car is guided.

Steering Neck: The vertical spindle carried by the steering yoke. It is the pivot of the bell crank by which the wheel is turned.

Steering Pillar: See "Steering Post".

Steering Post: The member through which the twist of the steering wheel is transmitted to the steering knuckle. The steering post often carries the spark and throttle levers also.

Steering Rod: The rod which connects the steering gear with the bell cranks or pivot arms, by means of which the motor car is guided.

Steering Wheel: The wheel by which the driver of a motor car guides it.

Steering Yoke: The Y-shaped piece in which the front axle terminates. The yoke carries the vertical steering spindle or steering neck.

Stephenson Link Motion: A reversing gear in which the ends of the two eccentric rods are connected by a link or quadrant sliding over a block at the end of the valve spindle.

Step-Up Coil: A coil used to transform low- into high-tension current.

Storage Battery: See "Accumulator".

Stroke: See "Piston Stroke".

Strainer, Gasoline: A wire netting for preventing impurities entering the gasoline feed system.

Strangle Tube: The narrowing of the throat of the carbureter just above the air inlets in order to increase the speed of the air, and thus increase the proportion of gas which will be picked up.

Stroke: The distance of travel of a piston from its point of farthest travel at one end of the cylinder to its point of farthest travel at the other end. Two strokes of the piston take place to every revolution of the crankshaft.

Stud Plate: The plate or frame in a planetary transmission system carrying studs upon which the central pinions revolve.

Suction Valve: The type of admission valve on an internal combustion engine which is opened by the suction of the piston within the cylinder and admits the mixture. The valve is normally held to its seat by a spring.

Sulphating of Battery: The formation of an inactive coating of lead sulphate on the surface of the plates of a storage battery. It is a source of loss in the battery.

Superheated Steam: Steam which has been still further heated after reaching the point of saturation.

Supplementary Air Valve: See "Auxiliary Air Valve".

Swivel Joint: The joint for connecting the steering arm of the wheel or lever-steering mechanism to the arms on the steering wheel. Also called *knuckle joint.*

T

Tachometer: An instrument for indicating the number of revolutions made by a machine in a unit of time.

Tandem Engine: A compound engine having two or more cylinders in a line, one behind the other, and with pistons attached to the same piston rod.

Tank Gage: See "Fuel-Level Indicator".

Tappet Rod: See "Push Rod".

Taxicab: A public motor-driven vehicle in which the fare is automatically registered by the taximeter.

Taximeter: An instrument in a public vehicle for mechanically indicating the fare charged.

Terminals: The connecting posts of electrical devices, as batteries or coils.

Thermal Unit: Usually called the *British Thermal Unit,* or *B. t. u.* A measure of mechanical work equal to the energy required to raise one pound of water one degree Fahrenheit.

Thermostat: An instrument to automatically regulate the temperature.

Thermosiphon Cooling: A method of cooling the cylinder of a gas engine. The water rises from the jackets and siphons into a radiator from whence it returns to the supply tank, doing away with the necessity for a circulating pump.

Three-Point Suspension: A method used for suspending motor car units, such as the motor, on three points.

Throttle: A valve placed in the admission pipe between the carbureter and the admission valve of the motor to control the speed and power of the motor by varying the supply of the mixture.

Throttle, Foot: See "Accelerator".

Throttle, Lever: A lever on the steering wheel which operates the carbureter throttle. See "Throttle".

Throttling: The act of closing the admission pipe of the engine so that the gas or steam is admitted to the cylinder less rapidly, thus cutting down the speed and power of the engine.

Thrust Bearing: A bearing which takes loads parallel with the axis of rotation of the shaft upon which it is fitted.

Tickler: A pin in a carbureter arranged to hold down the float in priming, also called *flushing pin* and *primer.*

Timer, Ignition: An ignition commutator.

Timing Gears: The gears which operate the camshaft and magneto shaft. The camshaft gear is twice as large as the crankshaft gear.

Timing Lever: A lever fitted to gas engines by means of which the time of ignition is changed. Also called *spark lever.*

Timing Valve: In a gas engine using float-tube ignition, a valve controlling the opening between the combustion space and the igniter.

Tip, Burner: A small earthen, aluminum, or platinum cover for the end of the burner tube of an acetylene lamp. It is usually provided with two holes, so placed that the jets from them meet and spread out in a fan shape.

Tire, Airless: See "Airless Tire".

Tire, Clincher: A type of pneumatic tire which is held to a clincher.

Tire, Cushion: Vehicle tire having a very thick rubber casing and very small air space. It is non-puncturable and does not have to be inflated, but is not as resilient as a pneumatic tire.

Tire, Non-Deflatable: See "Tire, Non-Puncturable".

Tire, Non Puncturable: A tire so constructed that it cannot be easily punctured or will not become deflated when punctured.

Tire, Punctures in: Holes or leaks in pneumatic tires caused by foreign substances penetrating the inner tube and allowing the air to escape.

Tire, Single-Tube: A pneumatic tire in which the inner and outer tubes are combined.

Tire, Solid: A tire made of solid, or nearly solid rubber.

Tire Band: A band to protect or repair a damaged pneumatic tire. See "Tire Protector".

Tire Bead: Lower edges of a pneumatic tire which grip the curved portion of a rim.

Tire Case: (1) A leather or metal case for carrying spare tire; same as *tire holder*. (2) The outer tube.

Tire Chain: See "Anti-Skid Device".

Tire Filling: Material to be introduced into the tire to take the place of air and do away with puncture troubles

Tire Gage: Gage used for measuring the air pressure in a pneumatic tire.

Tire Holder: A metal or leather case for carrying spare tires.

Tire-Inflating Tank: A tank containing compressed air or gas for inflating the tires.

Tire Inflater, Mechanical: A small mechanical pump for inflating pneumatic tires.

Tire Patch: See "Patch, Tire Repair".

Tire-Pressure Gage: A pressure gage to indicate the pressure of air in the tire.

Tire Protector: The sleeve or band placed over a tire to protect it from road wear.

Tire Pump: A pump for furnishing air under pressure to the tire, may be either hand- or power-operated.

Tire Sleeve: A sleeve to protect the injured part of a pneumatic tire. It is a tire protector which covers more of the circumference of the wheel than a tire band. See "Tire Protector".

Tire Tape: Adhesive tape used to bind the outer tube to the rim in repairing tires.

Tire Tool: Tool used to apply and remove a tire.

Tire Valve: A small valve in the inner tube to allow air to be pumped into the tube without permitting it to escape.

Tires, Creeping of: See "Creeping of Tires".

Tonneau: The rear seats of a motor car. Literally, the word means a round tank or water barrel.

Torque: Turning effort, or twisting effort of a rotating part.

Torque Rod: A rod attached at one end to the rear axle and at the other to the frame; used to prevent twisting of the rear-axle housing.

Torsion Rod: The shaft that transmits the turning impulse from the change gears to the rear axle. Usually spoken of as the *shaft.*

Touch Spark: See "Wipe Spark".

Tourabout: A light type of touring car.

Touring Car: A car with no removable rear seats, and a carrying capacity of four to seven persons.

Town Car: A car having the rear seats enclosed but the driver exposed.

Traction: The act of drawing or state of being drawn. The pull (or push) of wheels.

Tractor: A self propelled vehicle for hauling other vehicles or implements; a traction engine.

Transmission, Individual Clutch: A transmission consisting of a set of spur gears on parallel shafts which are always in mesh, different trains being picked up with a separate clutch for each set.

Transmission, Planetary: A transmission system in which a number of pinions revolve about a central pinion in a manner similar to the revolution of the planets about the sun; usual type consists of a central pinion surrounded by three or more pinions and an internal gear.

Transmission, Sliding Gear: A transmission system in which sliding change-speed gears are used.

Transmission Brake: Brake operating on the gearset shaft or end of the propeller shaft.

Transmission Gears: A set of gears by which power is transmitted. In automobiles, usually called *change-speed gears.*

Transmission Ratio: The ratio of the speed of the crankshaft to the speed of the transmission shaft or driving shaft.

Tread: That part of a wheel which comes in contact with the road.

Tread, Detachable: A tire covering to protect the outer tube, which may be taken off or replaced.

Trembler: The vibrating spring actuated by the induction coil magnet which rapidly connects and disconnects the primary circuit in connection with jump-spark ignition.

Truck: (1) A strong, comparatively slow-speed vehicle, designed for transporting heavy loads. (2) A swiveling carriage having small wheels, which may be placed under the wheels of a car.

Try Cock: A faucet or valve which may be opened by hand to ascertain the height of water in the boiler.

Tube Case: See "Tire Case".

Tube Ignition: See "Hot-Tube Ignition".

Tubing, Flexible: See "Flexible Tubing".

Tubular Radiator: An automobile radiator in which the jacket water circulates in a series of tubes.

Tungsten Lamp: Incandescent bulb with the filament made of tungsten wire.

Turning Moment: See "Torque".

Turning Radius: The radius of a circle which the wheels of a car describe in making its shortest turn.

Turntable: Device installed in the floor of a garage and used for turning motor cars around.

Two-Cycle or Two-Stroke Cycle Engine: An internal-combustion engine in which an impulse occurs at the beginning of every revolution, that is, at the beginning of every downward stroke of the piston.

Two-to-One Gear: The system of gearing in a four-cycle gas engine for driving the camshaft, which must revolve once to every two revolutions of the crankshaft.

GLOSSARY

U

Under Frame: The main frame of the chassis or running gear of a motor vehicle.

Unit-Power Plant: A power system consisting of a motor, gearset, and clutch which may be removed from the motor car as a unit.

Universal Joint: A mechanism for endwise connection of two shafts so that rotary motion may be transmitted when one shaft is at an angle with the other. Also called *universal coupling, flexible coupling, Cardan joint* and *Hooke's joint.*

Upkeep: The expenditure for maintenance or expenditure required to keep a vehicle in good condition and repair.

V

Vacuum Fuel Feed: A system of feeding the gasoline from a tank at the rear of an automobile by maintaining a partial vacuum at some point in the system, usually at the dash, the fuel flowing from this point by gravity to the carbureter.

Vacuum Line: In an indicator diagram, the line of absolute vacuum. It is at a distance corresponding to 14.7 pounds below the atmospheric line.

Valve: A device in a passage by which the flow of liquids or gases may be permitted or stopped.

Valve, Admission: The valve in the admission pipe of the engine leading from the carbureter to the cylinder by which the supply of fuel may be cut off.

Valve, Automatic: See "Automatic Valve".

Valve, Inlet: See "Inlet Valve".

Valve, Mixing: See "Mixing Valve".

Valve, Muffler Cut-Out: See "Cut-Out, Muffler".

Valve, Overhead: See "Overhead Valve".

Valve, Poppet: See "Poppet Valve".

Valve, Rotary: See "Motor, Rotary Valve".

Valve, Suction: An admission valve which is opened by the difference between the pressures in the atmosphere and in the cylinder.

Valve Cage: A valve-retaining pocket which is attached to the cylinder.

Valve Clearance: The clearance of play between the valve stem and the tappet.

Valve Gear: The mechanism by which the motion of the admission or exhaust valve is controlled.

Valve Grinding: The act of removing marks of corrosion, pitting, etc., from the seats and faces of poppet or disk valves. The surfaces to be ground are rotated in contact with each other, an abrasive having been supplied.

Valve Lift: See "Lift".

Valve Lifter: A device for raising a poppet valve from its seat.

Valve Seat: (1) That portion of the engine upon which the valve rests when it is closed. (2) The portion upon which the face of a valve is in contact when closed.

Valve Setting: The operation of adjusting the valves of an engine so that the events of the cycle occur at the proper time. Also called *valve timing.*

Valve Spring: The spring which is around the valve stem and is used to return the valve to closed position after it has been opened by the cam:

Valve Stem: The rod-like portion of a poppet valve.

Valve Timing: See "Valve Setting".

Vaporizer: A device to vaporize the fuel for an oil engine. In starting it is necessary to heat the vaporizer, but the exhaust gases afterwards keep it at the proper temperature. The carbureter of the gas engine properly belongs under the general head of *vaporizer,* but the term has become restricted to the vaporizer for oil engines.

Variable-Speed Device: See "Gear, Change-Speed".

Vertical Motor: An upright engine whose piston travel is in a vertical plane.

Vibrator: The part of the primary circuit of a jump-spark ignition system by which the circuit is rapidly interrupted to give a transformer effect in the coil.

Vibrator, Master: See "Master Vibrator".

Volatile: Passing easily from a liquid to a gaseous state, in opposition to *fixed.*

Volatilization: Evaporation of liquids upon exposure to the air at ordinary temperatures.

Volt: Practical unit of electromotive force; such an electromotive force as would cause a current of one ampere to flow through a resistance of one ohm.

Voltammeter: A voltmeter and an ammeter combined; sometimes refers to wattmeter.

Voltmeter: An instrument for measuring the difference of electric potential between the terminals of an electric circuit. It registers the electric pressure in volts.

Vulcanization: The operation of combining sulphur with rubber at a high temperature, either to make it soft, pliable, and elastic, or to harden it.

Vulcanizer: A furnace for the vulcanization of rubber.

W

Walking Beam: See "Rocker Arm".

Water Cooling: Method of removing the heat of an internal-combustion motor from the cylinders by means of a circulation of water between the cylinders and the outer casing.

Water Gage: An instrument used to indicate the height of water within a boiler or other water system. It consists of a glass tube connected at its upper and lower ends with the water system.

Water Jacket: A casing placed about the cylinder of an internal-combustion engine to permit a current of water to flow around it for cooling purposes.

Watt: The unit of electric power. It is the product of the current in amperes flowing in a circuit by the pressure in volts. It is $\frac{1}{746}$ of a horsepower.

Watt Hour: The unit of electrical energy. The given watt-hour capacity of a battery, for instance, means the ability of a battery to furnish one watt for the given number of hours or the given number of watts for one hour, or a number of watts for a number of hours such that their product will be the given watt hours.

Welding, Autogeneous: A method of joining two pieces of metal by melting by means of a blow torch burning acetylene in an atmosphere of oxygen. This melts the ends of the parts and these are then run together.

Wheel, Artillery: A wood-spoked wheel whose spokes are in line with a line drawn vertically through the hub.

Wheel, Dished: A wheel made concave or convex so that the hub is inside or outside as compared with the rim. This is to counteract the outward inclination of the wheel due to the fact that the spindle is tapered and that its outward center is lower than its inner center.

Wheel, Double-Interacting: The mechanism by which two wheels are hung on one hub or axle, the outer being shod with an ordinary solid tire and the inner with a pneumatic tire, so that the weight of the vehicle bears against the lowest point of the pneumatic tire of the inner wheel to give the durability and tractive properties of a solid tire with the resiliency of a pneumatic.

Wheel, Spare: See "Spare Wheel".

Wheel Steering: See "Steering Wheel".

Wheel, Wire: A wheel with spokes made of wire.

Wheel Puller: A device used for pulling automobile wheels from their axles.

Wheel Steer: A method of guiding a car by means of a hand wheel.

Wheel, Steering Angle for: The angle which the steering column makes with the horizontal. It varies from 90° to 30° or less.

Wheelbase: The distance between the road contact of one rear wheel with the point of road contact of the front wheel on the same side.

Wheels, Driving on All Four: The method of using all four wheels of an automobile as the driving wheels.

Wheels, Driving on Front: The method of using the two front wheels as the drivers.

Wheels, Steering on Rear: Method of guiding the vehicle by turning the rear wheels.

Whistle: An automobile accessory consisting of a signalling apparatus giving a loud or harsh sound. Also called a *horn*.

Wind Guard: See "Wind Shield".

Wind Shield: A glass front placed upright on the dash to protect the occupants of the car from the wind.

Wipe Spark: Form of primary sparking device in which a spark is produced by a moving terminal sliding over another terminal, the break thus made causing a spark. Also called *touch spark*.

Wipe-Spark Coil: A primary spark coil with which the spark is made by wiping contact.

Wire Drawing: The effect of steam passing through a partially closed valve or other constricted opening; so called from the thinness of the indicator diagram.

Working Pressure: The safe working pressure of a boiler, usually estimated as $\frac{1}{5}$ of the pressure at which a boiler will burst.

Worm: A helical screw thread.

Worm and Sector: A worm gear in which the worm wheel is not complete but is only a sector. Used especially in steering devices.

Worm Drive: A form of drive using worm gears. See "Gears, Worm".

Worm Gear: The spiral gear in which a worm or screw is used to rotate a wheel

Worm Wheel: A wheel rotated by a worm.

Wrist Pin: See "Piston Pin".

X

X Spring: A vehicle spring composed of two laminated springs so placed one upon the other that they form the letter X.

Y

Yoke, Steering: See "Steering Yoke".

Started in Volume 1

Also from CGR Publishing

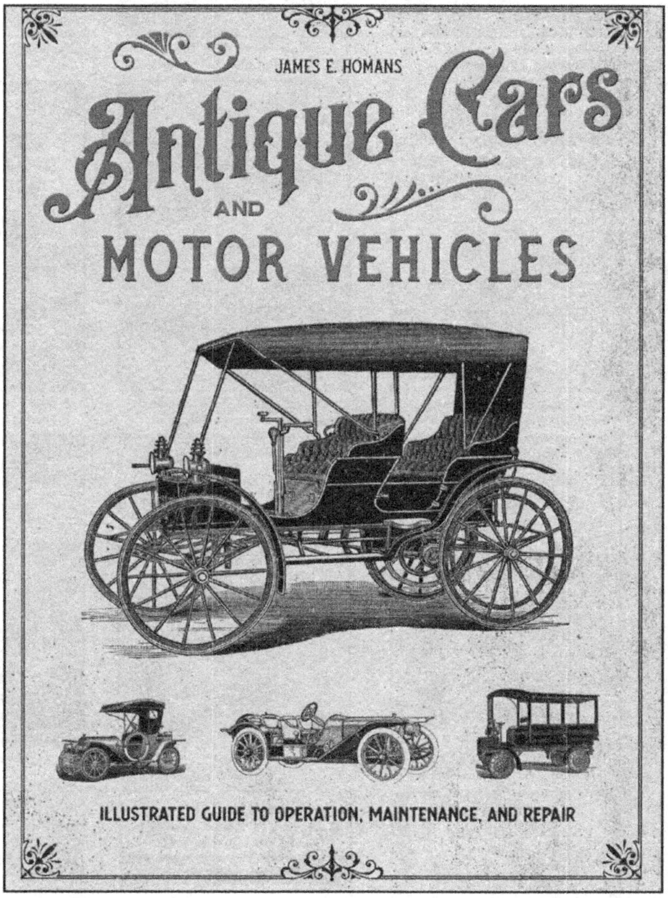

OTHER BOOKS FROM CGR PUBLISHING AT CGRPUBLISHING.COM

1939 New York World's Fair: The World of Tomorrow in Photographs

San Francisco 1915 World's Fair: The Panama-Pacific International Expo.

1904 St. Louis World's Fair: The Louisiana Purchase Exposition in Photographs

Chicago 1933 World's Fair: A Century of Progress in Photographs

19th Century New York: A Dramatic Collection of Images

The American Railway: The Trains, Railroads, and People Who Ran the Rails

The Aeroplane Speaks: Illustrated Historical Guide to Airplanes

The World's Fair of 1893 Ultra Massive Photographic Adventure Vol. 1

The World's Fair of 1893 Ultra Massive Photographic Adventure Vol. 2

The World's Fair of 1893 Ultra Massive Photographic Adventure Vol. 3

Henry Ford: My Life and Work - Enlarged Special Edition

Magnum Skywolf #1

Ethel the Cyborg Ninja Book 1

The Complete Ford Model T Guide: Enlarged Illustrated Special Edition

How To Draw Digital by Mark Bussler

Best of Gustave Doré Volume 1: Illustrations from History's Most Versatile...

Made in United States
Orlando, FL
10 January 2026

76596440R00240